Microspheres: Medical and Biological Applications

Editors

Alan Rembaum, Ph.D.
Jet Propulsion Laboratory
California Institute of Technology
Hydrocarbon Institute
Department of Chemistry
University of Southern California
Los Angeles, California

Zoltán A. Tökés, Ph.D.
Associate Professor of Biochemistry
School of Medicine and Comprehensive
Career Center
University of Southern California
Los Angeles, California

CRC Press
Taylor & Francis Group
Boca Raton London New York

CRC Press is an imprint of the
Taylor & Francis Group, an **informa** business

First published 1988 by CRC Press
Taylor & Francis Group
6000 Broken Sound Parkway NW, Suite 300
Boca Raton, FL 33487-2742

Reissued 2018 by CRC Press

A Library of Congress record exists under LC control number: 88007343

ISBN 13: 978-1-138-50620-6 (hbk)
ISBN 13: 978-0-203-71133-0 (ebk)

Visit the Taylor & Francis Web site at http://www.taylorandfrancis.com and the
CRC Press Web site at http://www.crcpress.com

THE EDITORS

Alan Rembaun, Ph.D., was born in 1916 in Ciechanow, Poland. In 1936, he went to Paris to study at the University of Sorbonne, and later received his "License es Sciences" from University of Lyon, France. During the Nazi occupation of France, he had to escape to Spain. He was arrested there and later escaped to England where he joined the Polish Liberation Forces. After the war, he came to the U.S. where he received his Ph.D. degree in Chemistry from the Syracuse University of New York in 1956. He joined the faculty at Akron University in Ohio and established himself as a recognized polymer chemist by investigating the polymerization of methyl methacrylate, the chemistry of methyl radicals, polymerization of methyl styrenes, and the electron transfer of siloxanes. In 1959, he went to Princeton University, New Jersey, to study the decomposition of ethyllithium in tetrahydrofuran and the cesium initiated diene polymerization and copolymerization. Two years later, at the Jet Propulsion Laboratories, California Institute of Technology, he developed a new class of block polymers, studied the kinetics of anionic polymerization of acenaphthylene, and investigated polymeric semiconductors. In collaboration with Drs. Felix Gutman and Allen Hermann, he devised a miniature battery which is now widely used in cardiac pacemakers. In 1970, he started to devote most of his investigation to novel polyelectrolytes, microspheres, and to their application to biological problems. He and Dr. William J. Dreyer of California Institute of Technology were involved in the application of latex spheres to study cell surfaces and in the development of immunolatex microspheres for scanning electron microscopy. His early work on the synthesis of magnetic polymeric microspheres of small and uniform size was not only meritorious in its own right in the field of polymer synthesis, but also provided opportunities for interactions with the disciplines of biology and medicine.

Dr. Rembaum has published more than 200 scientific papers. He co-authored or edited 5 books and was issued 56 patents. In 1968 and 1984, he received the NASA Gold Medal Award for exceptional scientific achievements and in 1986, after 24 years of service at the Jet Propulsion Laboratories, he received the NASA Career Award which recognized his outstanding scientific and technical contribution. He passed away in June 1986 and will be remembered as a stimulating, tireless, and inspiring scientist by his friends and colleagues.

Zoltán A. Tökés, Ph.D., is an Associate Professor in the Department of Biochemistry at the School of Medicine of the University of Southern California. He was born in 1940 in Budapest, Hungary. His initial encounter with biochemistry was at the Technical School for Fermentation and Food Preservation. In 1957, he came to the U.S. and worked in the field of mucopolysaccharide biosynthesis as a Research Assistant with Drs. Walter Marx and John W. Mehl. He received his B.Sc. degree in Chemistry, from the University of Southern California in 1964 and continued his graduate education at the California Institute of Technology, Pasadena. His research was on the cell surface changes with differentiation in Dr. William J. Dreyer's laboratory. After receiving his Ph.D. in Biochemistry, he joined the University of Malaya in Kuala Lumpur in 1970 as a lecturer, and conducted research on the digestive enzymes of carnivorous plants and on the erythrocyte abnormalities found among the aborigines. In 1972, he joined the Basel Institute for Immunology in Switzerland under the directorship of Dr. Neils K. Jerne. There he continued his research on the biochemistry of the cell surface and in collaboration with Dr. Hansruedi Kiefer he developed an assay to quantitate the cell surface associated proteolytic activity. He joined the University of Southern California's Comprehensive Cancer Center and the Department of Biochemistry in 1974. His investigation focuses on the molecular events that take place a the cell surface.

Dr. Tökés published more than 50 scientific papers, contriubted 18 books and recieved serveral U.S. patents. Dr. Tökés claims that there is no such thing as a boring moment in his laboratory. His laboratory and co-workers are credited with the successful encapsulation of anthracyclines in liposomes thereby reducing their undesirable host toxicity. They were successful in generating human monoclonal antibodies which preferentially bind to breast cancer cells and developing a membrane directed drug delivery with microspheres. His laboratory was also involved in the characterization of cells surface associated glycoproteins and proteinase inhibitors. Since 1975, Dr. Tökés served as the director of the Cell Culture Core Facility of the USC Comprehensive Cancer Center.

This book is gratefully dedicated by Alan Rembaum to his wife Danuta

CONTRIBUTORS

Wilhelm Ansorge, Ph.D.
Department of Instrumentation
European Molecular Biology Laboratory
Heidelberg, West Germany

Jon Olav Bjørgum, Cand.real.
Research Scientist
Division of Applied Chemistry
SINTEF
Trondheim, Norway

Robert A. Bloodgood, Ph.D.
Associate Professor
Department of Anatomy
University of Virginia School of Medicine
Charlottesville, Virginia

F. Chae
Memorial Sloan-Kettering Cancer Center
New York, New York

M. Chang
Jet Propulsion Laboratory
California Institute of Technology
Pasadena, California

N. Collins
Memorial Sloan-Kettering Cancer Center
New York, New York

Michael Colvin, Ph.D.
Chemist
Department of Applied Sciences
Jet Propulsion Laboratory
Pasadena, California

Håvard E. Danielsen, M.Sc.
Norsk Hydro's Institute for Cancer Research
The Norwegian Radium Hospital
Oslo, Norway

Turid Ellingsen, Dr.ing.
Department of Applied Chemistry
SINTEF
Trondheim, Norway

Emma Fernandez-Repollet
Department of Pharmacology
University of Puerto Rico School of Medicine
San Juan, Puerto Rico

R. M. Fitch
Louis Laboratory
S.C. Johnson & Sons, Inc.
Racine, Wisconsin

Steinar Funderud, Ph.D.
Biochemist
Apothekernes Laboratory for Special Preparaters
The Norwegian Radium Hospital
Oslo, Norway

Jean Gruenberg, Ph.D.
Department of Cell Biology
European Molecular Biology Laboratory
Heidelberg, West Germany

Kathryn E. Howell, Ph.D.
Department of Cell Biology
European Molecular Biology Laboratory
Heidelberg, West Germany

N. Kernan
Memorial Sloan-Kettering Cancer Center
New York, New York

J. Laver
Memorial Sloan-Kettering Cancer Center
New York, New York

K. M. Scholsky
Louis Laboratory
S.C. Johnson & Sons, Inc.
Racine, Wisconsin

A. Schwartz
Flow Cytometry Standards Corporation
Research Triangle Park, North Carolina

Tor Lea
Department of Immunology and Rheumatology
Rikshospitalet University Hospital
Oslo, Norway

Kjell Nustad, Ph.D.
Consultant
Central Laboratory
The Norwegian Radium Hospital
Oslo, Norway

R. O'Reilly
Memorial Sloan-Kettering Cancer Center
New York, New York

Chris D. Platsoucas, Ph.D.
Department of Immunology
M.D. Anderson Hospital
Houston, Texas

Albrecht Reith, Ph.D.
Head of Laboratory of Electron Microscopy and Morphometry
Department of Pathology
Norsk Hydro's Institute for Cancer Research
The Norwegian Radium Hospital
Oslo, Norway

Alan Rembaum, Ph.D.
Jet Propulsion Laboratory
California Institute of Technology
Hydrocarbon Institute
Department of Chemistry
University of Southern California
Los Angeles, California

Kevin L. Ross, Ph.D.
Research Fellow
Department of Medical Oncology
City of Hope National Medical Center
Duarte, California

Per Stenstad, Dr.ing.
Research Scientist
Department of Industrial Chemistry
University of Trondheim
Trondheim, Norway

Sven-Erik Strand, Ph.D.
Associate Professor
Department of Radiation Physics
Lund University
Lund, Sweden

Jerome A. Streifel
Molecular Biosystems, Inc.
San Diego, California

Zoltán A. Tökés, Ph.D.
Associate Professor of Biochemistry
Comprehensive Cancer Center
University of Southern California School of Medicine
Los Angeles, California

John Ugelstad, Ph.D.
Professor
Department of Industrial Chemistry
University of Trondheim
Trondheim, Norway

Frode Vartdal, Ph.D.
Senior Registrar
Institute of Immunology and Rheumatology
Rikshospitalet University Hospital
Oslo, Norway

TABLE OF CONTENTS

Chapter 1

THE COVALENT BINDING OF ENZYMES AND IMMUNOGLOBULINS TO HYDROPHILIC MICROSPHERES

M. Colvin, A. Smolka, A. Rembaum, and M. Chang

TABLE OF CONTENTS

ABSTRACT

A series of 1 μm hydrophilic microspheres consisting of poly-2-hydroxyethyl methacrylate (poly-HEMA), and a copolymer sequence of HEMA-acrolein microspheres were prepared by means of ionizing radiation. The enzymes alkaline phosphatase (AP), horseradish peroxidase (HRP), and a monoclonal antibody directed against a gastric ATPase were covalently bound to these microspheres. The resulting microsphere-protein conjugates were shown to retain their biological activity, as shown by the substrate turnover in the case of the bound enzymes, or by the competitive displacement of antigen in the case of immobilized antibody. The binding of AP and HRP to the microspheres was studied as a function of both time and pH in order to determine the optimal reaction conditions. HRP was also bound to polystyrene and preformed polystyrene microspheres with polyacrolein grafted on their surface in order to compare their enzymatic activity with HRP bound to the poly-HEMA and HEMA-acrolein microspheres.

I. INTRODUCTION

We have previously described the synthesis of both magnetic and nonmagnetic polymeric microspheres,[6] and we have reported their application in removal of neuroblastoma cells from normal cells in bone marrow,[14] in the separation of red blood cells,[10] and the separation of human B and T lymphocytes.[3] These applications were based on the formation of "immunomicrospheres", that is, polymeric microspheres to which either monoclonal or polyclonal antibodies were coupled. The antibody-coated microspheres then reacted specifically with the antigen or receptor present on the cell membrane, and thereby attached themselves to the cell surface.

The antibody was coupled to the microspheres either covalently or noncovalently, that is by hydrophobic interactions between the antibody and the microsphere resulting in surface adsorption. The difficulty in the noncovalently bound case was the gradual leakage of antibody from the immunomicrospheres.[6] Of similar importance was the problem associated with the noncovalent attachment of the microspheres to hydrophobic surfaces, i.e., cell membranes. In an earlier study, we reported the immobilization of protein and antibody on acrolein[5] and acrolein-grafted polystyrene microspheres,[4] but found a gradual leakage of protein over time from these particles. We, therefore, directed our attention to the synthesis and study of microspheres which were highly hydrophilic, to which protein (especially antibody) could be covalently bound with little or no physical adsorption of the protein on the microsphere surface.

In the present study, we describe a series of hydrophilic microspheres consisting of (1) poly-HEMA and (2) a copolymer sequence of HEMA-acrolein microspheres. Poly-HEMA was chosen because it has been found to be a very hydrophilic and biocompatible polymer; e.g., it has been used in soft contact lenses, burn dressings, and in drug-delivery matrices. In order to study the immobilization of protein or antibody onto the surface of the poly-HEMA microspheres, they were activated with 1,1-carbonyldiimidazole (CDI) prior to incubation with protein. The HEMA-acrolein microspheres did not require any prior activation for protein binding.

Two model proteins were chosen to study the immobilization of protein on these microsphers: HRP and AP, both of which are widely used in enzyme-linked immunosorbent assays (ELISA). To determine the optimum reaction conditions for HRP and AP binding, the enzymes were coupled at a fixed concentration with the CDI-activated poly-HEMA and HEMA-acrolein copolymer microspheres as a function of time and pH. HRP was also immobilized onto polystyrene- and acrolein-grafted polystyrene microspheres for comparison with the CDI-activated poly-HEMA and the HEMA-acrolein copolymer microspheres.

II. METHODS

Materials — 2-Hydroxyethyl methacrylate (HEMA, bp 67° C), acrolein (bp 2.5° C), and methacrylic acid (MAA, bp 63° C), all from Aldrich, were distilled at reduced pressure (3.5 mm, 100 mm, and 12 mmHg, respectively). A suspension of polystyrene microspheres of diameter 1 μm were obtained from Eastman Kodak and were washed several times by repeated centrifugation from distilled water. Electrophoresis grade *N,N'*-bisacrylamide (BIS) and acrylamide were from Biorad; 1,1,1-trimethylol propane triacrylate (TMPTA) from Celanese; and CDI from Sigma were used as received. Tetrahydrofuran (THF) was freshly distilled from sodium metal and benzophenone. Polyethelene oxide (PEO) (mol wt 100,000) was from Polysciences; HRP from Boehringer-Mannheim and AP (Sigma) were used as received. The dose rate of the cobalt gamma radiation source was 0.1 Mrad/hr.

Preparation of hydrophilic microspheres — poly-HEMA based copolymer microspheres activated by CDI and HEMA-acrolein copolymer microsphers with varying amounts of acrolein were prepared as follows:

Poly-HEMA based microspheres — 350 mℓ of a 0.4% w/v aqueous solution of PEO was purged with nitrogen gas for 15 min to remove dissolved oxygen. HEMA (4.7 g) and 0.67 g each of acrylamide, BIS, and MAA were added, and the solution purged for an additional 5 min with nitrogen, sealed and sonicated for 15 min, and placed in the cobalt gamma source for 22 hr at room temperature. The resulting microsphere suspension of diameter 1 μm was washed several times by centrifugation with distilled water.

HEMA-acrolein copolymer microspheres — To 15 mℓ 0.4% w/v PEO solution purged as above were added 1.4 g of combined acrolein and HEMA monomers, with acrolein amounts varying from 5% to 50% by weight, 0.76 g of MAA, and 0.83 g of TMPTA. After purging with nitrogen gas for 5 min, and sonication for 15 min, the solution was irradiated for 18 hr, the resulting microsphere suspension of diameter 1 μm was washed as described above.

CDI activation of the poly-HEMA based microspheres — Poly-HEMA microspheres were solvent exchanged from aqueous solution to anhydrous THF. The THF suspension was sealed and agitated for 24 hr and washed three times with anhydrous THF. The microspheres were then resuspended at 6.6% w/w in anhydrous THF. CDI (6.3 g) was added to the suspension, and the reaction carried out with rotation at 4°C for 48 hr in the dark.

Acrolein-grafted polystyrene microspheres — The preformed polystyrene microspheres were acrolein grafted according to the procedure of Rembaum et al.[4] The resulting microspheres were washed repeatedly by centrifugation from distilled water.

Binding of HRP to CDI-activated poly-HEMA microspheres — Aliquots (10 mg) of poly-HEMA microspheres were added to aliquots of HRP (0.5 mg) in 5 mℓ of either 0.1 *M* phosphate buffer pH 7.5, or 0.1 *M* phosphate buffer, pH 8.2, or 0.1 *M* carbonate buffer pH 9.6. The temperature was maintained at 4°C and at varying times up to 48 hr, the microspheres were recovered from the samples by centrifugation, and 5 mℓ 0.1 *M* 2-aminoethanol in 0.1 *M* phosphate buffer pH 7.4 added to the pellet to quench the reaction. Samples were sonicated for 15 min and stored in a thermostatted rotator at 4°C.

Binding of HRP to HEMA-acrolein copolymer microspheres — Polyglutaraldehyde and polyacrolein in the form of insoluble microspheres have been shown to bind protein via the aldehyde functional groups with the formation of Schiff bases.[3] This reaction was used to bind HRP to HEMA-acrolein microspheres in a protocol similar to that used for HEMA microspheres, with two modifications: buffers were prepared with 0.1% Tween 20 to reduce hydrophobic adsorption, and an additional 0.1 *M* acetate buffer pH 5.3 was used. The reactions were quenched with 5 mℓ 1% BSA and 0.1 *M* NH_2OH, in 0.1 *M* phosphate buffer, pH 7.4. Quenched samples were sonicated for 15 min and stored at 4° C.

Binding of HRP to polystyrene and acrolein-grafted polystyrene microspheres — HRP was immobilized on the polystyrene and acrolein-grafted polystyrene microspheres in a procedure similar to that used for the CDI-activated HEMA microspheres, at pH 9.6 with 0.1 *M* carbonate buffer. After an incubation period of 48 hr, the microspheres were washed five times with distilled water and stored at 4° C.

Assay of HRP binding to the microspheres — Aliquots of microspheres (10 mg) were washed three times in distilled water and resuspended at 4° C in 2 mℓ 0.25 *M* NaCl, 0.05 *M* acetate buffer pH 5.3. Two mℓ of substrate solution (2 mℓ 1 mg/mℓ O-phenylenediamine, 0.3 mℓ 3% H_2O_2, and 45 mℓ H_2O) was added, followed after an appropriate time interval by 1 mℓ 2.5 *N* H_2SO_4. The samples were centrifuged and the absorption of the supernatants measured at 490 nm.

Covalent binding of alkaline phosphatase to the microspheres — Poly-HEMA and HEMA-acrolein copolymer microsphere aliquots (2.5 mg) were washed three times in one of three buffers at pH 5.3, pH 7.4, or pH 9.6. Alkaline phosphatase (1 mg) dissolved in 500 μℓ of appropriate pH buffer was added to each of the microsphere pellets, and incubated overnight at 4° C with rotation. Microsphere samples were washed three times and resuspended in 200 μℓ substrate buffer (1 *M* diethanolamine, 2 m*M* $MgCl_2$, pH 9.6). To doubling dilutions of the microspheres in a 96-well microtitration plate were added 100 μℓ of 1 mg/mℓ *p*-nitrophenyl phosphate in substrate buffer, and absorption measurements at 410 nm made every 3 min using a Titertek Multiskan spectrophotometer.

Monoclonal antibodies — The antibodies were prepared as described.[12] Briefly, BALB/CJ mice were immunized with purified vesicular pig $H^+ + K^+$ ATPase obtained from gastric mucosal homogenates by differential and sucrose-gradient centrifugation followed by free-flow electrophoresis.[7] When circulating anti-ATPase activity was confirmed by ELISA, the mice were sacrificed and their lymph node cells fused with the nonsecreting myeloma cell line P3.X63.Ag8.653 using polyethylene glycol. Hybrids secreting anti-ATPase antibodies were cloned by limiting dilution and expanded to mass culture in vitro or as ascites in pristane-primed mice. The antibodies were purified by precipitation with ammonium sulfate and protein A-affinity chromatography.[9]

Biosynthetic labeling of antibodies — Hybridoma cells in log-phase growth (10^6 cells, hybridoma clones 4.4 and 7.22) were washed once in serum-free medium, and resuspended in 1 mℓ sterile-filtered leucine-free RPMI 1640 medium made 10% in fetal calf serum previously dialyzed against phosphate-buffered saline. Five μCi ^{14}C-leucine was added and the cells were incubated at 37° C for 24 hr with 6% CO_2. Cells were pelleted by centrifugation, and the supernatant applied to a protein A-Sepharose affinity column (2 mℓ bed volume). Radiolabeled monoclonal antibody was eluted with 2 column volumes 0.1 *M* citrate pH 3.5, neutralized with 1 *M* Tris, and dialyzed against PBS. The specific activity of antibodies 4.4 and 7.22 was 6 μCi/mg and 5 μCi/mg, respectively.

Binding of antibodies to microspheres — The microspheres were diluted to approximately 125 μg/mℓ. Fifty-milliliter aliquots were centrifuged, and to the pellets were added 50 μℓ of various dilutions of ^{14}C-monoclonal antibody. The suspensions were incubated with rotation at 4° C overnight, and then vacuum filtered through 0.3 μm Millipore membranes previously soaked in 1% bovine serum albumin (BSA) to prevent nonspecific retention of antibody. The filters were washed three times *in situ* with PBS, placed into 10 mℓ Aquasol liquid scintillation fluid and counted for 60 sec. The binding data were expressed as dp retained on the filters as a function of the amount of ^{14}C-antibody incubated with the microspheres.

Competitive binding of antigens to HEMA-acrolein immunomicrospheres — Microspheres were reacted with monoclonal antibody 7.22 as described above. Immunomicrosphere aliquots (300 μg/50 μℓ 1% BSA/PBS) were incubated overnight at 4° C with 100 fmol (10 μℓ) ^{125}I-ATPase and varying amounts of unlabeled ATPase up to 100 fmol. The immunomicrospheres were washed three times by centrifugation, and bound radioactivity measured by gamma counting.

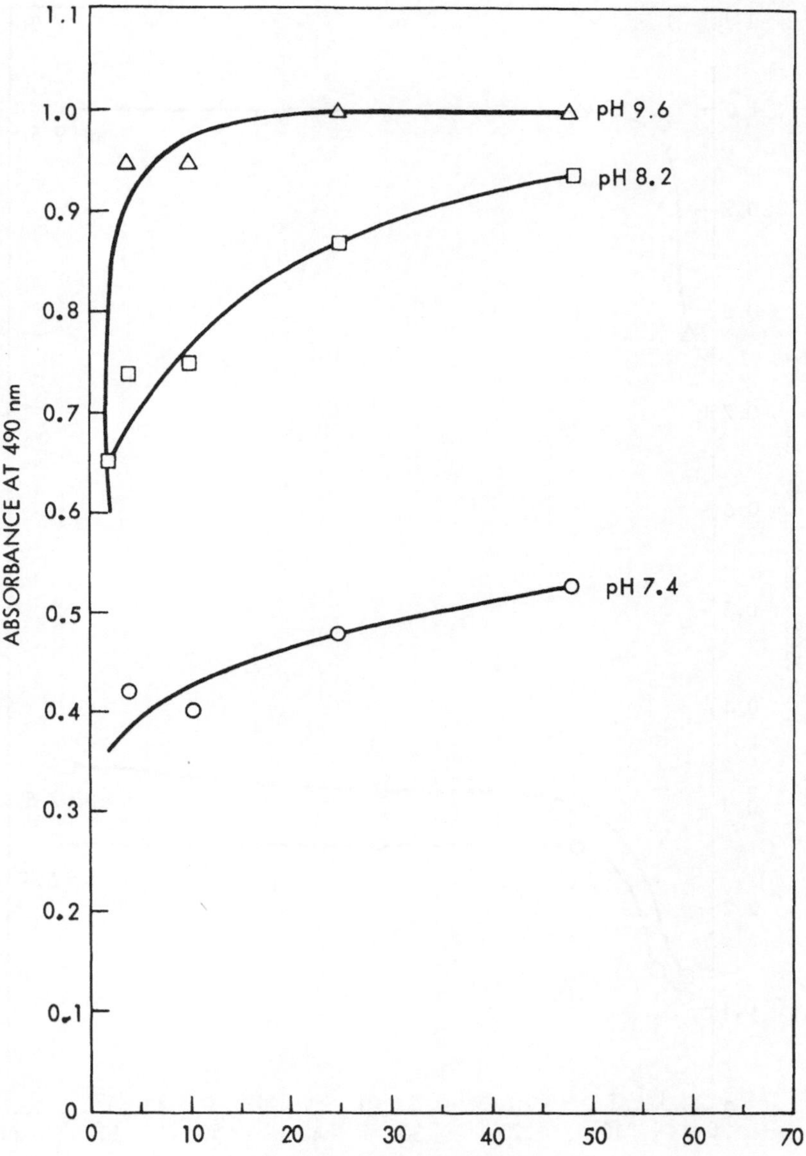

FIGURE 1. Enzymatic activity of HRP-conjugated poly-HEMA microspheres. The turnover of phenylenediamine was measured at 490 nm as a function of incubation time of HRP with CDI-activated poly-HEMA microspheres.

III. RESULTS

The reactivity of poly-HEMA and HEMA-acrolein microspheres with proteins was assessed by incubating the microspheres with HRP and AP, followed by measurements of the bound-enzyme activity.

Figure 1 represents the absorption at 490 nm of the reaction product of microsphere-bound HRP as a function of incubation time of the enzyme with CDI-activated poly-HEMA microspheres. Incubations were carried out at pH 7.4, pH 8.2, and pH 9.6. Alkaline pH clearly favored an increase in the enzymatic activity with 100% saturation after 10 hr at pH 9.4

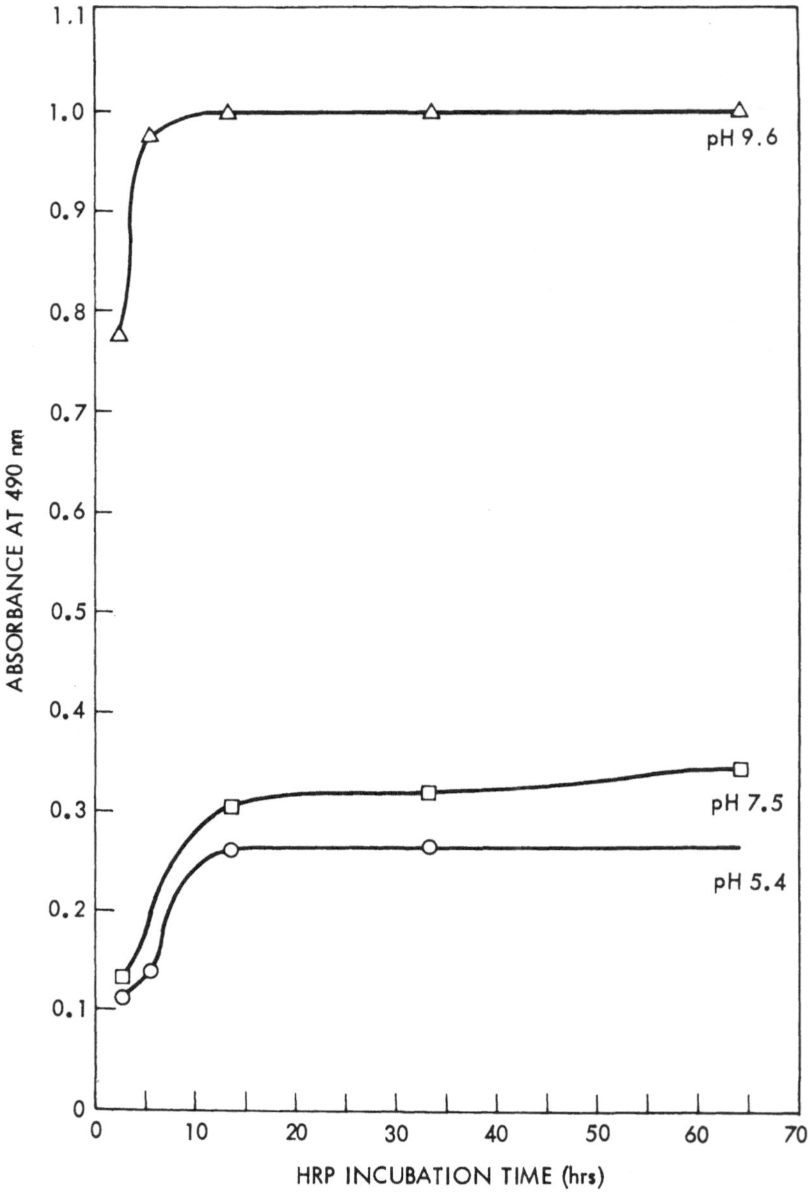

FIGURE 2. Enzymatic activity of HRP-conjugated HEMA-acrolein microspheres. The turnover of phenylenediamine was measured at 490 nm as a function of incubation time of HRP with HEMA-acrolein microspheres.

and only 40% and 75% saturation after 10 hr at pH 7.4 and pH 8.2, respectively. Figure 2 shows analogous data for HEMA-acrolein microspheres. In contrast to the poly-HEMA microspheres, the enzymatic activity was not found to be pH dependent, saturation occurring after 10 hr at pH 5.4, pH 7.5, and pH 9.6. The amount of enzymatic activity, however, was significantly affected by the pH, with four times as much activity found at pH 9.6 than at pH 5.4. At pH 9.6, HEMA-acrolein microspheres exhibited 50% less enzymatic activity than the poly-HEMA beads at the same pH.

Figure 3 shows the enzymatic activity of HRP bound to HEMA-acrolein microspheres as a function of acrolein content, measured at a binding pH of 9.6. The enzymatic activity was

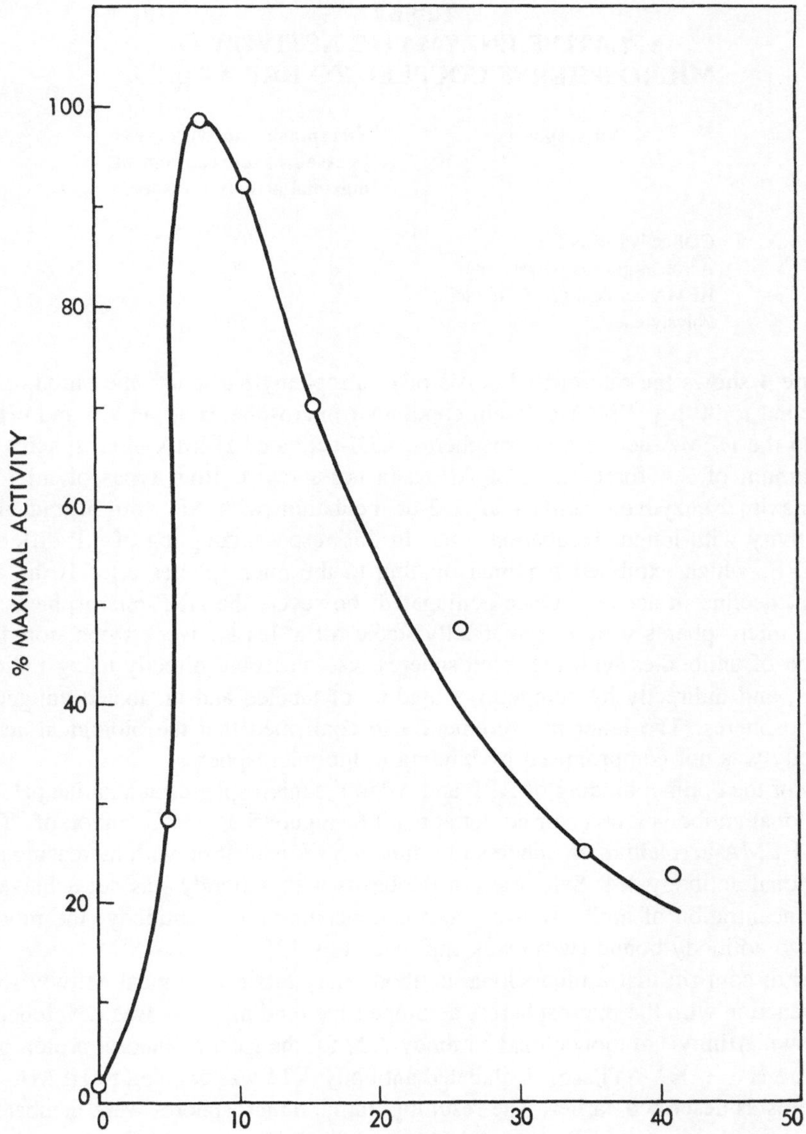

FIGURE 3. Enzymatic activity of HEMA-acrolein microspheres conjugated to HRP as a function of acrolein content. The enzymatic activity of the HEMA-acrolein microspheres conjugated with HRP was determined as a function of acrolein content at a binding pH of 9.6.

found to increase as the acrolein content rose to 7.5% by weight. Further increases of acrolein content, however, significantly lowered the enzymatic activity with 75% less enzymatic activity at 20% acrolein than at 7.5% acrolein.

Table 1 compares the relative enzymatic activity of 10-mg aliquots of the four different types of microspheres coupled to HRP at pH 9.6. With an optimal acrolein content of 7.5% by weight, the HEMA-acrolein microspheres were half as active as CDI-activated HEMA and acrolein-grafted polystyrene microspheres, while immodified polystyrene microspheres exhibited only one tenth of the maximal enzymatic activity.

Binding of AP to HEMA-acrolein microspheres was more efficient at pH 9.6 than at pH

Table 1
RELATIVE ENZYMATIC ACTIVITY OF MICROSPHERES COUPLED TO HRP AT pH 9.6

Microspheres	Enzymatic activity (expressed as percentage of maximal activity measured)
CDI-activated HEMA	100
Acrolein-grafted polystyrene	90
HEMA-acrolein (7.5% acrolein)	50
Polystyrene	10

5.3. Figure 4 shows the rate of hydrolysis of *p*-nitrophenylphosphate measured at 410 nm by AP bound to 30 μg HEMA-acrolein copolymer microspheres at pH 9.6 and pH 5.3. In contrast to the HEMA-acrolein microspheres, CDI-activated HEMA microspheres showed a pH optimum of 7.4 for binding of AP (data not shown). Both types of microspheres showed maximal enzymatic activity after 2 hr incubation with AP, with a gradual loss of bound activity with longer incubation time. In this respect, coupling of AP differed from that of HRP, which exhibited maximal binding to the microspheres after 10 hr, with no subsequent decline in activity. Once conjugated, however, the HRP-microspheres and the AP-bound microspheres were enzymatically stable for at least 1 week when stored at 4°C.

Reaction of antibodies with the microspheres was measured directly using radiolabeled antibodies, and indirectly by competitive binding of labeled and unlabeled antigen to immunomicrospheres. The latter measurement also confirmed that the biological activity of the antibody was not compromised by binding to the microspheres.

In view of the optimal binding of HRP and AP to the microspheres at alkaline pH, binding of monoclonal antibodies was carried out at pH 9.6. Figure 5 depicts retention of ^{14}C counts (dpm) by HEMA-acrolein microspheres as a function of incubation with increasing amounts of monoclonal antibody 4.4. Saturation of the beads with antibody was not achieved at the highest concentration of antibody used; at this concentration of antibody, the ratio of microsphere to antibody bound (w/w) was approximately 125.

In order to confirm that a monoclonal antibody may retain biological activity following covalent reaction with the microspheres, a competitive binding assay was developed, based on the known affinity[13] of monoclonal antibody 7.22 for the gastric mucosal proton-pumping enzyme, the $H^+ + K^+$ ATPase.[8] Unlabeled antibody 7.22 was coupled to HEMA-acrolein microspheres as described earlier. The resulting immunomicrospheres were incubated overnight with a fixed amount of ^{125}I-labeled ATPase, and varying amounts of unlabeled ATPase. Sampling of the immunomicrospheres revealed an inverse relationship between bound ^{125}I counts (dpm) and the amount of unlabeled ATPase, as shown in Figure 6. Such a relationship was not observed when antibody-deficient hybridoma supernatant replaced antibody 7.22 in the incubation with microspheres.

IV. DISCUSSION

The present study demonstrates covalent coupling of two enzymes, HRP and AP, and a monoclonal antibody directed against a gastric ATPase, to either poly-HEMA microspheres or HEMA-acrolein copolymer microspheres. The resulting conjugates were shown to retain the biological activities of the bound proteins, as shown by substrate turnover in the case of the bound enzymes, and by competitive displacement of antigen in the case of the immobilized monoclonal antibody.

Microspheres of diameter 1 μm were chosen because of their high surface area per unit volume, slow sedimentation rate, and their stability against aggregation. The poly-HEMA

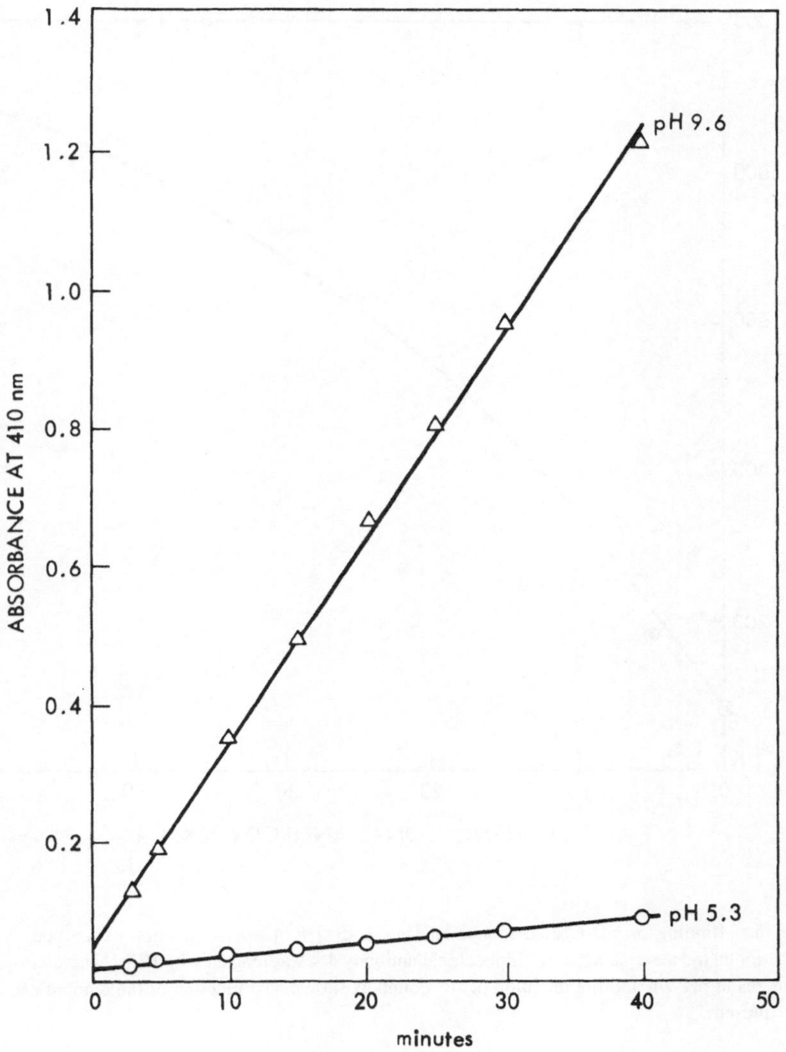

FIGURE 4. The rate of hydrolysis of *p*-nitrophenylphosphate by AP immobilized on HEMA-acrolein microspheres. The rate of hydrolysis of *p*-nitrophenylphosphate measured at 410 nm by AP bound to 30 μg HEMA-acrolein copolymer microspheres at pH 5.3 and pH 9.6.

and HEMA-acrolein copolymer microspheres had a relatively narrow size distribution and were quite pure since their preparation was achieved in the absence of free-radical initiator and other additives commonly used in emulsion polymerization. Both types of microspheres were very hydrophilic and therefore exhibited a low degree of hydrophobic adsorption of reactants or impurities in the samples being assayed, greatly increasing the sensitivity of the assay.

The binding of enzyme or antibody to the poly-HEMA microspheres required the microspheres to be activated prior to the incubation step, for this study CDI was chosen to activate the poly-HEMA microspheres. No protein binding was observed on the microspheres without prior activation (data not shown) which illustrates the absence of hydrophobic adsorption of protein on the microspheres.

CDI is suitable for the activation of substrates[1] and microspheres[2] containing hydroxyl groups. These activated substrates and microspheres can subsequently be coupled with protein or other compounds containing free-amino groups.

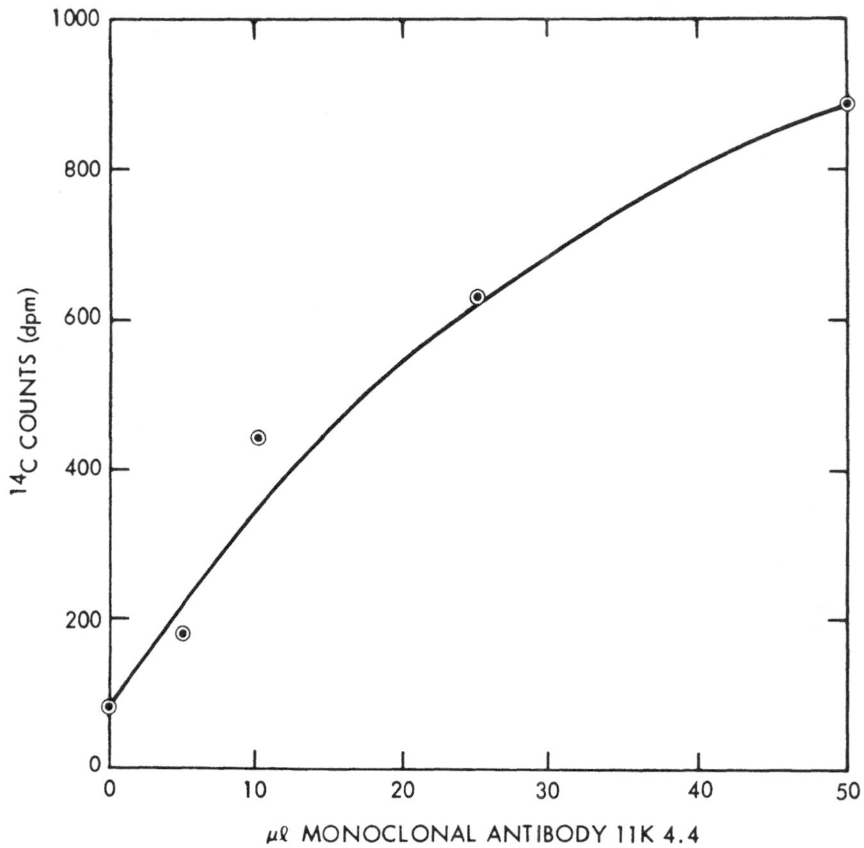

FIGURE 5. Binding of ^{14}C-labeled monoclonal antibody HK 4.4 as a function of antibody concentration in the reaction mixture. Monoclonal antibody 4.4 was reacted with HEMA-acrolein microspheres at pH 9.6 and the retention of ^{14}C counts is shown as a function of the amount of antibody present.

In the first step, the microspheres are incubated with CDI under anhydrous conditions. The CDI then undergoes nucleophilic attack by a hydroxyl group displacing an imidazole ring forming the microsphere-carbamate complex. In the second step, this complex undergoes a nucleophilic attack from an amino group of a protein molecule displacing the second imidazole ring, coupling the protein to the microsphere.

The results obtained for the coupling of HRP to the CDI-activated poly-HEMA microspheres agree with the results of Bethell et al.,[1] who found the optimum pH for the binding of protein to CDI-activated agarose to be between pH 9 to 10.

In contrast, the enzymatic activity of HRP coupled to HEMA-acrolein microspheres was maximal after 10 hr at each incubation pH, indicating a pH-dependent equilibrium of HRP with the microspheres. The composition of the HRP-bound HEMA-acrolein microspheres which exhibited maximum enzymatic activity were those with 7.5% acrolein as illustrated in Figure 3. Further increases in acrolein content resulted in a gradual decrease in the relative enzymatic activity. This inhibition may be caused by the multiple attachment of the enzyme with increasing acrolein content which results in a distortion or blocking of the active site of the bound enzyme.

The decline in the enzymatic activity of AP after 2 hr incubation time with both the CDI-

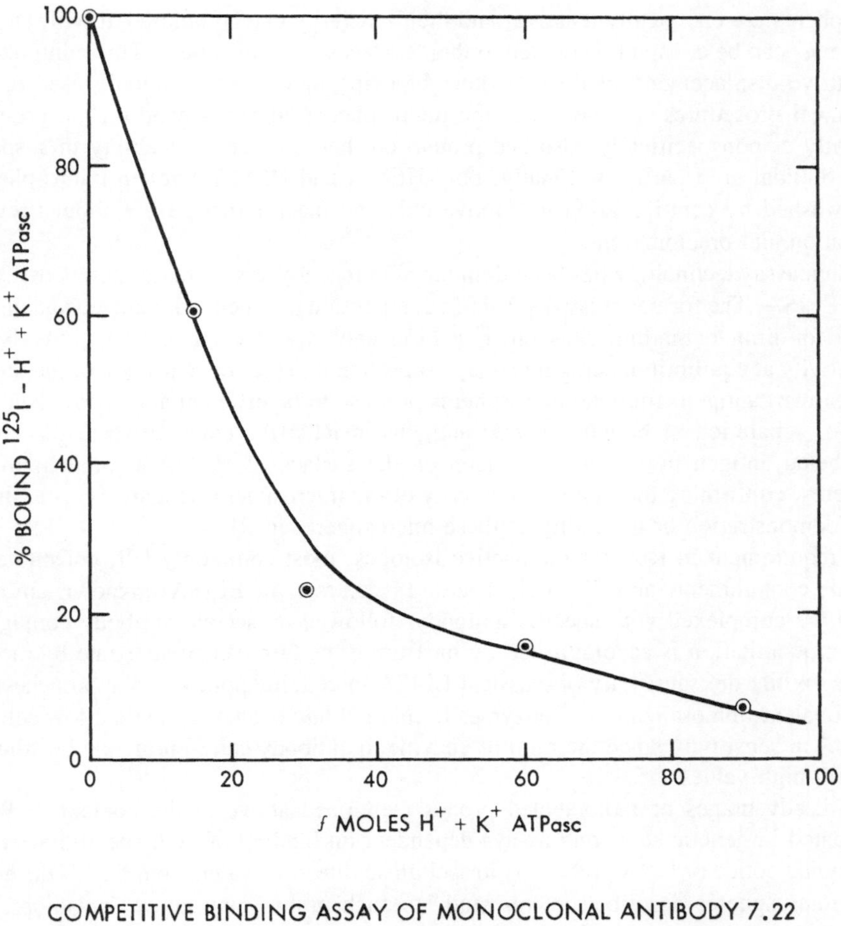

FIGURE 6. Competitive binding of monoclonal antibody 7.22 using HEMA-acrolein immunomicrospheres. HEMA-acrolein microspheres with antibody 7.22 immobilized on their surface were incubated with a fixed amount of ^{125}I-labeled ATPase, and varying amounts of unlabeled ATPase. The retention of ^{125}I counts is shown as a function of the amount of unlabeled ATPase.

activated HEMA microspheres and the HEMA-acrolein microspheres could be caused by interchain steric hindrance between enzyme molecules bound to the same microsphere. Although longer incubation periods probably result in a greater amount of bound enzyme, this is offset by the blocking or distortion of the active site due to the close proximity of an adjacent enzyme molecule.

The potential applications of of immobilized enzymes and antibodies encompass the fields of immunoassay, drug delivery, genetic screening, selective plasmaphoresis, biochemical synthesis, electron microscopy, and cell separation. The use of immunomicrospheres in cell separation has already been described.[3,10] Particulate markers are widely used in electron microscopy to visualize and localize target molecules in cells and tissues. Targets are recognized by virtue of their interaction with antibodies, lectins, substrates or ligands, which are in turn covalently or noncovalently conjugated to relatively electron-dense particles. Ferritin, imposil, hemocyanin, viruses, and most recently colloidal gold, have all been used as markers in transmission and scanning electron microscopy, with varying degrees of success. Colloidal gold with immobilized antibodies has offered outstanding specificity of recognition of cellular antigens, but is susceptible to nonspecific binding of macromolecules, and aggregation induced by centrifugal washing. Since poly-HEMA and HEMA-acrolein microspheres are chemically reactive, and their reactivity can be tailored at will, antibodies or enzymes can be covalently coupled to their surfaces, as shown here. This minimizes their competitive displacement by BSA or other blocking molecules commonly used in immunochemical procedures. Furthermore, the microspheres do not depend on the presence of covalently or nonspecifically adsorbed protein on their surface for stability in suspension, as do colloidal gold particles. Finally, poly-HEMA and HEMA-acrolein microspheres are easily washed by centrifugation to remove unbound macromolecules, without irreversible aggregation and precipitation.

Immunoassay technology has been dominated in recent years by radioimmunoassay (RIA) and by ELISA. The former assay depends on competition between radiolabeled and unlabeled antigen for limited binding sites on a specific antibody, the amount of antibody-bound radioactivity at equilibrium being inversely proportional to the concentration of the unlabeled (or unknown) antigen. Immunomicrospheres promise to be efficient and convenient precipitants for separation of bound and free antigens in RIA: the competitive displacement of radiolabeled antigen by unlabeled antigen on the surface of HEMA-acrolein immunomicrospheres, confirming the biological activity of the microspheres (Figure 6), is at the same time a demonstration of the utility of these microspheres in RIA.

The requirement in RIA for radioactive isotopes, most commonly ^{125}I, constitutes a biological, containment, and disposal hazard. In contrast, in ELISA unknown amounts of antigen are complexed with specific antibody, followed by second antibody conjugated to enzyme; quantitation is accomplished by measuring the turnover of substrate by the bound enzyme. While the sensitivity of classical ELISA in certain applications approaches that of RIA, covalent immobilization of enzymes to immunomicrospheres should allow substantial increases in sensitivity since the ratio of enzyme to antibody can in principle be adjusted to relatively high values.

The disadvantages of radiolabeled probes mentioned above in the context of RIA are exacerbated in genetic screening assays dependent on binding of high specific-activity oligodeoxynucleotide or DNA probes to nitrocellulose filters. Covalent coupling of the enzyme-linked microspheres described in this study to such probes may allow nonisotopic, high-sensitivity detection of cloned DNA sequences.

ACKNOWLEDGMENTS

This report represents one phase of research carried on at the Jet Propulsion Laboratory, California Institute of Technology, under Contract NAS7-918, sponsored by the National Aeronautics and Space Administration, and by Grant No. IR01-CA20668-05 awarded by the National Cancer Institute, DHEW, and Grant No. 70-1859 awarded by the Defense Advanced Research Project Agency.

REFERENCES

1. **Bethell, G. S., Ayers, J. S., and Hancock, W. S.,** *J. Biol. Chem.*, 254(8), 2372, 1979.
2. **Fornusek, L., Vetvicka, V., Kopecek, J., Zidkova, J., and Rossman, P.,** International Union of Pure and Applied Chemistry, 26th Microsymposium on Macromolecules, Prague, 1984.
3. **Kempner, D. H., Smolka, A., and Rembaum, A.,** *Electrophoresis*, 3, 109, 1982.
4. **Rembaum, A., Yen, R. C. K., Kempner, D. H., and Ugelstad, J.,** *J. Immun. Meth.*, 52, 341, 1982.
5. **Rembaum, A., Chang, M., Richards, G., and Li, M.,** *J. Polym. Sci.*, 22, 3609, 1984.
6. **Rembaum, A., Ugelstad, J., Kemstead, J. T., Chang, M., and Richards, G.,** in *Cell Labeling and Separation by Means of Monodisperse Magnetic and Non-Magnetic Microspheres*, Davis, S. S., Illum, L., McVie, J. G., and Tomlison, E., Eds., Elsevier, Amsterdam, 1984.

7. **Saccomani, G., Stewart, H. B., Shaw, D., Lewin, M., and Sachs, G.,** *Biochim. Biophys. Acta,* 465, 311, 1977.
8. **Sachs, G. H.,** *Rev. Physiol. Biochem. Pharmacol.,* 79, 33, 1977.
9. **Seppala, I., Sarvas, H., Peterfy, F., and Mabila, O.,** *Scand. J. Immunol.,* 14, 335, 1981.
10. **Smolka, A., Margel, S., Nerren, B. H., and Rembaum, A.,** *Biochim. Biophys. Acta,* 588, 246, 1979.
11. **Smolka, A., Kempner, D., and Rembaum, A.,** *Electrophoresis,* 3, 300, 1982.
12. **Smolka, A., Helander, H., and Sachs, G.,** *Am. J. Physiol.,* 245, G5899, 1983.
13. **Smolka, A. and Weinstein, W. M.,** *Gastroenterology,* 90(2), 532, 1986.
14. **Treleaven, J. G., Gibson, F. M., Ugelstad, J., Rembaum, A., Philip, T., Caine, G. D., and Kemshead, J. T.,** *Lancet,* 1, 8368, 1984.

Chapter 2

APPLICATION OF MICROSPHERES TO MEASURE CELL SURFACE-ASSOCIATED PROTEOLYTIC ENZYME ACTIVITIES

Z. A. Tökés

TABLE OF CONTENTS

I. INTRODUCTION

Proteolytic enzymes which are restricted to the topography of the cell surface deserve attention since the activity of such enzymes may be crucial for the alteration of external cellular environments, modification of the cell's own surface constituents, and for the selective alteration of adjacent cells. In addition, plasma membrane-bound proteases may be involved in target cell destruction.[1]

The mobility of a membrane-bound enzyme is restricted to the lipid bilayer. Detection of its activity in the external environment of the cell suggests that the active site of the enzyme faces outward. In this position, the enzyme is restricted from attacking the components of the membrane unless the cleavable entities are placed in front of the active site. Figure 1 illustrates the most likely configurations of such a hypothetical enzyme. The active site would also be available to external or membrane-bound inhibitors which may regulate its activity.[2] Such surface-bound enzymes may selectively modify other cells when close contact occurs between cells.

Ordinary assays for proteases are not applicable in measuring selectively the activity of surface-bound enzymes. The development of an assay to measure such an enzyme activity requires two criteria. The first is that the substrate should be accessible only to the outside of the cells and should not be internalized by endocytosis. The second criterion is that the substrate should be available in two configurations, either in direct contact with the cell surface, or adjacent to the cell without contact. In the first configuration, the substrate can be digested by the membrane-bound and released enzymes. In the no-contact configurations only released enzymes can cleave the substrate. The increase in proteolysis from the no-contact to contact configurations measures the available cell surface-associated enzyme activity.

An approach is formulated in this chapter that meets the two criteria outlined above in measuring proteolytic activity available at the cell surface. The application of microspheres allows the substrate to be brought into contact with the surface of viable cells, and conditions are provided whereby microspheres present the substrate to the vicinity of cells without any cellular contact. The substrates are proteinaceous molecules, previously tagged either with radioisotopes or fluorescent markers, and subsequently linked to the microspheres by covalent bonds. The rate of substrate cleavage from the solid support is proportional to the enzyme activity. Illustrations are provided on how the assay can be used to measure enzymatic activity on various cell lines grown as monolayers and on lymphocytes in suspension. The method can be useful to assess the effect of biological events involving membrane perturbations. Furthermore, the approach is not limited to the quantitation of proteolytic enzymes and with modifications, it can be used to measure a wide variety of enzyme activities such as esterases, glycosidases, or glycosyltransferases.

II. OUTLINES OF THE ASSAY

A. Choice of Microspheres as Substrate Carriers

The selection of the best microspheres as substrate carriers for the assay is based on five criteria.

1. A suspension of microspheres should be accurately pipetable. This requires that the beads do not sediment too rapidly. Particles with larger than 50 μm diameters tend to sediment, a condition making it difficult to portion out the substrate evenly to each assay sample.
2. Microspheres should be too large for endocytosis in order to avoid the internalization of the substrate. Beads with diameter larger that 5 μm are excluded by most cells

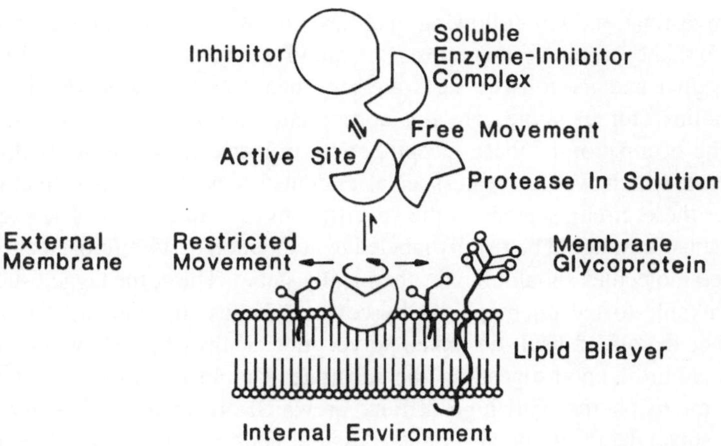

FIGURE 1. A hypothetical cell-surface associated protease is attached to the plasma membrane's lipid bilayer. The active site is facing toward the outside and is available to inhibitors. The lipid bilayer restricts the mobility and the movement of this enzyme and also inhibits the digestion of cell components which are integral parts of the plasma membrane of the same cell. Upon detachment the enzyme may form dimers via the lipophilic portion of the molecule. Serum protease inhibitors, such as α_1-protease inhibitor, or α_1-antichymotrypsin, may compete for the released enzyme in the serum and form inactive, soluble enzyme-inhibitor complexes.

under normal physiological conditions. However, certain cells like macrophages are particularly active in phagocytosis and require special selection of microspheres which are not engulfed.

3. The beads should not be porous, since small molecular-weight peptides produced by enzyme cleavage might become trapped inside their matrix.
4. The microspheres alone should not deliver a false cellular signal when in contact with the cell surface. Polycationic surfaces, for example, may cause an extensive cell-surface rearrangement or agglutination of lymphocytes.
5. The carrier beads should have chemical side groups available for the covalent linkage of the substrate.

The experience summarized in this chapter was generated with 26 μm diameter polystyrene beads chemically modified to introduce amino groups on their surfaces. The procedure of Filippusson and Hornby served for the chemical modification of polystyrene microspheres.[3] Briefly, 2.5 mℓ of Dow latex microspheres with 26 μm diameter (10% solid v/v) was stirred with chilled (0° C) mixture of concentrated H_2SO_4 and concentrated HNO_3 (2:1) for 10 min. The suspension was then centrifuged at approximately 2500 g for 5 min. The acid was removed and the beads were washed three times with chilled-distilled water by centrifugation. The beads may change to a light cream color upon acid treatment, but extensive discoloration should be avoided. For the reduction, 11 g of $SnCl_2$ (Merck, analytical grade) was dissolved in 10 mℓ of concentrated HCl and added to the washed nitrated beads. The suspension was stirred for 1 hr at room temperature. At the end of reduction, the beads were centrifuged and washed successively with 0.1 *N* HCl, H_2O, 0.1 *N* NaOH, and H_2O until the pH returned to neutral. Before each coupling, the microspheres were extensively washed by repeated centrifugation.

B. Choice of Substrates to Measure Proteolytic Activities

The choice of substrate is determined by the type of proteolytic activity and by the cell type under investigation. The following qualities are essential for a molecule to function as substrate. First, the proteinaceous substrate should not give a specific cell-surface signal unless that signal and its consequences are the objectives for the study. For this reason, immunoglobulins, for example, are not appropriate substrates for lymphocytes with Fc receptors. The occupation of these receptors may trigger cellular events such as lysosomal enzyme release which would create special circumstances. Some investigators, however, may consider these circumstances as the specific aim of their studies. The second criterion was that the substrate should be readily labeled by iodination or with fluorescent dye markers, and the tagged molecules should remain chemically stable. Third, the tagged substrate should be readily linkable to the microspheres by covalent bonds. In general, it is preferred that the substrate be devoid of disulfide-linked polypeptide chains which show unusual sensitivity to reduction. Fourth, upon digestion, the substrates should not yield peptides which bind back to the microspheres with high affinity or which are readily taken up by the cells. Finally, the substrate itself, or any of its possible contaminants, should be devoid of all proteolytic enzyme activity. To insure this condition, the practice was adopted to treat all substrates with diisopropylfluorophosphate (DIFP) to irreversibly inhibit any contaminating neutral serine proteases.

Experiments were done and results obtained in our laboratory using as substrates radioiodinated bovine casein, fetuin, immunoglobulins, heavy chains of immunoglobulins, and erythrocyte-derived glycophorin. This review will summarize the data obtained with casein, unless specified otherwise. The following labeling procedure was adapted for the assay.

Ten milligrams of casein (DIFCO Laboratories, Detroit, Mich.) was dissolved in 1.5 mℓ of 8 *M* urea at 40° C and diluted to 10 mℓ with 0.05 *M* phosphate buffer pH 7.2. Twenty microliters of DIFP was added three times at 10-min intervals and the mixture was dialyzed overnight against phosphate-buffered saline (PBS). Aliquots containing 0.5 to 1.0 mg casein were usually labeled with ^{125}I using the lactoperoxidase method.[4] More recently the Iodogen method was adapted. In this procedure, 250 μg of casein was reacted with 200 μg chloroglycouril (Pierce Chemicals, Rockford, Ill.) and 1 mCi Na ^{125}I (Amersham) was added for 15 min at room temperature.[5] The iodinated protein was dialyzed against (PBS) for 24 hr.

The labeled proteins were coupled to the chemically modified microspheres using 1-cyclohexyl-3-(2-morpholinoethyl)-carbodiimide metho-*p*-toluenesulfonate, (CDI) (Aldrich Chemicals).[6] In a typical coupling reaction, 3 to 7 $\times$ 10^6 microspheres were mixed with 50 to 100 μg of labeled casein in 2 mℓ of 0.1 *M* phosphate buffer, pH 7.2. Four milligrams of CDI was added to this reaction mixture, and the reaction was carried out for 20 min at room temperature. This was repeated with an additional 4 mg of CDI while the pH was adjusted to 7.2 with PBS. At the end of incubation, the beads were washed twice with PBS and centrifuged. After the second wash 5 mℓ of 6 *M* guanidine-HCl was added, and the labeled protein-microsphere complexes were allowed to react for 15 min. The mixture was centrifuged, and the top layer, containing the microspheres, was removed, diluted with PBS, and centrifuged repeatedly to remove all guanidine-HCl. This treatment resulted in the complete removal of all loosely bound casein from the microsphere surface. The resulting substrate-bead complexes were stored at 4° C in the presence of 1% sodium azide, and were used within 10 days of preparation after several washings. In general, 5% of the labeled casein remained covalently linked to the microspheres after treatment with the guanidine-HCl. The specific activity ranged from 1.5 to 5.0 dpm per bead. Assuming that the microspheres were perfect spheres, it was estimated that 1 to 3 radioiodinated protein molecules, with an average molecular weight of 50,000 daltons/100 Å^2 were necessary to obtain reliable results. For the assay, 50- to 100-μℓ suspensions of labeled beads were pipeted to each sample containing 5 $\times$ 10^5 to 2.5 $\times$ 10^6 beads/mℓ. No more than 250 to 400 dpm/hr should

FIGURE 2. Diagrammatic illustration of the rolling-beads assay (not drawn to scale). Cells are seeded only on the lower half of the dish. When the bead-substrate complex is introduced directly over the cell surface, contact occurs. In the no-contact position, the dishes are turned, the cells are on the upper portion of the slope, and the microsphere-substrate complex is introduced in the lower section only.[7]

be released from control beads which were incubated without any contact with cells or cell supernatants.

C. Procedures for Studying Monolayers of Cells

A method was designed to grow cells on an adjustable and tilted platform using 35-mm plastic culture dishes (Falcon Plastics, Oxnard, Calif.). One half of the petri dish surface was seeded with cells which were allowed to attach, spread out, and grow while the dish was positioned on the tilted platform. Growth media covered only the lower half of the petri dish surface; the upper half of the surface remained free of cells. Figure 2 illustrates the arrangement.

Three sets of cultures were set up in quadruplicates on the tilted platform. Each sample received 1.9 mℓ of Eagle's minimum essential media (MEM) and 0.1 mℓ suspensions of labeled substrate microspheres. The first set, designated as "control", consisted of previously washed petri dishes with culture medium and microspheres only. Experience showed that three rinses with culture media with 10-min incubations at 37°C between every rinse, significantly reduced artifacts due to surfactants used by manufacturers. In the second set of cultures, the cells remained in the lower half of the dish and the bead-substrate complexes were introduced directly over the viable cell surfaces. These sets of samples were designated as "contact" assay points. For the third set of samples, designated as "no-contact", the culture dishes were turned 180° and placed on the tilted platform in such a way that the cells were on the upper slope, and the bead-substrate complexes were rolled over the cell-free surface area of the dish.[7]

Incubations were carried out in a Hotpack humidified incubator at 37°C in the presence of 5% CO_2/air mixture. In order to maintain motion for the microspheres, a Virtex test-tube shaker was kept operational at a position near the incubator which caused sufficient vibration to gently roll the beads.

In a typical assay, the petri dishes were seeded with 1.0 mℓ of cell suspension. In cases where the cells were allowed to grow and cover half the surface area, approximately 5×10^4 cells per milliliter were added. Growth media containing gentamycin and amphotericin B was changed every second day until a density of 4 to 5×10^5 cells per dish was reached. In other assays, cells were allowed to attach for 12 hr prior to the assay and analyzed without actual growth in the petri dishes. In this case, the percent attachment of each cell line determined the actual number of cells used for seeding. Depending on the cell shape and size, each assay sample contained between 100,000 to 500,000 cells. Attention was given to keep nearly identical number of cells for each quadruplicate sample. Corrections were applied at the end of the assay when the number of cells could be accurately counted after trypsinization. To quantitate the amount of labeled peptides removed from the microspheres after a given length of incubation, an aliquot of 50 to 100 μℓ was withdrawn from each sample carefully to avoid the sedimented beads. The aliquot was mixed with 1 mℓ of ice-cold 1% human serum albumin (DIFP treated) in saline, stirred, and rapidly centrifuged at 2000 × *g* to sediment any microspheres that may have been accidentally removed. Samples of 1 mℓ each were withdrawn from the centrifuge tubes, and the amount of radioactivity was measured in a Packard Autogamma scintillation spectrometer. Each petri dish could be sampled three to four times, after increasing lengths of incubation with the substrate. For the purpose of comparing various cell lines, the proteolytic activity (percent substrate released from microspheres per unit time) for 10^5 cells was established. Greater than 95% viability was necessary during the whole assay period.

D. Procedures for Studying Suspensions of Cells

Lymphocytes were extensively investigated for the availability of surface-associated proteolytic activity. In these studies, suspensions of cells were cosedimented with the substrate-microsphere complexes, and the rates of substrate release were measured in their supernatants. Cosedimentation provided the contact configuration for the assay. In order to assure maximum interaction between microspheres and cells, the samples were gently stirred and resedimented at regular time intervals. For the no-contact configuration, cells were treated in an identical manner in the absence of microspheres or in contact with substrate-free beads, and the supernatants of these samples were assayed subsequently with labeled substrate-bead complexes.[2]

In a typical assay, 2 to 5 × 10^5 cells were incubated in serum-free Medium 199 with 1 to 2 × 10^5 beads carrying 2 to 4 × 10^5 dpm of labeled substrate. Sample volumes ranged from 0.5 to 1.5 mℓ. For controls, cells were replaced with 0.1 mℓ of 1% human or bovine serum albumin (BSA) in PBS, previously treated with DIFP. All incubations were done in triplicate. Cosedimentation was achieved by gentle centrifugation at 100 to 500 × *g*. Microspheres and cells were gently stirred and resedimented at every 30 min to achieve greater interaction. For the no-contact samples, identical incubations were set up without labeled substrate-bead complexes. Supernatants from these samples were withdrawn at various time intervals and incubated with the labeled substrate-microsphere complexes. The amount of substrate cleaved from the microspheres was assessed by withdrawing 50 to 100 μℓ of the supernatant and mixing it with 1 mℓ of albumin solution, followed by centrifugation, and by counting the radioactivity in the bead-free supernatants using a Packard Autogamma spectrometer. Proteolytic activity was expressed as the percent of radioactivity released from the beads per unit time by 10^5 cells.

E. Types of Cells Investigated

More than 20 different cell types were investigated in our laboratory. Some of these cells were further investigated after various treatments of their plasma membranes. This chapter will summarize results obtained with suspensions of erythrocytes, human and murine lymphocytes, and with monolayers of endothelial cells, myocytes, normal and transformed hepatomas, fibroblasts, and epithelial cells.

Erythrocyte preparation — Fresh human venous blood was heparinized and macrophages, granulocytes, some lymphocytes, and platelets were adsorbed onto cotton wool. Erythrocytes and a fraction of lymphocytes were washed off with Hanks' medium. This cell suspension was layered over 8.25% Ficoll (Pharmacia, Uppsala, Sweden), specific gravity 1.077 and centrifuged at 1000 × *g* for 20 min. Erythrocytes packed in the Ficoll layer were collected and washed with PBS. The fractionation through Ficoll was repeated, and the cells were washed free of Ficoll.

Human leukocytes — Suspensions of leukocytes in a serum-gelatin mixture were prepared from normal adult donors as previously described.[8,9] One milliliter aliquots of leukocyte suspension, containing 3.2 × 10^6 mononuclear cells, were incubated overnight in glass culture tubes. After rinsing the tubes with Medium 199, 3.2 × 10^5 adherent cells remained. These cells were designated either as macrophages or as glass-adherent leukocytes (GAL), and were assayed in the same tubes with labeled substrate-microsphere complexes. Part of the serum-gelatin leukocyte suspension was used for the preparation of T-enriched lymphocytes (TEL). Red cells, thrombocytes, and polymorphonuclear leukocytes were removed by centrifugation on a step gradient of Ficoll-Paque (Pharmacia, Piscataway, N.J.). Adherent and phagocytic cells were then removed by a magnet after carbonyl-ion ingestion and by adherence to glass bottles and to Sephadex G-25. A portion of the cells purified by a Ficoll-Paque gradient were spared further purification and were designated as unfractionated leukocytes.

Murine lymphocytes — Rat (DA strain) lymph node cells were obtained from axillary, brachial, inguinal, and mesenteric nodes. The nodes were removed and placed in Hank's medium at room temperature. Single-cell suspension was prepared under sterile conditions by teasing the nodes with forceps and Dispo pipettes. The cells were washed by repeated sedimentation. These preparations were 99% viable as estimated by trypan-blue exclusion and contained less than 0.1% granulocytes. Over 80% of these cells were capable of transforming to blast cells by pokeweed mitogen. Lymph node cells from Balb/C mice were prepared similarly. Spleen lymphocytes were prepared by removing a fresh spleen and gently teasing it in cold Medium 199 containing 1% antibiotic-antimycotic mixture and 2% heat-inactivated fetal bovine serum (FBS). The single-cell suspension was layered on Ficoll-Urovision, density 1.077, and centrifuged for 15 min at 1500 rpm in MSE-Mistral Centrifuge. The cell layer at interface was recovered and diluted twofold with cold Medium 199. The cells were washed free of Ficoll by repeated centrifugation. Viability was 95 to 99%. These preparations usually contained 2 to 8% granulocytes.[10]

Endothelial cells — Primary calf aorta endothelial cells were obtained according to the method of Eisenstein.[11] Cells, identified as endothelial by the presence of Weibel-Palade bodies, were grown and subcultured in RPMI 1640 medium supplemented with 10 to 20% FBS and 20 m*M* Hepes buffer. Cells were plated on 35-mm plastic culture dishes, using 3 × 10^4 cells per milliliter. Cultures from the second, fourth, and sixth passages were used for the protease assay.[7]

Rat liver cell lines — Investigation included three different cell lines and their various passages. Detailed descriptions of these cells have been published.[12] Cells designated as RL-34, maintained typical epithelial morphology and had diploid number of chromosomes. These cells have not acquired tumorigenicity, as tested by inoculation into immune-suppressed host animals and did not grow in colonies in soft agar culture. With these characteristics the cells resembled normal cells even though they adapted to cell-culture conditions. The second cell line was derived from RL-34 after carcinogen (4-nitroquinoline 1-oxide, 4NQO) treatment. These cells were investigated using five different passages. Between passages 51 to 88, they began to express tumorigenicity. Chromosome analysis revealed a gradual shift toward tetraploid range. Cells formed aggregates, three-dimensional clusters, and pile ups. The third cell line with tetraploid chromosomes, RL-T, was spontaneously transformed and expressed highly tumorigenic properties.

Quail embryo fibroblasts and tumor cell lines — Carcinogen-induced quail tumor cell lines were obtained from Dr. Peter Vogt, University of Southern California. Their characterization has been published earlier.[13] Investigation included five cell lines from carcinogen-induced tumors with different tumorigenic properties. For normal cells, chick or quail embryo fibroblasts in their early passages were used.

Human breast and prostate adenocarcinoma cells — The human breast adenocarcinoma cell line MCF-7, passages 6 to 14 and 78 to 88, and HBL-100 were obtained from Dr.

David Kingsbury at Naval Biosciences Laboratory, Oakland, and from the Mason Research Institute, Worcester, Mass. Prostate carcinoma cell line PC-3 was established from a poorly differentiated human prostate adenocarcinoma which had metastasized to bone.[14]

Myocyte cultures — Primary cultures of neonatal rat heart cells were established according to previously published techniques.[15] These nondividing cells were studied 48 hr after attachment to the plastic petri dishes. Nonattached cells were discarded after 6, 12, and 24 hr of incubation.

Murine embryo fibroblasts — The C3H/10T-1/2 mouse-embryo fibroblast cells can be oncogenically transformed with a variety of carcinogenic chemicals and radiations. The establishment of these cells and relevant culturing techniques have been published.[16] MCA is a cell line derived from these embryo fibroblasts after transformation by chemical carcinogenesis.

F. Sensitivity of the Assay

Soluble trypsin was used to gauge the sensitivity of the assay. Porcine trypsin (Worthington) was dissolved in cold MEM cell culture media. Incubations were set up consisting of 0.1 mℓ substrate-microsphere suspension, 0.1 mℓ enzyme solution of varying concentrations, and 0.8 mℓ of MEM. The samples were incubated at 37°C with gentle agitation, and at various time intervals, 0.1-mℓ aliquots were removed. These sample aliquots were mixed with 1.0 mℓ of 1% human serum albumin at 0°C, rapidly centrifuged, and then 0.5 mℓ of the supernatnat, free of microspheres, was counted in a Packard autogamma spectrometer. In a typical assay 100,000 microspheres were used to which 200,000 dpm of ^{125}I-casein were covalently attached. The lowest amount of detectable enzyme activity was 5 ng of trypsin which released approximately 0.5% of the substrate per hour. The amount of substrate released was proportional to the length of incubation and to the amount of the enzyme until 15 to 20% of casein was removed. Consequently, most assays were set up to result in a substrate release of 1 to 10% during 90-min incubations, an amount which represents approximately 10 to 200 ng of soluble trypsin-equivalent activity. Since considerable variations were obtained with different batches of substrate-microsphere complexes, it became essential to use an identical batch when different sets of cells were compared.

The amount of substrate released from the microspheres was proportional to the cell number. For the investigation of each cell type, it was essential to establish the ratio of microspheres to cells which provides sufficient excess of substrate for the available enzyme activity and assures an increase in the amount of digested substrate in proportion to time and cell number. For cells in suspension, the assay in general required a minimum of 50,000 cells. This number of human peripheral lymphocytes, depleted of macrophages and granulocytes, released 0.5% of the substrate from 50,000 microspheres in 60 min. Doubling the cell number resulted in 1.0% release; 200,000 cells released 2.0% of the substrate. Further increase in cell number failed to result in proportional increase of substrate release due to insufficient contact with microspheres. When 75,000 lymphocytes were incubated with equal number, or with 1.5×10^5 and 3×10^5 microspheres, approximately the same amount of ^{125}I-peptides were released in 60 min (4200 ± 850 dpm). This indicates that the substrate was present in sufficient excess to reveal the maximum enzyme velocity under the assay conditions.

The sensitivity of the assay with monolayers of cells was similar to the one with suspensions of cells. The lowest number of cells assayable was, in general, 40,000. This number of cells shows less than 50% confluency over the one-half surface area of the petri dish. Results with limiting number of cells per assay varied somewhat as a function of cell size and surface area. Most reliable values for the rate of substrate release were obtained with a range of 50,000 to 400,000 cells incubated with 75,000 to 325,000 microspheres. For example, linear increase in the rate of substrate removal during a 120-min incubation was obtained

Table 1
RELEASE OF ^{125}I-PEPTIDES FROM MICROSPHERES BY RAT LIVER EPITHELIAL CELLS

Cell lines	Contact	No-contact	Increased release due to contact
90-Min Incubation			
RL-34	2320 ± 250	1930 ± 220	390
RL-4NQO	3380 ± 405	1780 ± 292	1600
RL-T	5760 ± 440	4860 ± 430	900
120-Min Incubation			
RL-34	2480 ± 320	2020 ± 230	460
RL-4NQO	4090 ± 360	1890 ± 240	2100
RL-T	7110 ± 310	5210 ± 470	1900

Note: All assay points were performed in triplicate and the whole assay was repeated four times with similar results. These results are expressed as dpm released and were obtained by incubating 400,000 cells with 150,000 microspheres to which 377,000 dpm ^{125}I-casein was attached.

in the range of 50,000 to 200,000 cells using rat liver epithelial or calf aorta endothelial cells. The first statistically significant increase in substrate release occurred in 15- to 30-min incubation.

III. RESULTS WITH MONOLAYERS OF CELLS

A. Rat Liver Epithelial Cells

The initial objective of the study was to compare proteolytic activity of closely related cell lines exhibiting different growing characteristics and stages of malignant transformations. In the first sets of experiments three rat liver epithelial cell lines were investigated.[12] Results are summarized in Table 1.

RL-34 cells exhibited normal cellular growth patterns. These cells grew to confluency and did not pile up on each other. They failed to grow in soft agar and to develop tumor in the immune-suppressed hosts. The lowest rate of peptide release in contact configuration was observed with these cells. Both the transformed, RL-T, and the carcinogen-treated RL-4NQO cells released significantly higher amounts of peptides from microspheres. However, when the released enzyme activity was assayed in no-contact configurations, they were indistinguishable from the RL-4NQO cells (passages 51 to 55). Only the RL-T cells showed consistently higher activity. After passages 79 and 85 of RL-4NQO cells, a significant increase in proteolysis was observed both in contact and no-contact configurations. Cells from these passages began to show low levels of tumor induction in immune-suppressed hosts.

Both RL-T and RL-4NQO cells showed increased substrate digestion (as compared to RL-34 cells) due to contact. The increase due to contact with RL-34 cells was statistically significant since similar values were obtained in four out of four experiments. Contact-related proteolysis was particularly significant with the RL-T cell line since these cells tend to pile up on each other, and their surface area available for interaction with microspheres is lower.

B. Quail Embryo Fibroblasts and Fibrosarcoma Cell Lines

In order to further test closely related normal and transformed cell lines with varying

Table 2
RELEASE OF ^{125}I-PEPTIDES FROM MICROSPHERES BY QUAIL FIBROSARCOMA CELL LINES (IN DPM)

Cell lines	Contact	No-contact	Increased release due to contact
60-Min Incubation			
QT- 2	1320	1050	270
QT- 4	1320	1010	310
QT- 7	1290	1005	285
QT-29	1170	1050	120
QT-46	1350	1160	190
QEF	805	710	95
120-Min Incubation			
QT- 2	2060	1210	850
QT- 4	2180	1240	940
QT- 7	1990	1180	810
QT-29	1825	1570	255
QT-46	2060	1740	320
QEF	1240	995	245
180-Min Incubation			
QT- 2	3420	1690	1730
QT- 4	3750	1830	1920
QT- 7	3310	1760	1550
QT-29	2300	1820	480
QT-46	2520	1930	590
QEF	1670	1205	465

Note: Each petri dish contained between 320,000 to 440,000 cells and 150,000 microspheres carrying 175,000 dpm of ^{125}I-casein. The values were adjusted to indicate the release rate by 100,000 cells. These values are the average from three samples. The average standard deviation was ±7% of the indicated values.

growth properties, five fibrosarcoma and a primary quail embryo fibroblast (QEF) cell lines were examined for proteolytic activity. The fibrosarcomas (QT) were derived from carcinogen-induced tumors in Japanese quail and have been extensively characterized previously.[13] They exhibit different morphological growth patterns and shapes. QT-2 and QT-7 had round shapes with a saturation density of 1.4 and 1.5 × 10^3 cells per square milliliter. QT-29 cells were shaped polygonal with saturation density of 1.1 × 10^3 cells per square milliliter, and QT-4 and QT-46 cells were fibroblastic with 1.4 and 2.7 × 10^3 cells per square milliliter, respectively. Normal QEF cells were fibroblastic with a saturation density of 0.5 × 10^3 cells per square milliliter. Two cell lines did not require trypsin for transfer (QT-2 and QT-7) indicating a loose attachment to plastic dishes. All the other cells had to be trypsin treated for detachment and subsequent passaging.

Four experiments were performed with quail cells. The results of a representative study is summarized in Table 2. The lowest activity in contact and no-contact configurations was obtained with the normal QET cells. All fibrosarcoma cells showed higher activity, although the activities of QT-29 and QT-46 cells closely resembled that of QET cells. The highest activity consistently observed was with QT-4 cells followed by QT-2 and QT-7 cells.

Tumorigenicity, as determined by the number of developing tumors in animals after inoculating them with 500,000 cells,[13] did not correlate with proteolytic activity. The most tumorigenic cells, QT-2 and QT-29, which were capable of inducing tumors in 87% and 82% of the host animals, had medium-high and low proteolytic activity. Conversely, the lowest tumorigenic cells, QT-4 (30% tumor induction), had the highest proteolytic activity.

Table 3
RELEASE OF ^{125}I-PEPTIDES FROM MICROSPHERES BY 10T-1/2 MURINE FIBROBLASTS AND THEIR TRANSFORMED CELLS, MCA

Cell line	Contact	No-contact	Increased release due to contact
90-Min Incubation			
10T-1/2	1660 ± 66	1720 ± 73	0
MCA	1760 ± 129	1680 ± 98	80
180-Min Incubation			
10T-1/2	2580 ± 168	2270 ± 186	310
MCA	2870 ± 104	2340 ± 92	530
360-Min Incubation			
10T-1/2	3480 ± 159	3060 ± 170	420
MCA	5480 ± 270	4010 ± 182	1470

Note: The results are for 400,000 cells at 90% confluency incubated with 100,000 microspheres to which 250,000 dpm ^{125}I-casein was coupled. The results are expressed as dpm released.

Cell shape did not seem to have an effect since QT-4, QT-46, and QET cells were all predominantly fibroblastic and had vastly different enzyme activities. Similarly, no correlation was found between the saturation growth density of cells per unit area and the observed enzyme activities.

Cell lines were ranked according to their metabolic activities as determined by hexose uptake.[13] The following ranking was obtained using the uptake of isotope-labeled deoxy-D-glucose: QT-4, QT-2, QT-46, QT-29, and QEF. There appears to be a correlation between hexose uptake and proteolysis which suggests that this relationship deserves further detailed investigation.

In separate sets of experiments, rat liver epithelial cells were compared with quail fibrosarcoma cell lines using the same batch of ^{125}I-casein-microsphere complexes. RL-T and RL-4NQO cells had comparable proteolytic activity to QT-2 and QT-7 cells. RL-34 cells had an only slightly higher rate of (60 to 80%) substrate digestion than QEF cells. It was interesting to observe, however, that increased adaptation to cell culture conditions by QEF cells resulted in progressive increase proteolysis. Thus, after 3 to 5 passages of the primary QEF cells, a 30% increase was observed, and after 7 to 9 passages, 50% higher activity was noted. These cells from later passages became indistinguishable from the RL-34 cells in terms of their proteolytic activities.

C. Murine Embryo Fibroblasts, C3H/10T-1/2

The usefulness of these cells for the study of in vitro carcinogenesis and differentiation is well established. A comparison was made between the MCA cells, transformed by chemical carcinogen, and the parental cells, 10T-1/2. The results are summarized in Table 3. Although during the first 30 min of incubation substantial peptide release (0.7%) was observed both under contact and no-contact assay configurations, no significant increase was detectable due to contact. Only after 180- and 360-min incubations did the activity increase with contact, and the MCA cells began to show significantly higher activity (Table 3). The activity of 10T-1/2 cells was similar to the QEF and RL-34 cells with the exception of the released enzyme activity which appeared to be higher. Further examination of enzymatic changes of these cells during the in vitro process of transformation and differentiation would be of considerable interest.

Table 4
RELEASE OF ^{125}I-PEPTIDES FROM GLYCOPHORIN-MICROSPHERES BY HUMAN BREAST ADENOCARCINOMA CELLS

Cell line	Contact	No-contact
60-min incubation		
MCF-7	2850 ± 200	2550 ± 180
HBL-100	2600 ± 210	3250 ± 250
180-min incubation		
MCF-7	3550 ± 190	3650 ± 240
HBL-100	2900 ± 165	3850 ± 210
240-min incubation		
MCF-7	3750 ± 220	4400 ± 280
HBL-100	3200 ± 150	4550 ± 240

Note: 400,000 cells per dish were assayed. Each dish received 150,000 microspheres coupled to 1.9 × 10^6 dpm of labeled glycoprotein. Specific activity of ^{125}I-glycophorin was 4.5 × 10^6 dpm/μg. Results are expressed as dpm released per 100,000 cells.

D. Human Breast and Prostate Adenocarcinoma Cells

Three human adenocarcinoma cells were investigated with the microsphere assay. All of them had higher proteolytic activity in contact and no-contact assay configurations than the QET cells. The most activity was shown by HBL-100 cells. MCF-7 and PC-3 cells were as active as bovine endothelial cells assayed at 70% confluency.

HBL-100 and MCF-7 cells were further tested with microspheres carrying ^{125}I-labeled glycophorin. The question was raised whether integral membrane components are also substrates for the cell-surface associated proteolytic activity. Glycophorin is a major erythrocyte membrane constituent which contains a high number of negatively charged sialic acid residues. This glycoprotein was purified to homogeneity from human erythrocyte stroma using lithium diiodosalicylate.[17] It was treated with DIFP and iodinated the same way as casein. The ^{125}I-labeled glycophorin was coupled to the microspheres and its rate of digestion while in contact or without contact with cells was measured. The results are summarized in Table 4. The unique finding is that, at all times, the amount of peptides released in no-contact configuration was indistinguishable or higher than in-contact configuration. Three interpretations can be proposed. The first one is the possibility that different types of enzymes are assayed for in-contact and no-contact positions. Thus, the enzymes retained at the cell surface are less effective in cleaving glycophorin than the released enzyme.

The second possibility is that the glycophorin molecules, which are covered with sialic acid residues, hinder the interaction with a potential enzyme-active site, since the cell surface is covered with numerous glycoproteins. Such restrictions would not exist in solution or in the no-contact configuration. The third interpretation is that the presence of membrane-bound proteinase inhibitors selectively inhibit those proteases which would cleave glycophorin at the cell surface. If these inhibitors are retained at the plasma membrane, and the enzymes are secreted in an uninhibited form, a higher rate of proteolysis would be observed in no-contact configuration. The existence of a membrane-bound and active proteinase inhibitor,

Table 5
RELEASE OF ^{125}I-LABELED PEPTIDES FROM MICROSPHERES BY CALF AORTA ENDOTHELIAL CELLS

Incubation time (min)	Contact	No-contact	Increased release due to contact
60	1870 ± 160	1420 ± 51	450
180	3100 ± 140	2110 ± 95	990
240	4650 ± 186	2550 ± 140	2100

Note: Experiments were performed with 400,000 cells incubated with 150,000 microspheres carrying 110,000 dpm of ^{125}I-casein. The cells were from their third passage at 70 to 80% confluency. Results are expressed as dpm released per 100,000 cells. The results are expressed as dpm released.

alpha-1-antichymotrypsin, on MCF-7 cells has been reported recently.[18] Deciphering the three alternatives presents an enticing challenge.

E. Endothelial Cells

Endothelial cells serve as a model for regulated tissue invasion. During wound healing these cells are capable of invading a given tissue in a similar manner to malignant cells. However, their migration is controlled and stops with the completion of the microvasculature. Therefore, it was of interest to examine these cells for their proteolytic activity. The results, using casein as substrate, are summarized in Table 5. Endothelial cells exhibited levels of proteolytic activity which are comparable to those obtained with malignant cells MCA, RL-T, and QT-2. The activity was due to enzymes which were produced by the endothelial cells and not absorbed from the fetal bovine serum, since the cells had the same levels of proteolytic activity after passage in media supplemented with 5% protease-free BSA. These findings established that normal cells, capable of controlled tissue invasion, possess similar proteolytic capabilities to some of the transformed malignant cell lines.[7]

IV. RESULTS WITH CELL SUSPENSIONS

The usefulness of microsphere-substrate complexes to detect contact mediated proteolysis can be illustrated with lymphocytes. The beads cosediment with lymphocytes during gentle centrifugation. Supernatants from incubations of unlabeled casein microspheres with lymphocytes can be analyzed for the released enzyme activity (no-contact configuration). Typical results obtained with various subpopulations of human peripheral leukocytes and murine lymphocytes are summarized in Table 6.

Erythrocytes had no detectable proteolytic activity while remaining intact. Upon lysis (not illustrated here), however, a limited proteolytic activity was detectable. The next lowest activity was observed with purified and unstimulated (resting) T lymphocytes both in contact and no-contact configurations. It is not certain whether the detected proteolysis comes from T lymphocytes or from a very small number of contaminating granulocytes. The highest activity was observed with unfractionated murine spleen cells. If granulocytes and macrophages were removed, however, the remaining cells behaved similarly to lymph node cells, which indicate that the majority of peptidase activity was produced by granulocytes. Macrophages were always higher than B-enriched lymphocytes but not as active as granulocytes (not illustrated here). Granulocytes had such high activity, that, under similar assay conditions, 48% of the ^{125}I-casein would be released from the microspheres.[19] Proteolytic activity

Table 6
RELEASE OF ^{125}I-PEPTIDES FROM CASEIN MICROSPHERES BY HUMAN PERIPHERAL LEUKOCYTES AND MURINE LYMPHOCYTES DURING INCUBATIONS FOR 100 MIN

Cell types	Contact	No-contact	Increased release due to contact
No cells, beads only	710 ± 140	—	—
Human			
Erythrocytes	755 ± 90	720 ± 110	35
Leukocytes	2150 ± 320	810 ± 100	1340
T-Enriched lymphocytes	1200 ± 210	790 ± 110	410
Macrophages	3400 ± 650	910 ± 130	2490
Murine			
Spleen cells	16000 ± 800	9800 ± 750	6200
B-Enriched lymphocytes	2600 ± 280	1100 ± 140	1500
Lymph node cells	1950 ± 210	960 ± 90	990
T-Enriched lymphocytes	940 ± 130	860 ± 110	80

Note: The incubations were carried out in serum-free Medium 199 with 300,000 cells and using 100,000 beads carrying 160,000 dpm of labeled casein. All samples were run in triplicate. Glass-adherent cells are designated as macrophages. The results are expressed as dpm released.

in the supernatants was 95% inhibited by 1% FBS and DIFP.[10] However, in contact configuration these inhibitions were only 50%.

Investigations of subclasses of lymphocytes require a careful selection of substrates. Cytotoxic lymphocytes or natural killer cells may not degrade casein and the assay may not detect the newly discovered T cell-specific proteases present in these cells.[20]

V. MEMBRANE PERTURBATIONS AND THEIR EFFECT ON ENZYME ACTIVITIES

The microsphere assay is particularly revealing when it is used to monitor events at the cell surface. Signals for proliferation and differentiation can be introduced during the assay and the consequent changes in enzyme activity are measured simultaneously. To illustrate this application, the results of three lymphocyte treatments are summarized. These treatments include two mitogens, lipopolysaccharides (LPS) and conconavalin A (ConA), and the oxidative signal from $NaIO_4$.[8]

LPS is a polyclonal B-lymphocyte mitogen which binds to the plasma membrane and subsequently leads to B-cell proliferation and maturation. It has no similar effects on T lymphocytes. Lymph nodes are rich in T lymphocytes and also contain glass-adherent macrophages. Lymph node cells respond to ConA as demonstrated by increased thymidine uptake followed by T-cell proliferation.[2] The lack of response of purified T lymphocytes, suggests the involvement of adherent cells in the mitogen-stimulated signal process. To highlight the utility of the microsphere enzyme assay, results of mitogen-stimulated proteolytic activity are illustrated. In Figure 3, the rate of peptide release is shown by the rate of lymph node cells released after treatment by ConA and LPS. Cells were incubated simultaneously for 100 min with varying concentrations of mitogen and ^{125}I-casein microspheres in contact configuration. The amount of labeled-peptide release did not change with LPS. With ConA, significant increase was seen already at 1 μg/mℓ, and at 50 μg/mℓ the proteolytic activity was doubled. When the glass-adherent cells were removed from the lymph node cells, no significant stimulation was detectable (not illustrated).

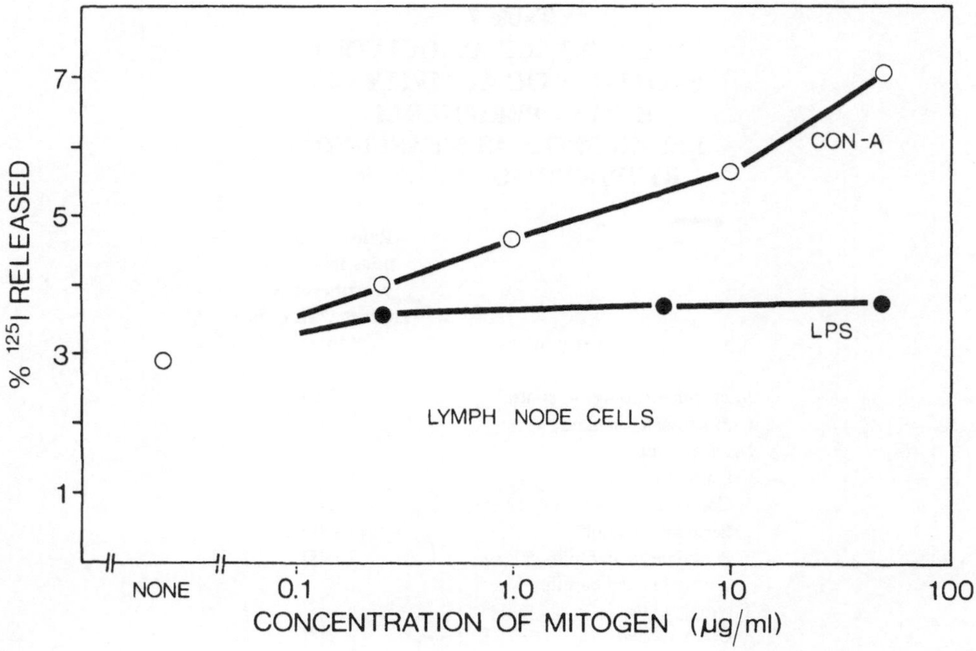

FIGURE 3. Peptide release by rat lymph node cells treated with mitogens is shown. Duplicate samples of 300,000 cells and 100,000 microspheres carrying 200,000 dpm of ^{125}I-labeled casein were incubated at 37°C for 100 min. Increasing concentrations of mitogens were given and the cells were incubated with the beads in contact configuration.

Proteolytic activities of human peripheral lymphocytes after mitogenic stimulation are summarized in Table 7. Untreated macrophages demonstrate substantial activity which can be further stimulated (60 to 70%) by ConA. The largest increase (fivefold) was observed after periodate treatment resulting in aldehyde residues capable of cross-linking by Schiffs base condensation to other membrane components possessing amino groups. Similar activities were obtained when neuraminidase was used to remove terminal sialic acid residues followed by a galactose-oxidase treatment of the exposed galactose residues. This oxidase treatment also resulted in aldehyde groups which cross-link with other membrane constituents (not illustrated).

In one set of experiments, periodate-treated macrophages were allowed to come into contact with the bead-substrate complex and 50 μg of unlabeled casein was introduced per assay. This amount of unlabeled substrate represented approximately a 1000-fold increase in substrate concentration, and in soluble enzyme-substrate assays, it would create a substrate competition whereby the amount of labeled casein digested would be significantly diminished. As indicated in Table 7, only 15% decrease in digestion was observed, signifying that soluble molecules may be excluded during the intimate interaction between microspheres and cells. Plasma proteinase-inhibitors and "innocent bystanding molecules" in solution may also be excluded in surface-to-surface proteolytic cleavage. The biological relevance of this reaction can be illustrated by replacing the microsphere-casein complexes with cell surface-iodinated syngeneic T lymphocytes. Under identical assay conditions, in contact with macrophages, 75% of labeled surface components were removed from the T cells in 100 min.[2] The addition of 1% serum did not decrease this proteolytic surface modification (unpublished observation). Thus, the microspheres were excellent models for measuring events that take place during cell-cell interactions. Furthermore, incubation of T lymphocytes in contact with macrophages led to increased ^{3}H-thymidine incorporation, signifying that a normal signal for proliferation was initiated.

Table 7
CELL-SURFACE ASSOCIATED PROTEOLYTIC ACTIVITY OF HUMAN PERIPHERAL LEUKOCYTES AS MEASURED BY ^{125}I-PEPTIDE RELEASE

Cell types and treatment	Released peptides from microspheres in contact with cells (dpm)
Microspheres only — control	740
Unfractionated leukocytes	2100
Macrophages	
Untreated	3300
ConA treated	5500
Periodate treated	18600
Periodate treated plus 50 μg of unlabeled casein	15800
T lymphocytes	
ConA treated	800
LPS treated	750

Note: Each sample contained 300,000 cells and 100,000 beads carrying 220,000 dpm of ^{125}I-labeled casein. Incubations were carried out in duplicate at 37°C for 100 min. Periodate treatment was performed prior to adding the beads as described previously.[8]

VI. DISCUSSION

The purpose of this chapter is to summarize the experience obtained with the rolling-microsphere assay and to illustrate its usefulness to measure enzyme activity at the cell surface. The term "cell surface-associated proteolytic activity" is used instead of "membrane-bound activity" to emphasize that the assay does not prove the existence of an enzyme which is an integral part of the plasma membrane. Soluble lysosomal enzymes could be released at cell surface resulting in a transient higher concentration at the vicinity of microspheres during contact configuration. However, such an enzyme would readily compete for soluble substrates and would be inhibited by plasma protease inhibitors. Certainly, these alternatives can be tested with the assay. Evidence for membrane-bound enzymes is now available for a large variety of cells, including macrophages, endothelial cells (plasminogen activator), neurons, and fibrosarcomas. The rolling-beads assay provides data consistent with membrane-bound enzymes. In combination with a variety of specific substrates and inhibitors, it helps to reveal the biological significance of these enzymes.

The assay is carried out under nontoxic conditions. None of the cells studied with microspheres lost their viability during incubations as determined with labeled chromium release. Furthermore, the plain microspheres alone did not induce or stimulate the surface proteolytic activity. This was demonstrated by pre-incubating monolayers of cells with increasing numbers of modified polystyrene beads, removing them, and re-introducing the substrate-microsphere complexes to the cells. In no case was there any detectable change in proteolysis due to treatment with plain microspheres.

The method will be particularly useful to assay protease inhibitors which bind to the cell surface. For example, alpha-1-antichymotrypsin, an active inhibitor of cathepsin G is a major glycoprotein synthesized by normal human breast epithelial cells and by a number of established adenocarcinoma cells. Immunofluorescence reveals the presence of this inhibitor on the surface of the cells.[18] Its biological function is not yet elucidated, but with the application of this assay one may probe its effectiveness of inhibiting a variety of enzymes introduced from the outside to the cells. The availability of these inhibitors as a function of cell cycle, cell density, supporting matrix, and membrane turnover can now be investigated.

Microspheres are also adaptable to study other enzymes which may play a role in cell-cell interactions. Synthetic peptides carrying labeled carbohydrate side chains can be used as substrates for glycosidase enzymes. The release of labeled phosphorylated or sulfated entities can be measured as well as the activity of RNAases and DNAases. Another approach for measuring surface enzymes would utilize the microspheres as acceptors. For example, we are investigating sialyl-transferase activity by incubating cells with labeled CMP-sialic acid and microspheres carrying covalently linked desialylated fetuin as an acceptor for the activated sialyl residue. Alternatively, a substrate and a second type of molecule capable of delivering a signal for proliferation or maturation may be introduced to the cells on the same microsphere. The effect of the signal molecule on the digestion of different substrates can thus be measured. All of these combinations offer enticing opportunities for cell biology.

REFERENCES

1. **DiStefano, J. F.,** *Cancer Res.,* 46, 1114, 1986.
2. **Tökés, Z. A.,** Cell surface-associated protease activity in lymphocyte interactions. The Immune System, Vol. 2, Steinberg, C. M. and Lefkovits, I., S. Karger, Basel, 1981, 249.
3. **Filippusson, H. and Hornby, W. H.,** *Biochem. J.,* 120, 215, 1970.
4. **Marchalonis, J. J., Cone, R. E., and Santer, V.,** *Biochem. J.,* 124, 921, 1971.
5. **Fraker, P. J. and Speck, J. C.,** *Biochem. Biophys. Res. Commun.,* 80, 429, 1978.
6. **Tökés, Z. A. and Chambers, S. M.,** *Biochem. Biophys. Acta,* 389, 325, 1975.
7. **Tökés, Z. A. and Sorgente, N.,** *Biochem. Biophys. Res. Commun.,* 73, 965, 1976.
8. **Tökés, Z. A., Bruszewski, W. B., and O'Brien, R. L.,** *Birth Defects: Original Article Series,* Vol. XIV, (No. 2), 1978, 195.
9. **Parker, J. W. and Lukes, R. T.,** *Am. J. Clin. Pathol.,* 56, 174, 1971.
10. **Tökés, Z. A. and Kiefer, H.,** *J. Supramol. Struct.,* 4, 507, 1976.
11. **Eisenstein, R., Kuettner, K. E., Neopolitan, C., Soble, L. W., and Sorgente, N.,** *Am. J. Pathol.,* 81, 337, 1975.
12. **Tökés, Z. A., Sorgente, N., and Okigaki, T.,** *Prog. Clin. Biol. Res.,* 17, 615, 1977.
13. **Moscovici, C., Moscovici, G., Jimenez, H., Lai, M. M. C., Hayman, M. T., and Vogt, P. K.,** *Cell,* 11, 95, 1977.
14. **Kaighn, M. E., Narayan, S., Ohnuki, K., Lechner, J. F., and Jones, L. W.,** *Invest. Urol.,* 17, 16, 1979.
15. **Simpson, P. and Savion, S.,** *Circ. Res.,* 50, 101, 1982.
16. **Mondal, S., Brankow, D. W., and Heidelberger, C.,** *Cancer Res.,* 36, 2254, 1976.
17. **Marchesi, V. T. and Andrews, E. P.,** *Science,* 174, 1247, 1971.
18. **Gendler, S. J. and Tökés, Z. A.,** *Biochim. Biophys. Acta,* 882, 242, 1986.
19. **Trinchieri, G., Bauman, P., DeMarchi, M., and Tökés, Z. A.,** *J. Immunol.,* 115, 249, 1975.
20. **Gershenfeld, H. K. and Weissman, I. L.,** *Science,* 232, 854, 1986.

Chapter 3

MAGNETIC FREE-FLOW IMMUNOISOLATION SYSTEM DESIGNED FOR SUBCELLULAR FRACTIONATION

Kathryn E. Howell, Wilhelm Ansorge, and Jean Gruenberg

TABLE OF CONTENTS

I. INTRODUCTION

Immunoreagents have become increasingly valuable for cell and molecular biology because of their specificity and flexibility. They are providing a means of addressing a number of problems that have been experimentally inaccessible. One very important technology, that of sorting either cells[1-7] or subcellular compartments[8-14] is now being approached using the selectivity and sensitivity of antibodies. Parallel advances have been made recently in the field of polymer chemistry so that optimal solid supports are being produced for coupling of the antibodies.[15-20] These fields, immunochemistry and polymer chemistry, are being combined to produce separation systems effective in isolation of both cells and subcellular compartments.

We present in this chapter our efforts to develop an immunoisolation system for organelle fractionation. The antibody provides the specificity of the isolation as it will recognize its antigen present on the exposed surface of the compartment to be isolated. The unique feature of our system is that the immunoisolation is carried out in free flow. A magnetic solid support to which specific antibodies are attached, is maintained in suspension within a magnetic field. The advantages of the free-flow system are low contamination by nonspecific components and limited stress applied to the cellular material due to the gentle washing conditions in the free-flow chamber.

After homogenization of a cell many of the organelles are in the form of membrane vesicles. For example, vesicles derived from the plasma membrane, the Golgi complex, the endoplasmic reticulum, and the endosomal system are the major components of what is termed a microsomal fraction. Particulate organelles, like the nucleus, mitochondria, lysosomes, and peroxisomes usually remain intact after homogenization. However, even under ideal circumstances, these organelles vesiculate to some extent. In earlier immunoisolation experiments, membrane vesicles bound to a solid support were washed by pelleting and resuspension which resulted in further homogenization of the isolated vesicles. Thus, much of the specifically isolated membrane is lost during the washes since the membrane vesicles continuously break and reseal into many smaller vesicles. Yet the washes are essential to reduce the nonspecific binding. It is for these reasons that we undertook the development of the free-flow immunoisolation system for organelle fractionation. The simplicity of the system and the gentle conditions give it widespread application.

II. MODEL ANTIGEN/ANTIBODY COUPLE

In order to systematically study and optimize the immunoisolation system and its component parts, we have established a model system. A viral particle, vesicular stomatitis virus (VSV), was selected for isolation. This was because we are using the VSV surface glycoprotein-G (VSV-G) — following implantation in the membrane — as antigen in our experiments on the immunoisolation of compartments of the endocytic pathway.[21,22] In addition, large amounts of the virus are easily prepared and radioactively labeled.

The VSV is a bullet-shaped particle of 170 × 65 nm. Its external envelope is a typical membrane that reflects the lipid composition of the host cell plasma membrane and contains a single membrane-spanning glycoprotein, G.[23] We raised antibodies against this protein for immunoisolation.

Components of the immunoisolation system are listed below. Each of these will be discussed in detail after introduction of a standard assay and two novel evaluation criteria.

1. Immunoadsorbent (ImAd)
    - Solid support
    - Linker molecule
    - Specific antibody

2. Input fraction
   - Specific component
   - Nonspecific component
3. Optimal conditions of immunoisolation
4. Instrumentation, magnetic free-flow chamber

III. EVALUATION CRITERIA AND STANDARD ASSAY

Subcellular fractionation is monitored using the evaluation criteria of yield and relative specific activity (RSA) calculated from a balance sheet of the fractionation. These two parameters were defined for subcellular fractionation by de Duve and his colleagues[24] (for a detailed review see Beaufay and Amar-Costesec[25]).

$$\text{Yield} = \frac{(\text{amount of specific component})\text{Isolated}}{(\text{amount of specific component})\text{Input}}$$

$$\text{Relative Specific Activity} = \frac{\dfrac{(\text{activity of specific component})\text{Isolated}}{(\text{activity of specific component})\text{Input}}}{\dfrac{(\text{amount of component uniformly distributed})\text{Isolated}}{(\text{amount of component uniformly distributed})\text{Input}}}$$

In the experiments reported here, the input is defined as the fraction presented to the ImAd and protein or radioactive label is used as the component uniformly distributed.

Immunoisolation has the advantage that values for two additional evaluation criteria can be calculated which provide additional information: enrichment and efficiency index. Enrichment is generally used to express the extent of purification. The immunoisolation protocol provides a unique means to distinguish between the amount of the specific and the nonspecific (contamination) binding of a uniformly distributed component (protein). Evaluation of experiments using this criteria will be presented in Table 1.

$$\text{Enrichment} = \frac{\dfrac{\text{amount of protein-specific component}}{\text{amount of protein — other components}}\text{Isolated}}{\dfrac{\text{amount of protein-specific component}}{\text{amount of protein — other components}}\text{Input}}$$

The numerator is directly calculated from the experimental data with a specific (SP) and a control (C) ImAd: [(SP-ImAd) − (C-ImAd)]/C-ImAd. This expression provides valuable information on its own. The values for the denominator must come from coordinate biochemical and morphological analysis, and may be difficult to obtain for subcellular fractionation studies.

Efficiency is usually a broadly defined term which expresses some combination of yield and purity. Since immunoisolation is based on an antigen/antibody interaction, it is essential to clearly define an efficiency term (E) that also reflects the specificity of this interaction. The term E is obtained from an efficiency index (EI), which is an expression of the amount of input required to achieve saturation of the ImAd and is calculated for both the specific and control ImAds.

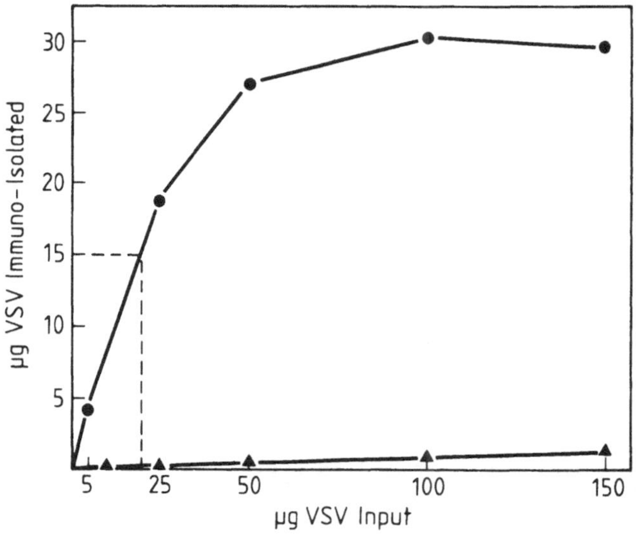

FIGURE 1. Immunoisolation of VSV using a specific (●---●---●) and a nonspecific (control) (▲---▲---▲) ImAd. 1 mg ImAd was used per point. Specific ImAd = M-267 beads/AP S αM Fc/M α VSV-G. Control ImAd = M-267 beads/AP S α M Fc/M IgG. The input fraction contained 5 to 150 μg, 3 × 10^3 cpm [^{35}S]/μg in 1 mℓ PBS per 5 mg/mℓ BSA. Binding was for 2 hr, at 4°C, with mixing. The ImAd was pelleted and unbound fraction was removed. Then the ImAd was washed three times by pelleting and resuspension with PBS-BSA. All fractions were counted and recoveries calculated. The dashed lines denote the half saturation point used in the efficiency calculation.

$$\text{Efficiency Index}_{\text{SP-ImAd}} = \frac{\text{amount of specific component isolated at saturation}}{\text{amount of input required for half saturation}}$$

$$\text{Efficiency Index}_{\text{C-ImAd}} = \frac{\text{amount of specific component bound at saturation}}{\text{amount of input required for half saturation}}$$

$$\text{Efficiency} = \frac{\text{efficiency index}_{\text{SP-ImAd}}}{\text{efficiency index}_{\text{C-ImAd}}}$$

The efficiency term, E, provides a valuable means to compare various ImAds and will be documented in Table 2.

An example of the standard assay using the VSV model system which permits the calculation of the isolation efficiency is shown in Figure 1. An increasing amount of the input fraction, 5 to 150 μg protein of the total VSV particle, is presented to a constant amount of ImAd, 1 mg. The ImAd consists of a solid support to which specific antibody is bound via a linker molecule. The isolation is easily monitored because the virus has been metabolically labeled with (^{35}S)-methionine. The reaction is carried out in an Eppendorf tube in 1 mℓ phosphate-buffered saline (PBS) and 5 mg/mℓ bovine serum albumin (BSA).

The amount of virus isolated is determined using the [^{35}S] label on an ImAd prepared with either specific antibody or a nonspecific antibody (Figure 1).

$$EI_{SP\text{-}ImAd} = \frac{30.0\ \mu g}{20.0\ \mu g} = 1.52 \qquad EI_{C\text{-}ImAd} = \frac{0.8\ \mu g}{50.0\ \mu g} = 0.02$$

therefore E = 1.52/0.02 = 76.0. In practice a system with E approaching 100 would be considered extremely efficient and one with E approaching 1 would be totally inefficient.

IV. IMMUNOISOLATION SYSTEM

A. Immunoadsorbent (ImAd)

1. Solid Support

A wide variety of solid supports have been used for the immunoisolation of cells and for cell labeling. Many of these are discussed in other contributions to this volume. Solid supports used for immunoisolation of organelles are more limited: (1) cellulose meshwork diazotized for chemical coupling of antibodies;[8,12,14] (2) polyacrylamide beads to which antibodies are covalently coupled,[9,10] and (3) fixed bacterial cells, *Staphylococcus aureus,*[11,13] which express protein A on their outer surface.

Our goal was to take the immunoisolation procedure one step further and carry it out in free flow. An ImAd prepared with a magnetic solid support is contained in a chamber within the magnetic field as a dilute, disperse suspension while the washing buffer flows through, providing effective and gentle removal of unbound components. This circumvents a major problem of immunoisolation on solid supports, the need to pellet and resuspend in order to efficiently remove the unbound and nonspecific components of the input fraction. With the free-flow system, more effective washing can be achieved resulting in lower nonspecific binding. More important, the gentle conditions prevent losses due to vesicularization of the specifically bound vesicles resulting in the maintenance of high-specific binding. The special instrumentation for this system is described in Section IV. D.

We have worked with a few of the many generations of solid supports developed by John Ugelstad and his collaborators at the University of Trondheim, Norway.[26] Those used are monodisperse, 3 μm diameter, polystyrene/divinyl benzene and contain 28% Fe. Introduction of the magnetic material in the solid support is achieved by an *in situ* process invented by Ugelstad which required the beads to be porous. Recently, the porosity has been successfully reduced from 160 (macroporous) to 4 m^2/g (microporous).[18]

The "beads" meet our general requirements for an appropriate solid support for immunoisolation:

1. Linker molecules and specific antibodies can be bound easily without significant loss of binding activity.
2. Nonspecific binding is within an acceptable range.
3. The beads are monodisperse and therefore, equally contained within the magnetic field.
4. The magnetic content is great enough that during washing at realistic flow rates the beads are maintained in suspension within the magnetic field.
5. The size of the bead, 3 μm, is appropriate for working with limited (μg) amounts of organelle fractions.
6. Organelle vesicles are restricted to the outer surface of the bead.
7. Beads are compatible with the processing steps required for transmission electron microscopy, which allows a morphological evaluation of an immunoisolation experiment.

We have been using beads which provide either hydrophobic binding or chemical coupling to attach the linker molecule. For the hydrophobic type the linker molecules are adsorbed onto the surface. The remaining hydrophobic sites are quenched subsequently with excess

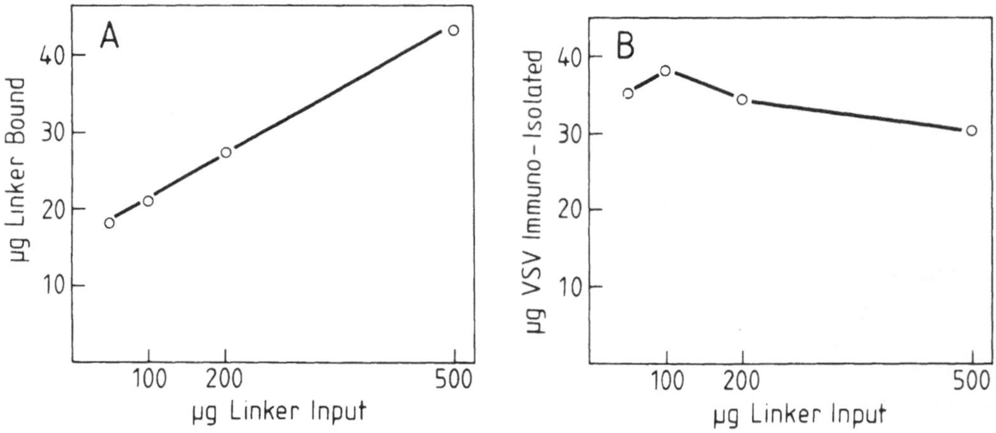

FIGURE 2. Determination of the amount of linker antibody required for maximal immunoisolation. (A) An increasing amount of AP S α R Fc, 5 to 500 μg plus trace amounts of [^{125}I]-AP S α R Fc, 2 × 10^4 cpm per tube, were reacted with 1 mg M-261 beads in 0.1 *M* potassium phosphate buffer pH 7.8 for 4 hr, at RT, with mixing. The unbound fraction was removed and the beads were washed three times with 0.05 *M* Tris/HCl buffer pH 7.5, 0.1 *M* NaCl, and 0.01% BSA; and then two times with PBS-BSA. (B) each sample (1 mg beads plus bound-linker antibody) was reacted with 100 μg AP R α VSV-G (see Figure 4), washed three times with PBS-BSA and then reacted with 50 μg [^{35}S]-VSV as in Figure 1.

protein, 5 mg/mℓ BSA. Examples of this bead type which have been used are M-97 and M-261, in order of reduced porosity. For the hydrophilic type the linker molecules are covalently coupled to free polyethylene oxide (PEO) chains extending from a layer of cross-linked acrylates covering the particle, for example M-267. The coupling procedure for the hydrophobic type is described by Ugelstad et al.[18] and for the hydrophilic type by Nustad et al.[27,28]

2. Linker Molecule

The procedures available to couple proteins to solid supports are usually inefficient. Only a small percentage of the input molecules actually bind, some are inactivated by the procedure, and others are bound in an unfavorable orientation. An easy and economical way to limit waste of specific antibody is to use a linker molecule placed between the solid support and the specific antibody. An additional advantage of a linker molecule is to provide flexibility which gives the antibody a better chance to find its antigen.[29] Use of generic antibodies raised against the Fc region of the IgG molecule produces an effective ImAd. The specific antibody is properly oriented and a decreased amount of specific antibody is required for an experiment. When working with rabbit polyclonal antibodies this may be an important criterion. An IgG fraction contains approximately 10% specific antibody. Therefore, in order to obtain a high enough density of specific antibody, affinity-purified antibodies must be used. When working with monoclonal antibodies use of an antimouse IgG linker will provide the affinity purification. Thus hybridoma supernatant or ascities fluid can be used directly. When unlimited quantities of affinity-purified specific polyclonal antibody and/or IgG fraction of specific monoclonal antibodies are available, the linker molecule may be omitted.

The amount of linker required for maximum isolation efficiency must be tested for each antigen/antibody couple and each type of solid support used. An example is shown in Figure 2. The amount of linker antibody (50 to 500 μg input) which binds to 1 mg M-261 is shown in Figure 2A. At each point the sample is then reacted with 100 μg specific antibody. The ImAds produced were used to immunoisolate an input fraction of 50 μg VSV (Figure 2B).

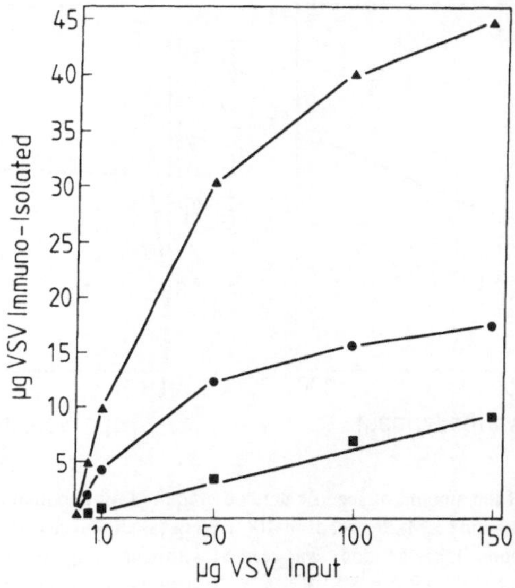

FIGURE 3. Comparison of linker antibodies for immunoisolation. ImAds were prepared using M-97 beads and AP S α R IgG (●---●---●) or AP S α R Fc as linker (▲---▲---▲) and AP R α VSVG, as specific antibody; a control ImAd was AP S α R IgG/control R IgG (■---■---■). Experimental conditions are as in Figure 1.

With the M-261, 50 μg linker per milligram of solid support already provides maximum isolation.

The significance of orientation of the specific antibody is shown in Figure 3. The use of linker antibody raised against only the Fc region of the rabbit IgG molecule doubled the isolation efficiency when compared to linker antibody raised against the entire rabbit IgG molecule. This experiment was carried out with a hydrophobic, more porous solid support, M-97, and the nonspecific binding is considerably higher than obtained in the experiment in Figure 1.

3. Specific Antibody

The specific antibody is always the most important reagent of the immunoisolation system. The affinity constants of the antibody/antigen interaction determined in solution will not necessarily reflect the conditions of immunoisolation where the antigen is membrane bound and the antibody immobilized on a solid phase. The efficiency of the ImAd is determined by a combination of the affinity of the specific antibody for its antigen and the density of bound antibodies. For this reason a great deal of effort has been put into optimizing the coupling conditions to obtain a high density of active molecules.[18,20,27-30]

We have shown that efficient immunoisolation was obtained in the range of 800 to 1000 specific antibody molecules per square micrometer ImAd surface area.[21] This quantitation was possible with nonporous solid supports. To obtain this density all the bound antibody must be specific, i.e., affinity purified if a polyclonal is used, and available to bind its antigen, i.e., properly oriented with an anti-Fc linker.

Experimentally the capacity of each ImAd is obtained by determining the amount of specific antibody-bound (Figure 4A) and the corresponding amount of antigen immunoisolated (Figure 4B). The linker antibody is bound at optimal concentration (Figure 2) followed by inputs of specific antibody from 1 to 100 μg/mg solid support (Figure 4A). Immunoisolation of an input of 50 μg VSV was carried out using the ImAds produced and this data

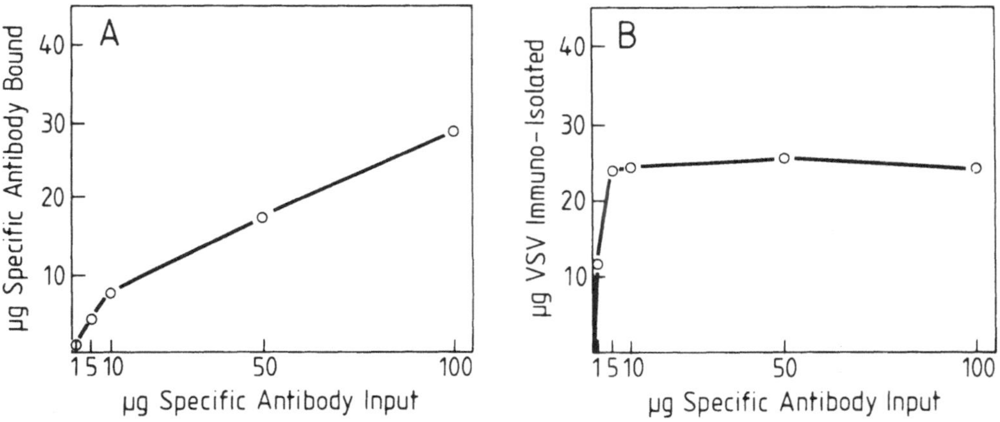

FIGURE 4. Determination of the amount of specific antibody required for maximal immunoisolation. (A) Linker antibody, AP S α R Fc, was bound to M-261 beads at 100 μg/mg (as established in experiment shown in Figure 2). Then 1 mg M-261 plus bound-linker antibody was reacted with increasing concentrations, 1 to 100 μg AP R α VSV-G plus trace amounts of [^{125}I]-AP R α VSV-G, 2 × 10^5 cpm per tube. The binding was overnight at 4°C, with mixing. The unbound fraction was removed and the beads were washed three times with PBS-BSA. All fractions were counted and recoveries were calculated. (B) each ImAd (1 mg M-261 plus bound-linker antibody plus bound-specific antibody) was reacted with 50 μg[^{35}S]-VSV. Experimental conditions are as in Figure 1.

is presented in Figure 4B. Saturation binding of VSV is achieved with an input of 5 μg Ap R α VSV-G,* where 4.4 μg of IgG are bound.

The ImAd efficiency (see Figure 1) provides information with respect to the specific/control ratio and permits direct comparison between various ImAds (see Table 1). This criteria was used in screening monoclonal antibodies against VSV-G. Only one of 12 positive clones produced an ImAd of equal isolation efficiency as the affinity-purified rabbit polyclonal antibody, Figure 5. In both examples, appropriate orientation was provided by the corresponding anti-Fc linker antibody.

B. Input Fraction

The input fraction varies considerably depending on the experimental system. The cellular component to be isolated (specific component) expresses a specific antigen on its outer surface, while the remaining cellular components (nonspecific) either lack the specific antigen or bear it in a site inaccessible to the antibody. When isolated, nonspecific components are defined as contamination. Specific and nonspecific components may be present in the input fraction in considerably different ratios.

For evaluation of subcellular fractionation studies at least three appropriate markers should be selected: one for the specific, another for the nonspecific, and a third which is uniformly distributed and includes both specific and nonspecific components. The ideal markers would be strict in definition and easy to assess. In general, the selection of the markers will determine the extent of information obtained from the experiment. Traditionally, for cell fractionation enzymatic activities are used to mark the various subcellular compartments.[24,25] However, they usually require fairly large samples and their specificities are not always well defined. Therefore, the trend is toward more specific markers such as those based on immunological specificity.

In the experiments reported here, each component has been metabolically labeled with a different isotope, and therefore we have used the distribution of the labels as markers for the specific and nonspecific components.

* See abbreviations.

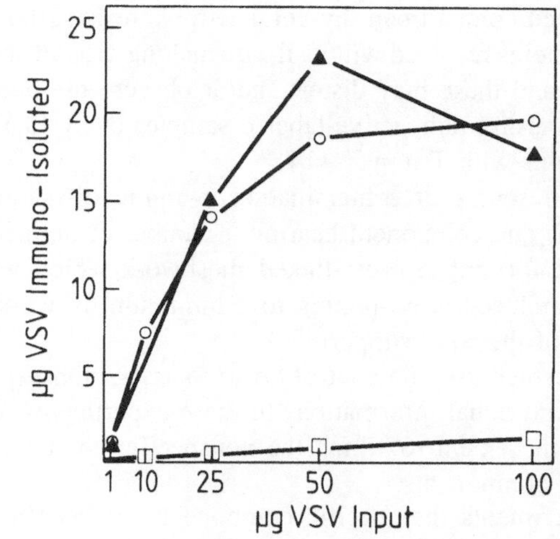

FIGURE 5. Immunoisolation of VSV using ImAds prepared with rabbit polyclonal antibody (○---○---○) (M-261/AP S α R Fc/AP R α VSV-G) and with a mouse monoclonal, clone 17.2.21.4.(▲---▲---▲) (M261/AP S α M Fc/M α VSV-G). Controls (□---□---□) (M261/AP S α M Fc/control M IgG and M-261/AP S α R Fc/control R IgG). Experimental conditions are as in Figure 1.

1. Specific Component

The density of antigen molecules accessible on the surface of the component to be isolated is another important parameter determining the success for immunoisolation. We have implanted increasing amounts of VSV-G into the plasma membrane of various cell types, and shown that there is a critical density of approximately 50 antigen molecules per square micrometer of membrane surface area for immunoisolation.[21] This corresponds to 0.2% of the total membrane protein, based on the estimate of 30,000 copies of protein per square micrometer of membrane surface area.[31] For example, erythrocyte vesicles with G protein implanted into the membrane at the critical density exhibit a EI_{SP} = 0.54 using the same ImAd as used in the experiment in Figure 1. With increased antigen densities both the rate of immunoisolation and the yield increase. This can be documented in the VSV model system, where the G protein is the only protein of the viral membrane and as expected a high EI_{SP} = 1.52 is obtained (see Figure 1); the maximum value would be 2.0.

2. Nonspecific Component

A major goal in immunoisolation experiments is to limit nonspecific binding or contamination to a minimum. Ideally, one would select an ImAd with a surface to which proteins could only bind via the antigen/antibody interaction. Using our evaluation criteria we are able to test our ImAds and select those with the best surfaces.

The first line of defense against nonspecific binding is to try to prevent it. Immunoisolation experiments are carried out in the presence of a high concentration of exogenous protein in an attempt to quench potential sites for nonspecific interactions. We use 5 mg/mℓ BSA, but many other proteins would serve equally well. Assay of the material isolated is usually expressed as a specific activity and amount of protein is most often selected as the component uniformly distributed in the input fraction. The addition of a quenching protein will affect this analysis. This problem can be overcome if the input fraction is prepared with a uniform radioactive label.[14]

Analysis of the isolated compartment by gel electrophoresis suffers from the same complication. The major proteins resolved will be the quenching protein and the immunoglobulin heavy and light chains and these may distort and/or obscure the proteins of interest. This can be circumvented by using radioactive-labeled samples or by first extracting the hydrophobic membrane proteins with Triton X-114.[32]

Entrapment is another source of contamination in immunoisolation experiments and is caused by cross-linking; one component bearing a number of antigen molecules will bind to more than one bead forming a cross-linked meshwork which will entrap nonspecific components. We have reduced cross-linking to a minimum by working in dilute solution and close to saturation of the solid support.

We have already presented use of a control ImAd to assess nonspecific binding. "Mixing experiments" provide additional information. In these experiments, one can evaluate both the contamination and the extent to which the nonspecific component interferes with the isolation of the specific components.

For our mixing experiments the specific component is [^{35}S]-VSV and the nonspecific component is a total microsomal fraction prepared from a cultured cell metabolically labeled with ^{3}H-leucine. This microsomal fraction contains all the cellular organelles and compartments except the nuclei (see Figure 10), and represents the input fraction for many immunoisolation experiments of subcellular compartments. The binding of the specific component is unaffected by increasing amounts of the microsomal fraction bound to the ImAd (Figure 6). The extent of the nonspecific binding with the microsomal fraction prepared from a cultured cell is greater than with the corresponding fraction prepared from rat liver, or erythrocyte membrane vesicles, or another virus, avian fowl-plague virus (FPV), (Figure 7).

C. Optimal Conditions of Immunoisolation

In the previous paragraphs we have described the important parameters for the preparation of the ImAd, and used a number of different solid supports in the illustrative experiments. In order to compare these ImAds we applied the evaluation criteria defined earlier: efficiency index and efficiency, and the values are presented in Table 1. The increased efficiency of the M-267 over the M-97 is a result of the very low nonspecific binding obtained with the M-267 beads. Since they were the most efficient of the ImAds tested, they were used for the free-flow experiments.

Other experimental conditions needed to be optimized: relationship between the concentration of input and ImAd, temperature, and time of incubation. The relationship between input and ImAd has been shown in Figure 1. Saturation of 1 mg ImAd corresponds to approximately 50 μg VSV bound and is achieved with 100 μg VSV input. When excess ImAd is required, contamination is likely to become a more serious problem — for example, when complete removal of the specific component from the input fraction is desired.

Temperature and time are obvious parameters to vary in the experimental protocol. Most antigen/antibody reactions are not significantly temperature dependent and are relatively rapid. There is no advantage to working at 20°C compared with 4°C (data not shown), and with biological samples the higher temperature may be a considerable disadvantage. In our standard system, optimal binding is attained in approximately 4 hr at 4°C, as shown in Figure 8. Prolonged incubation time may increase the contamination since the level of nonspecific binding increases with time.

All the experiments illustrated are carried out by reacting the input fraction with the completed ImAd, i.e., the specific antibody is immobilized on the ImAd prior to immunoisolation. An alternative sequence is first to bind the specific antibody to the input fraction and then to remove the excess unbound antibody. The input fraction plus bound antibody is then mixed with a solid support with linker antibody bound. A major advantage of binding

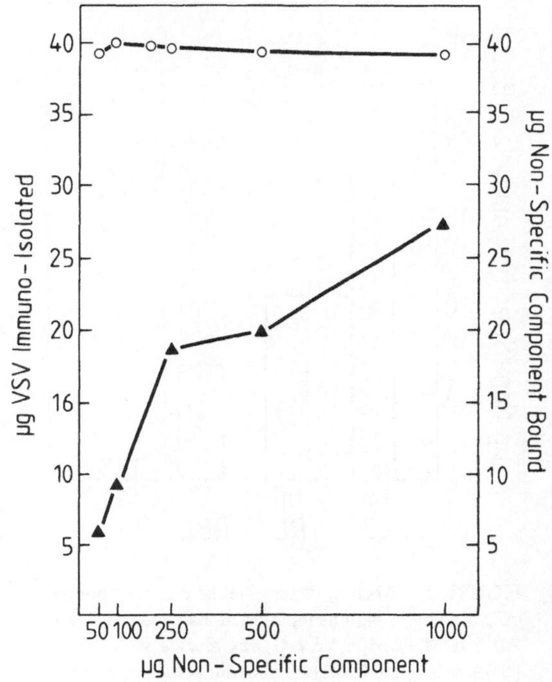

FIGURE 6. Mixing experiment to quantitate the nonspecific component bound and to determine the effect of the nonspecific component on the binding of the specific component. One mg M261/AP S α M Fc/M α VSV-G was used per point. The input fraction in 1 mℓ PBS-BSA consisted of: specific component, 50 µg VSV, 1×10^4 cpm [^{35}S] (○---○---○) (left ordinate); increasing concentration of nonspecific component, 50 to 1000 µg of [^{3}H]-leucine microsomal fraction, 1×10^3 cpm/µg (▲---▲---▲) (right ordinate). Experimental conditions are as in Figure 1.

the specific antibody directly to the input fraction is that the affinity purification of the specific antibody takes place directly on the component of interest, i.e., total antiserum can be used. To a large extent the choice between these two protocols depends on: (1) how easy it is to remove the free from the bound specific antibody; and (2) how the isolated component is affected by the combination of excess antibody and increased time of the isolation required in the alternative sequence. For example, if the antigen-containing component is large, like a cell, removal of the unbound antibody by centrifugation is simple and rapid. In the VSV isolation this is less convenient because an ultracentrifugation step is required.

We preferred to use the completed ImAd protocol because it was more convenient. The ImAd can be prepared in large batches and is stable when stored in PBS-BSA for more than a month (Figure 8) as compared with the same ImAd assayed on day 1 of preparation.

D. Instrumentation, Magnetic Free-Flow Chamber

As already mentioned, there are two major reasons for establishing an immunoisolation system in free flow: first, to limit the extent of stress on the isolated material and second, to decrease contamination. The instrumentation was designed to be versatile, so that it would be equally adaptable to the isolation of small particles, like viruses and lipoproteins, as well as subcellular compartments or cells. The process is rapid and easily maintained sterile. The instrumentation is inexpensive and requires minimal manipulation and maintenance to permit general use. The diagram shows the magnetic free-flow chamber for immunoisolation.

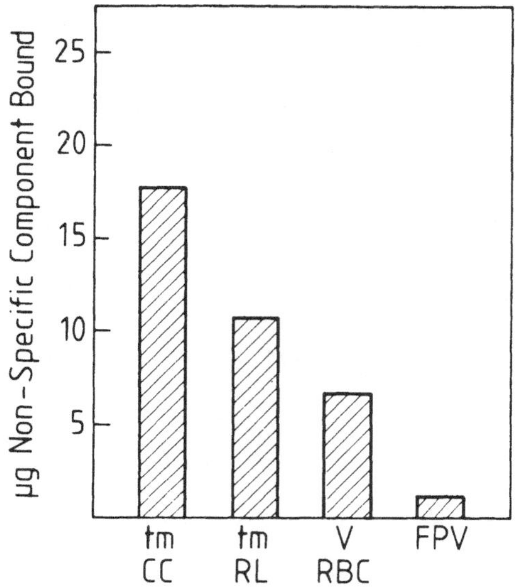

FIGURE 7. Assay to determine the extent of nonspecific binding of various input fractions. One mg M-261/AP S α M Fc/M α VSV-G was reacted with 250 µg [^{3}H]-labeled protein of: (1) microsomal fraction prepared from a cultured cell (TM, CC) (2) microsomal fraction prepared from rat liver (TM, RL); (3) vesicles prepared from human erythrocytes (V, RBC), and (4) avian fowl plague virus (FPV). Experimental conditions are as in Figure 1.

Table 1
IMMUNOISOLATION EFFICIENCIES

Solid support	Linker antibody[a]	αVSV-G specific antibody	Efficiency (%)
M-267	S α M Fc	Monoclonal	95.0
M-261	S α M Fc	Monoclonal	43.0
M-97	S α R Fc	AP	16.0
M-97	S α R IgG	AP	6.0

[a] See abbreviations.

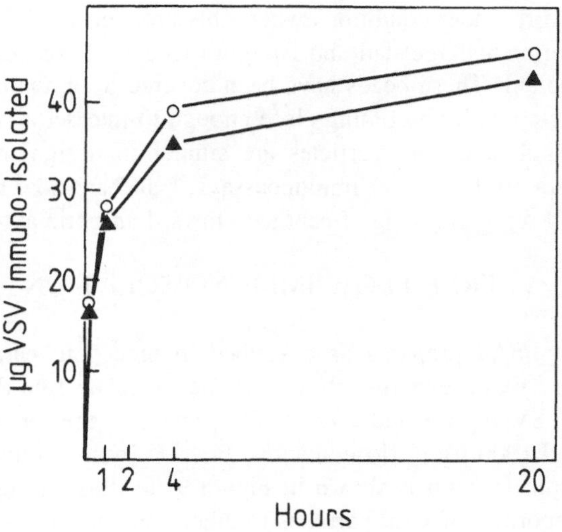

FIGURE 8. Time curve of immunoisolation. 1 mg M-261/AP S α M Fc/M α VSV-G is reacted with 50 µg VSV, 2 × 10^5cpm [^{35}S] in 1 mℓ PBS-BSA for 15 min to 20 hr. The experiment (○---○---○) was carried out with the ImAd on day 1. The experiment (▲---▲---▲) was carried out with the same ImAd after storage for 1 month in PSA-BSA. Experimental conditions, except for time, are as in Figure 1.

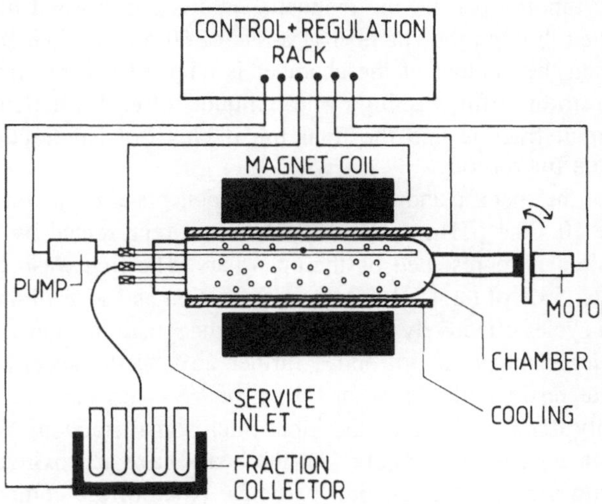

The isolation chamber, cooled to 4°C, is placed within the magnetic field generated by an electromagnet (B = 0.1 T). The shape of the magnetic field has been designed to permit the beads to be maintained in a disperse suspension and to be contained within the chamber while a buffer flows through the chamber. The flow rate of the buffer is regulated by a pump joined to the inlet, and a fraction collector is connected to the outlet. The chamber can be slowly rocked at variable speed to disperse both the input fraction and the ImAd. All functions of the isolation system can be automated and programmed via a microprocessor control rack. A more detailed report will be published elsewhere.

Aggregation of the beads may occur in the magnetic field. The extent of the aggregation depends on the magnetic moment of the bead, their composition, and the strength of the magnetic field. We work under conditions where this problem is minimized. Based on the results from our experiments Ugelstad and Bjørgum recently have designed solid supports to overcome this problem. These beads have been covered by a nonmagnetic layer, which prevents the magnetic cores from coming close enough to interact and induce aggregation. The new magnetic shell and core particles are similar in design to those produced by Ugelstad and collaborators for use in immunoassays,[28] and are now being tested. We anticipate that these will represent a significant step toward an optimal solid support.

V. FREE-FLOW IMMUNOISOLATION

The most efficient ImAd prepared as described in preceding paragraphs with M-267 particles was used in mixing experiments to test the magnetic free-flow isolation. In the example presented, VSV represented 10% of the protein of the input which consisted of 100 μg [^{35}S]-VSV and 1000 μg [^{3}H]-microsomal fraction from a cultured cell. An electron micrograph of the input fraction is shown in Figure 9, to illustrate the components being separated and the proportion of viral particles to other components. A very thin pellet was prepared so that the composition of the input fraction could be visualized in a single micrograph. At the top of the pellet there are mostly free ribosomes and polysomes while a little further, the vesicles derived from the rough endoplasmic reticulum become the major component. The viral particles band toward the center of the pellet, and toward the bottom, mitochondria and lysosomes are present.

This experiment was carried out in a 15-mℓ chamber, but any size may be substituted. The input fraction in PBS-BSA and the ImAd (2 mg M-267/AP S α M Fc/M α VSV-G) are added to the chamber which is then closed and rocked at four cycles per minute for 2 hr at 4°C. After the binding period, the magnetic field is applied and the washing buffer (PBS-BSA) is pumped through the chamber at a rate of 60 mℓ/hr. Four fractions of 15 mℓ are collected and then the content of the chamber is emptied and the ImAd plus isolated fraction separated from the fifth washing cycle. Aliquots of each fraction are counted and an aliquot of the input fraction and the final ImAd plus isolated fraction are fixed and processed for electron microscopy.

The distribution of the specific and nonspecific components after the immunoisolation are illustrated in Figure 10. The [^{3}H]-microsomal fraction is represented by the cross-hatched bars and the [^{35}S]-VSV is represented by the open bars. The first wash cycle contains the unbound components; 56% of the [^{3}H]-microsomal fraction and 35% of the [^{35}S]-VSV. The following four wash cycles effectively remove most of the remaining nonspecific component, 43% of the [^{3}H]-microsomal fraction, and a further 26% of the specific component, the [^{35}S]-VSV. The final result is that 39% or 39 μg of VSV has been immunoisolated and contamination is only 0.1% or 1 μg of the input microsomal fraction. When the ImAd is examined by electron microscopy (Figure 11), the beads have approximately 50% of their surface occupied with a monolayer of viral particles. None of the components of the microsomal fraction are obvious in the isolated fraction, consistent with the label distribution.

A striking result was obtained in the free-flow system when compared with a similar immunoisolation in the test tube (Figure 6). Quantitative comparison between the test tube assay (Figure 6) and the free-flow assay (Figure 10) using the evaluation criteria is presented in Table 2. The yield of the [^{35}S]-VSV is lower in the free-flow experiment. This can be explained by the lower concentration of the ImAd in the input fraction. However, a significantly lower nonspecific binding was achieved with the free-flow protocol, 1 vs. 20 μg [^{3}H]-microsomal fraction, resulting in an RSA value close to the theoretical maximum and a 20-fold increase in the enrichment over the test tube experiment. The total enrichment from the input fraction is 390-fold.

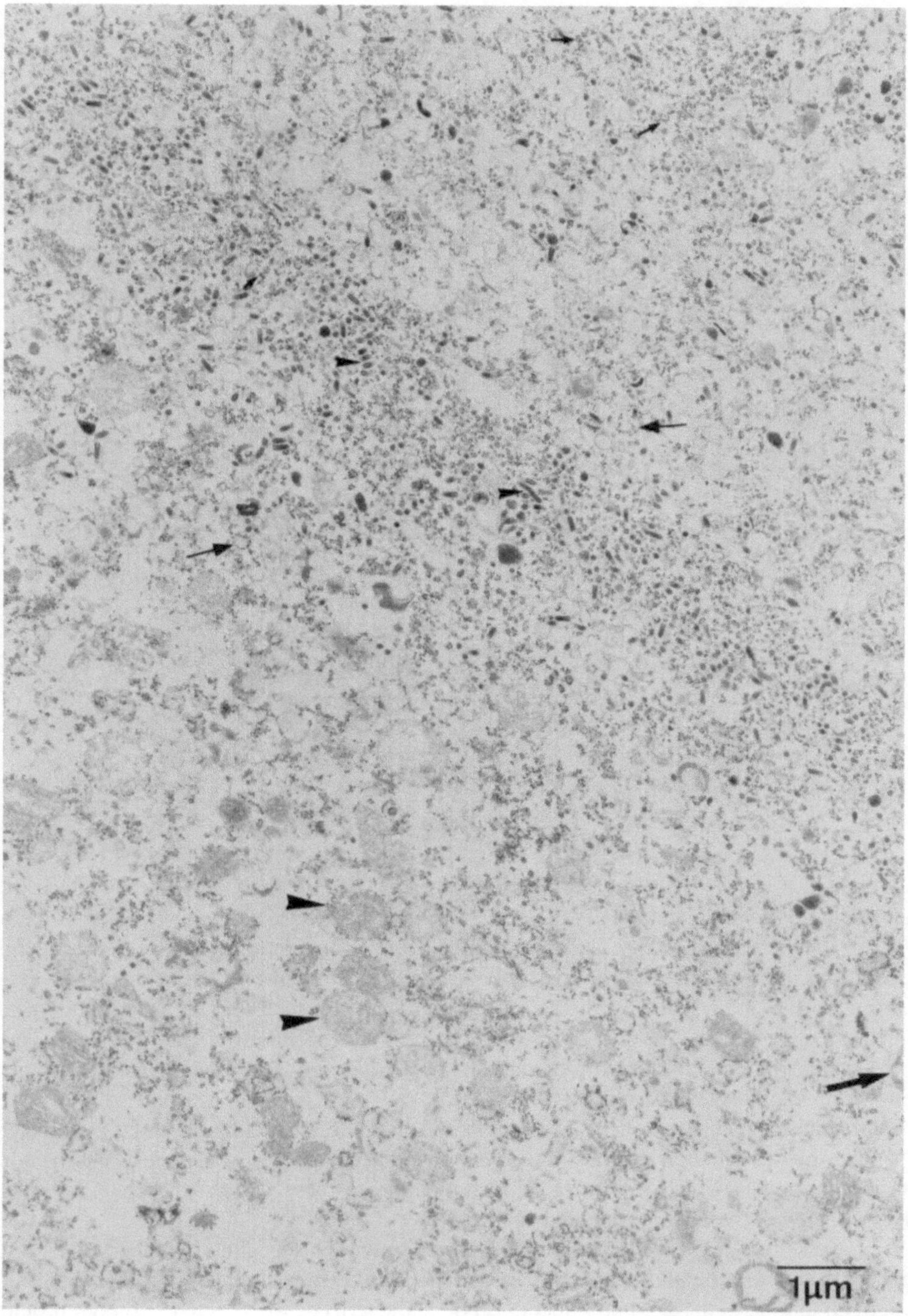

FIGURE 9. Electron micrograph of a thin section of the input fraction of the free-flow immunoisolation experiment. The input fraction, 100 μg VSV mixed with 1000 μg microsomal fraction prepared from a culture cell. An aliquot of 50 μg of the input fraction was fixed in suspension with 2% glutaraldehyde in 0.1 *M* cacodylate buffer pH 7.2 for 30 min and then centrifuged at 20,000 rpm for 20 min in 0.8-mℓ tubes which fit the SW 50 rotor with adaptors. The pellet was postfixed with 2.0% OsO_4 in 0.1 *M* cacodylate buffer pH 7.2 and then stained with 2% uranylacetate in H_2O for 1 hr, dehydrated and embedded in Spurr's resin. The micrograph is a field through the entire depth of the pellet. VSV, small arrow head; free ribosomes and polysomes, small arrow; vesicles of rough endoplasmic reticulum, intermediate arrow; mitochondria, large arrow head; and lysosomes, large arrow; the bar is equal to 1 μm. (Magnification × 12,150.)

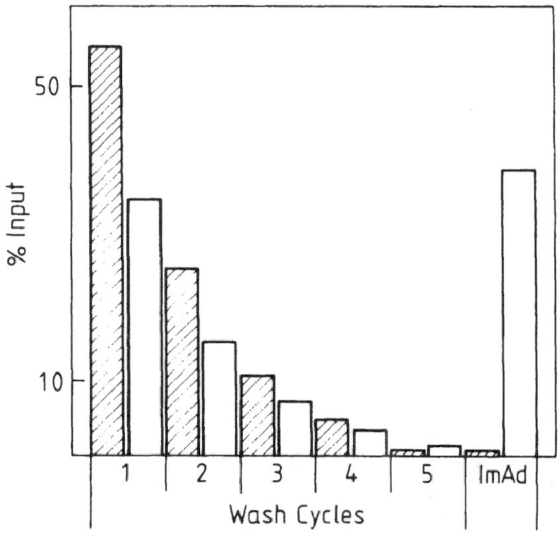

FIGURE 10. Distribution of the specific component, [^{35}S]-VSV, open bars, and the nonspecific component, [^{3}H]-TM, cross-hatched bars, in the free-flow immunoisolation experiment. After 2 mg ImAd (M-267/AP S α M Fc/M α VSV-G) is reacted with the input fraction, 100 μg [^{35}S]-VSV and 1000 μg [^{3}H]-microsomal fraction, the magnetic field is applied and PBS-BSA is pumped through the suspended ImAd. Fractions of 15 mℓ are collected and aliquots are counted. Data are expressed as percent of the input (ordinate) recovered in the wash cycles and on the final ImAd (abscissa).

VI. DISCUSSION AND PERSPECTIVES

Recent advances in the field of cell and molecular biology have begun to elucidate both the extent and the complexity of intracellular transport pathways. Definitive characterization of these pathways and functional studies using cell-free reconstitution systems require purified preparations of the compartments involved.[22,35] One of the major obstacles confronting those studying membrane traffic is the paucity of techniques available for the isolation of these compartments. This is due to the fact that during homogenization of the cell, these compartments form membrane vesicles of similar size and density which are difficult to separate using centrifugation techniques.

The most promising approach to the specific isolation of subcellular compartments takes advantage of the functional and biochemical differences between organelles. That is, one can use antibodies which recognize antigens unique to these compartments to positively select vesicles derived from a given organelle.

We have set out to design an immunoisolation system for subcellular compartments. Our aim is to isolate the compartments of the endocytic and secretory pathways. Since, at present, an antigen unique to these compartments is not characterized, we selected a foreign protein, the G-protein of VSV. It has been shown by a number of groups that this viral glyocoprotein can be introduced and selectively accumulated in the membrane of the various compartments along both the endocytic and secretory pathways (for a review see Reference 33). This makes it possible to isolate different compartments, using the same antigen/antibody couple.[21,22,35] This being our goal, the total virus becomes an advantageous model to optimize the immunoisolation system. It has a membrane similar to a cellular membrane and is in the same size range as vesicles derived from intracellular compartments following homogenization.

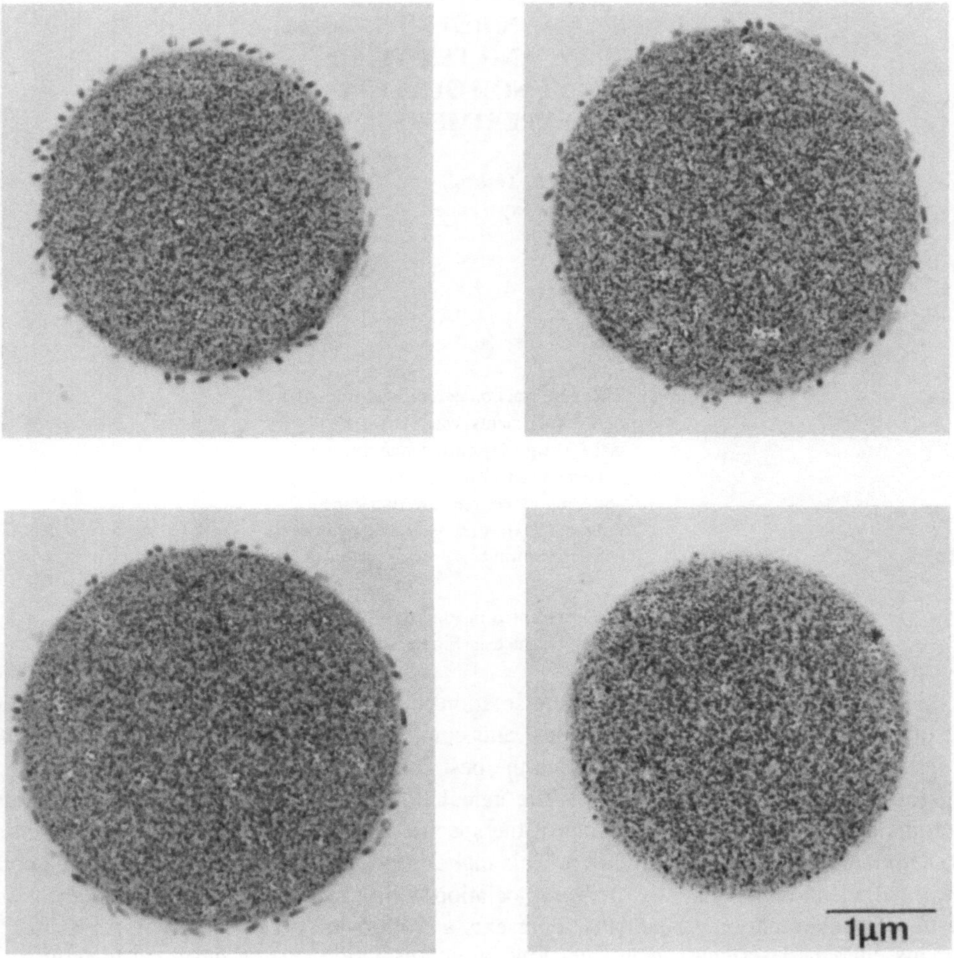

FIGURE 11. Electron micrographs of thin sections of the ImAd plus bound material after free-flow immunoisolation. Experiment with specific ImAd are shown in A, B, and C, and control ImAd in D. The ImAds plus bound material were mixed with 2% low melting agarose at 30°C and cooled on ice until the agarose solidified. The block was sliced into small cubes and fixed and processed as in Figure 9. The specific ImAd is covered with a monolayer of VSV particles. The control ImAd has no virus or components of the microsomal fraction bound to its surface. The bar is equal to 1 μm. (Magnification × 15,400.)

In the immunoisolation system presented here each step of the isolation protocol is discussed in detail, providing insight into the critical points and into the possibilities for variation and adaption to other needs. Our objective has been to provide a comprehensive study of the potential and practicalities of immunoisolation.

In earlier immunoisolation experiments, a wide variety of materials and conditions have been used; these have been referred to in the introduction to this chapter. However, it is difficult to make a decision as to the most appropriate materials and protocol for a given application. Comparisons are complicated by the lack of standard evaluation criteria.

In the subcellular fractionation field the standard criteria — the balance sheet, yield, and relative specific activity of selected markers[24,25] — are of enormous value to judge the success of the experiments. These criteria also allow comparison between the many contributions to the literature. To facilitate the future evaluation of immunoisolation experiments we have introduced two new criteria.

Table 2
COMPARISON BETWEEN FREE-FLOW AND TEST-TUBE IMMUNOISOLATION EXPERIMENTS

	Test-tube experiment[a]	Free-flow experiment[b]
Yield	80%	39%
RSA	7.3	10.7
RSA_{max}	11.0	11.0
Enrichment	20.0	390.0

Note: The input fraction per milligram ImAd for both experiments was 50 μg [^{35}S]-VSV and 500 μg [^{3}H]-microsomal fraction. One mg was used in the test-tube experiment and 2 mg in the free-flow experiment. Enrichment $_{max}$ has no real meaning because it reaches infinity.

[a] Data are presented in Figure 6.
[b] Data are presented in Figure 10.

Efficiency is used to optimize immunoisolation conditions in terms of selection and titration of linker molecules, specific antibodies, and comparison of solid supports. The efficiency criterion is a useful parameter to define the best conditions for high yield and low contamination in the experimental system. The calculation of the second criterion, enrichment, permits the evaluation of the amount of the specific compartment as well as the amount of contamination in the isolated fraction. This makes it possible to determine if a small amount of a marker known to have its major association with another compartment is present as a result of contamination, or actually represents a cross-over between two compartments.[9,34]

This chapter describes a new free-flow immunoisolation system designed to reduce the stress applied to the isolated material. The experiments documented here also demonstrate that high yields are achieved under conditions where nonspecific binding of a cellular fraction is significantly reduced over conventional immunoisolation protocols. The isolation of the minor cellular compartments involved in membrane biogenesis and sorting is now possible and offers great potential for cell biology.

ACKNOWLEDGMENTS

We are very grateful to John Ugelstad and his collaborators for supplying the magnetic solid supports used in this study and for the many hours of stimulating and informative discussion. We also thank Kell Nustad for sharing his expertise in working with the beads and coupling of antibodies to them.

We thank Annette Scharm, Ursula Reuter-Carlson, and Josef Stegemann for excellent technical assistance. We also thank Daniel Cutler, Eileen Devaney, Laura Roman, and Kai Simons for critically reading the manuscript and providing helpful suggestions.

Finally, thanks are due to Marianne Remy and Anne Walter for word processing the manuscript.

ABBREVIATIONS

Control, C; efficiency, E; efficiency index, E.I.; avian fowl plague virus, FPV; immunoadsorbent, ImAd; phosphate buffered saline, PBS, relative specific activity, RSA; room temperature, RT; specific, SP; total microsomes, TM; vesicular stomatitis virus, VSV; the glycoprotein-G of VSV, VSV-G.

The different linker antibodies are raised in sheep with:

1. IgG fraction prepared from rabbit serum
2. IgG fraction prepared from mouse serum
3. Fc constant fragment of rabbit IgG
4. Fc constant fragment of mouse IgG

The linker antibody is always affinity purified on a column with rabbit or mouse IgG coupled to CnBr-activated Sepharose 4B, (Pharmacia, Uppsala, Sweden).

Other abberviations for the antibodies are

affinity purified	= AP
mouse	= M
rabbit	= R
sheep	= S

- Antibodies produced in a sheep against (1) IgG fraction prepared from rabbit serum = S α R IgG
- Antibodies produced in a sheep against (2) IgG fraction prepared from mouse serum = S α M IgG
- Antibodies produced in a sheep against (3) Fc fragment of rabbit IgG = S α R Fc
- Antibodies produced in a sheep against (4) Fc fragment of mouse IgG = S α M Fc
- Antibodies produced in rabbit against a micellar form of glycoprotein G of the vesicular stomatitis virus = R α VSV-G
- Antibodies produced by mouse hybridoma clone No 17.2.21.4 = M α VSV-G. The mouse was immunized with the same antigen as the rabbit.

The notation used: type of bead/type of linker/specific antibody = ImAd.

REFERENCES

1. **Molday, R. S., Yen, S. P. S., and Rembaum, A.,** Application of magnetic microspheres in labelling and separation of cells, *Nature (London),* 268, 437, 1977.
2. **Kronick, P. L., Campbell, G. L., and Joseph, K.,** Magnetic microspheres prepared by redox polymerization used in cell separation based on gangliosides, *Science,* 200, 1047, 1978.
3. **Meier, D. H., Lagenaur, C., and Schachner, M.,** Immunoselection of oligodendrocytes by magnetic beads. I. Determination of antibody coupling parameters and cell binding conditions, *J. Neurosci. Res.,* 7, 119, 1982.
4. **Basch, R. S., Berman, J. W., and Lakow, E.,** Cell separation using positive immunoselective techniques, *J. Immunol. Methods,* 56, 269, 1983.
5. **Parks, D. R. and Herzenberg, L. A.,** Fluorescence-activated cell sorting: theory, experimental optimization, and applications in lymphoid cell biology, in *Methods in Enzymology,* DiSabato, G., Langone, J. J., and VanVunakis, H., Eds., Academic Press, New York, 108, 197, 1984.
6. **Treleaven, J. G., Gibson, F. M., Ugelstad, J., Rembaum, A., Philip, T., Caine, G. D., and Kemshead, J. T.,** *Lancet,* 1, 70, 1984.

7. **Wormmeester, J., Stiekema, F., and DeGroot, K.,** A simple method for immunoselective cell separation with the avidin-biotin system, *J. Immunol. Methods,* 67, 389, 1984.
8. **Luzio, J. P., Newby, A. C., and Hales, C. N.,** A rapid immunological procedure for the isolation of hormonally sensitive rat fat-cell plasma membrane, *Biochem. J.,* 154, 11, 1976.
9. **Ito, A. and Palade, G. E.,** Presence of NADPH-cytochrome P-450 reductase in rat liver Golgi membranes. Evidence obtained by immunoadsorption method, *J. Cell Biol.,* 79, 590, 1978.
10. **Miljanich, G. P., Brasier, A. R., and Kelly, R. B.,** Partial purification of presynaptic plasma membrane by immunoadsorption, *J. Cell Biol.,* 94, 88, 1982.
11. **Merisko, E. M., Farquhar, M. G., and Palade, G. E.,** Coated vesicle isolation by immunoadsorption on *Staphylococcus aureus* cells, *J. Cell Biol.,* 92, 846, 1982.
12. **Richardson, P. J., Siddle, K., and Luzio, J. P.,** Immunoaffinity purification of intact, metabolically active, cholinergic nerve terminals from mammalian brain, *Biochem. J.,* 219, 647, 1984.
13. **Roman, L. M. and Hubbard, A. L.,** A domain-specific marker for the hepatocyte plasma membrane. III. Isolation of bile canicular membrane by immunoadsorption, *J. Cell Biol.,* 98, 1497, 1984.
14. **Devaney, E. and Howell, K. E.,** Immuno-isolation of a plasma membrane fraction from the FAO cell, *EMBO J.,* 4, 3123, 1985.
15. **Margel, S., Beitler, U., and Ofarim, M.,** Polyacrolein microspheres as a new tool in cell biology, *J. Cell Sci.,* 56, 157, 1982.
16. **Molday, R. S. and Mackenzie, D.,** Immunospecific ferromagnetic iron-dextran reagents for the labeling and magnetic separation of cells, *J. Immunol. Methods,* 52, 353, 1982.
17. **Rembaum, A., Yen, R. C., Kempner, D. H., and Ugelstad, J.,** Cell labeling and magnetic separation by means of immunoreagents based on polyacrolein microspheres, *J. Immunol. Methods,* 53, 341, 1982.
18. **Ugelstad, J., Rembaum, A., Kemshead, J. T., Nustad, K., Funderud, S., and Schmid, R.,** Preparation and biomedical applications of monodisperse polymer particles, in *Microspheres and Drug Therapy, Immunological and Medical Aspects,* Davis, S. S., Illum, L., McVie, J. G., and Tomlinson, E., Eds., Elsevier, Amsterdam, 365, 1984.
19. **Rembaum, A., Ugelstad, J., Kemshead, J. T., Chang, M., and Richards, G.,** Cell labeling and separation by means of monodisperse magnetic and nonmagnetic microspheres, in *Microspheres and Drug Therapy, Immunological and Medical Aspects,* Davis, S. S., Illum, L., McVie, J. G., and Tomlinson, E., Eds., Elsevier, Amsterdam, 383, 1984.
20. **Rembaum, A.,** personal communication, 1985.
21. **Gruenberg, J. and Howell, K. E.,** Immuno-isolation of vesicles using antigenic sites either located on the cytoplasmic or on the exoplasmic domain of an implanted viral protein. A quantitative analysis, *Eur. J. Cell Biol.,* 38, 312, 1985.
22. **Gruenberg, J. and Howell, K. E.,** Reconstitution of vesicle fusions occurring in endocytosis with a cell-free system, *EMBO J.,* 5, 3091, 1986.
23. **Wagner, R. R.,** Reproduction of rhabdoviruses, in *Comprehensive Virology,* Vol. 4, Fraenkel-Conrat, H. and Wagner, R. R., Eds., Plenum Press, New York, 1975, 1.
24. **deDuve, C.,** The separation and characterization of subcellular particles, *Harvey Lect.,* 59, 49, 1965.
25. **Beaufay, H. and Amar-Costesec, A.,** Cell fractionation techniques, in *Methods in Membrane Biology,* Vol. 6, Korn, E. D., Ed., Plenum Press, New York, 1976, 1.
26. **Ugelstad, J., Söderberg, L., Berge, A., and Bergström, J.,** Monodisperse polymer particles — a step forward for chromatography, *Nature (London),* 305, 95, 1983.
27. **Nustad, K., Ugelstad, J., Berge, A., Ellingsen, T., Schmidt, R., Johansen, L., and Børmer, O.,** Use of monosized polymer particles in immunoassays, in *Radioimmunoassay and Related Procedures in Medicine,* IAEA-SM 259/19, 1982, 45.
28. **Nustad, K., Johansen, L., Ugelstad, J., Ellingsen, T., and Berge, A.,** Hydrophilic monodisperse particles as solid-phase material in immunoassays: comparison of shell-and-core particles with compact particles, *Eur. Surg. Res.,* 16 (Suppl. 2), 80, 1984.
29. **Cuatrecasas, P.,** Membrane receptors, *Annu. Rev. Biochem.,* 43, 169, 1974.
30. **Avrameas, S., Ternynck, T., and Guesdon, J.-L.,** Coupling of enzymes to antibodies and antigens, *Scand. J. Immunol.,* 8 (Suppl. 7), 7, 1978.
31. **Quinn, P., Griffiths, G., and Warren, G.,** Density of newly synthesized plasma membrane proteins in intracellular membranes. II. Biochemical studies, *J. Cell Biol.,* 98, 2142, 1984.
32. **Bordier, C.,** Phase separation of integral membrane proteins in Triton X-114 solution, *J. Biol. Chem.,* 256, 1604, 1981.
33. **Simons, K. and Warren, G.,** Semliki Forest Virus: a probe for membrane traffic in the animal cell, *Adv. Protein Chem.,* 36, 79, 1984.
34. **Howell, K. E., Ito, A., and Palade, G. E.,** Endoplasmic reticulum marker enzymes in Golgi fractions — what does this mean?, *J. Cell Biol.,* 79, 581, 1978.
35. **Gruenberg, J. and Howell, K. E.,** An internalized transmembrane protein residues in a fusion-competent endosome for less than 5 min, *Proc. Nat. Acad. Sci. U.S.A.,* 84, 5758, 1987.

Chapter 4

MONODISPERSE POLYMER PARTICLES IN IMMUNOASSAYS AND CELL SEPARATION

Kjell Nustad, Havard Danielsen, Albrecht Reith, Steinar Funderud, Tor Lea, Frode Vartdal, and John Ugelstad

TABLE OF CONTENTS

I. INTRODUCTION

A new two-step method for preparation of monodisperse particles was developed by Ugelstad et al. in 1980.[1] The procedure is suitable for making compact, porous, and core-and-shell particles from a wide variety of monomers in particle sizes ranging from 1 to 100 μm.[2] In this chapter we will review the results obtained so far with these particles in immunoassays and cell separation.

II. MONOSIZED PARTICLES IN IMMUNOASSAYS

A. Why Use Monosized Particles in Immunoassays?

Most immunoassays are dependent on the physical separation of excess labeled reagent from that bound to the specific antibody. Precipitation of the immune complexes with polyethylene glycol and an anti-immunoglobulin antibody is commonly used for this purpose. However, this method is prone to interference from hyperlipemia and other changes in the serum composition.[3,4] Such influence can readily be avoided using solid-phase antibodies as separation reagents.

The solid phase can either be stationary, such as antibodies bound to the wall of the tube, or mobile, like antibodies bound to small particles. The coated tube has many advantages such as ease of separation of free reagents from that bound to the wall, and nonspecific binding can be reduced by simple and rapid washing procedures. The main advantage of a particle-bound antibody is that the mobile particle will allow a rapid reaction with its antigen. For example, the time needed for a placental alkaline phosphatase antibody to bind its enzyme was only 5 min when the antibody was bound to particles. However, it took 24 hr to bind the same amount of enzyme when the same amount of antibody was bound to the wall of a microtiter well.[5]

The disadvantage of particles is the work involved in collecting and washing them using centrifugation. However, easier handling of particles can be accomplished by collecting the particles by filtration or by using magnetic particles. Magnetic particles with the desired properties for immunoassays are just now being developed by Ugelstad, and will not be discussed in this chapter. However, particles made by the procedure of Ugelstad et al.[1] are extremely uniform in size and should be ideal for collection by filtration. These particles should also be valuable in agglutination assays and particle-counting immunoassays.[6] The increasing use of antibody excess methods should also stimulate use of particle-bound antibodies since such preparations are stable for years, and it is easy to obtain true antibody excess conditions, thus creating rapid, robust immunometric assays.[7]

B. Essential Properties for Particles to be Used in Immunoassays

Particle diameter should be as small as possible for each analytical application. This will favor rapid Brownian movements of particles in suspension and give a maximum surface area with a minimum amount of solid material. Monosized particles are essential for particle-counting immunoassays. Equal size is a great advantage when particles are to be collected by filtration, and gives uniform reaction conditions during incubations in all analytical applications. The density of the particles should be close to that of water. However, for some applications particles should also be easy to collect by centrifugation. We find that a core-and-shell particle with a density of 1.07 g/cm^3 and a diameter of 3 μm is satisfactory for most applications. These particles have a core of cross-linked low-density polymer, and a cross-linked hydrophilic shell with hydroxyl groups stemming from hydroxy ethyl methacrylate.[8]

The particle surface should be smooth and allow a stable binding of antibodies. Porous particles should generally be avoided since antibodies inside pores would usually be less efficient than those on the particle surface. The surface should have low nonspecific binding of labeled analyte/antibodies. The core-and-shell particles described above are satisfactory in most instances.[8-10] However, a search should be made for more hydrophilic shells with even less nonspecific binding for the detection of analytes present in trace amounts in biological fluids.

C. Antibody Attachment to the Particles

1. Hydrophobic Adsorption of Antibodies

Hydrophobic particles usually made from styrene and divinylbenzene should first be treated with 50% ethanol in water to remove traces of detergent left over from the production process. The particles should then be dispersed in water to reduce particle-to-particle interactions. Antibody adsorption is initiated by adding the particles in water to the antibody in phosphate-buffered saline (PBS) (0.15 *M* NaCl, 0.01 *M* Na-phosphate, pH 7.5). The final salt concentration should be as low as possible; however, some affinity-purified antibodies will denature if the final concentration is below 0.04 *M* Na. A few minutes are allowed for the rapid adsorption of antibodies to the particles. Salt/buffer is now added to strengthen the hydrophobic binding and control the pH. We use the following recipe: three volumes of particles in water are added to one volume of antibody in PBS followed by one volume of 0.5 *M* borate buffer pH 9.5 or 0.5 *M* Na-phosphate buffer pH 7.4. Either buffer works equally well with our antibodies. A typical coupling would be conducted with a final particle concentration of 40 mg/mℓ and with 5 μg antibody per milligram of particles.

2. Chemical Coupling of Antibodies

Using particles with hydroxyl groups we find a striking difference when activating these particles with different methods. Our sheep antirabbit IgG antibody prefers coupling to sulfonyl chloride-activated particles, as shown in Figure 1A.[9] The ability of antirabbit IgG antibody (Ab_2) to bind its antigen was measured, as shown in Figure 1B. Affinity-purified Ab_2 served as a primary standard and the binding capacity of 1 mg of this antibody was defined as one unit.[9] Particle-bound Ab_2 gave a parallel dose-response curve and was used as a more convenient secondary standard. The specific binding capacity (U/mg Ab_2 bound) of a particle-bound Ab_2 should then be 1.0 as indicated by the dotted line in Figure 1A. However, Ab_2 coupled to sulfonyl chloride-activated particles was superior to soluble antibody, whereas Ab_2 coupled to carbonyl-di-imidazol-activated particles was inferior (Figure 1A). Both methods showed reduced specific binding capacity with increased density of antibodies on the particle surface (Figure 1A). It should be noted that an affinity-purified rabbit antikallikrein antibody did not show this preference for sulfonyl chloride-activated particles.[10]

Particles activated with *p*-toluene sulfonyl chloride (tosylated) are very stable and can be stored at 4°C in water for months with only minor loss in antibody-binding capacity. The actual coupling occurs in two steps. The antibody is first rapidly adsorbed to hydrophobic groups introduced by tosylation and a chemical coupling is completed only after 24 hr at pH 9.5 and room temperature, as shown in Figure 2.[8] The hydrophobic nature of the activated particles means that the actual coupling of antibodies should be performed exactly as described for hydrophobic adsorption above. The limitations of tosylation as an activation procedure are particle aggregation which is pronounced when using particles with a diameter under 2 μm. The pH, temperature, and the time needed for a chemical coupling to be completed are usually well tolerated by antibodies.[8] However, less stable ligands should be coupled using the more reactive trifluoroethane sulfonyl chloride, which will allow coupling at pH 7.5 and 4°C in 10 to 30 min.[8,11]

The long-term stability of antibodies on particles is excellent. Several preparations of chemically coupled antibodies have unchanged binding capacity for antigen after 1 to 2 yr

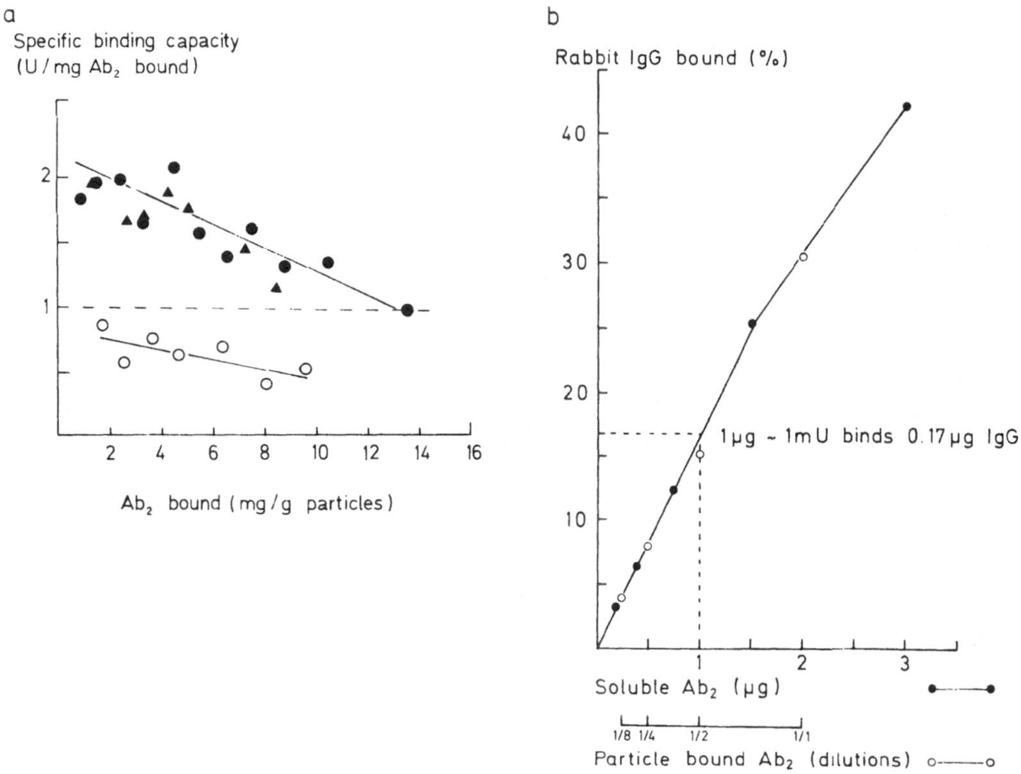

FIGURE 1. Comparison of methods for coupling of affinity-purified sheep antirabbit IgG (Ab_2) to monodisperse polymer particles and the measurement of its rabbit IgG-binding capacity. (A) Ab_2 was coupled to particles activated with *p*-toluene sulfonyl chloride (▲), trifluoroethane sulfonyl chloride (●), or carbonyl-di-imidazole (○). The relationship between Ab_2 bound to the particles and its rabbit IgG binding capacity is shown. The broken line (----) indicates the theoretical specific rabbit IgG-binding capacity using soluble Ab_2 as standard. 1 Unit = the rabbit IgG-binding capacity of 1 mg soluble Ab_2. (From Nustad, K., et al., *Radioimmunoassay and Related Procedures in Medicine 1982*, IAEA, Vienna, 1982, 45. With permission.) (B) Standard curve for the rabbit IgG binding assay. Purified rabbit IgG (1 µg), iodinated IgG (30,000 cpm, 1.2 ng) and 50 µℓ human serum in 100-µℓ assay buffer (0.05 *M* Tris, 0.1 *M* NaCl, 0.01 % BSA, 0.01% merthiolate, pH 7.4) were incubated with 1 mℓ Ab_2 in assay buffer for 2 hr at 20°C. Polyethylene glycol (500 µℓ of an 18% solution of MW 6000) was added, followed by whirl mixing and 30-min incubation at 20°C. After centrifugation (3000 × *g* for 15 min) and decantation of the supernatant fluid, the immune complexes formed were determined by counting the labeled IgG in the precipitate. Polyethylene glycol was omitted when particle-bound Ab_2 was assayed. An analogous assay was used to determine the mouse IgG-binding capacity of antimouse IgG antibodies.

storage at 4°C, whereas antibodies adsorbed on particles usually lose about 30 to 50% of their binding capacity during such storage.

No detrimental effects on antibody activity or particle dispersity were observed after freezing and thawing particles with antibodies seven times, nor when freeze drying such preparations.

3. Antibody Orientation on the Particle Surface

The ideal situation would be a firm attachment of the antibody to the particle through the carboxyl terminal end of the heavy chain of the immunoglobulin molecule with complete preservation of antigen binding properties, but such chemistry is not available to us. The closest approach would probably be a coupling through the sugar residues present in the hinge region of the antibody molecule.[12]

A similar orientation is probably obtained by hydrophobic adsorption since a major hy-

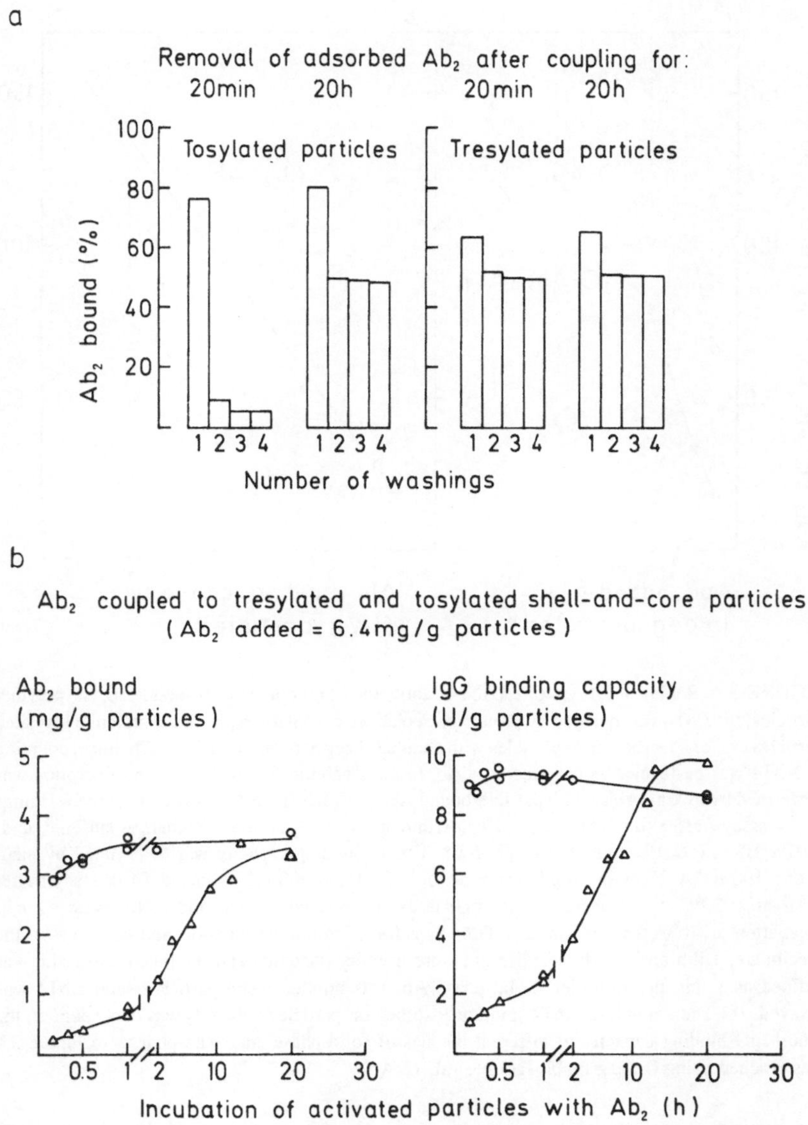

FIGURE 2. Coupling of affinity-purified sheep antirabbit IgG (Ab_2) to sulfonyl chloride-activated particles. (a) Removal of adsorbed Ab_2 from particles activated with *p*-toluene sulfonyl chloride (tosylated) and trifluoroethane sulfonyl chloride (tresylated). Particles and Ab_2 were mixed and rotated end-over-end for 20 min or 20 hr at 20°C, the particles were collected by centrifugation, the supernatant removed and the particle pellet washed briefly in 0.1 *M* borate buffer, pH 9.5 (washing 1). The particles were then washed by end-over-end rotation at 20°C in the following buffers: 1 *M* ethanolamine HCl, 0.1% Tween 20, pH 9.5 (washing 2); 0.05 *M* Tris, 0.1 *M* NaCl, 0.01% BSA, 0.01% merthiolate, 0.1% Tween 20, pH 7.4 (washing 3); and in the last buffer without detergent (washing 4). (b) Effect of incubation time on the chemical coupling of Ab_2 to activated particles. The Ab_2 was incubated with activated particles for the times indicated. The particles were washed as described above. Tosylated (△) and tresylated (○) monodisperse polymer particles XP 4101 from Dyno Industrier A.S., Oslo, Norway, were used. The amount of Ab_2 bound was estimated by addition of a trace amount of iodinated Ab_2. The rabbit IgG binding capacity was determined as described in the legend to Figure 1B. (From Nustad, K., et al., *Eur. Surg. Res.*, 16 (Suppl. 2), 80, 1984. With permission.)

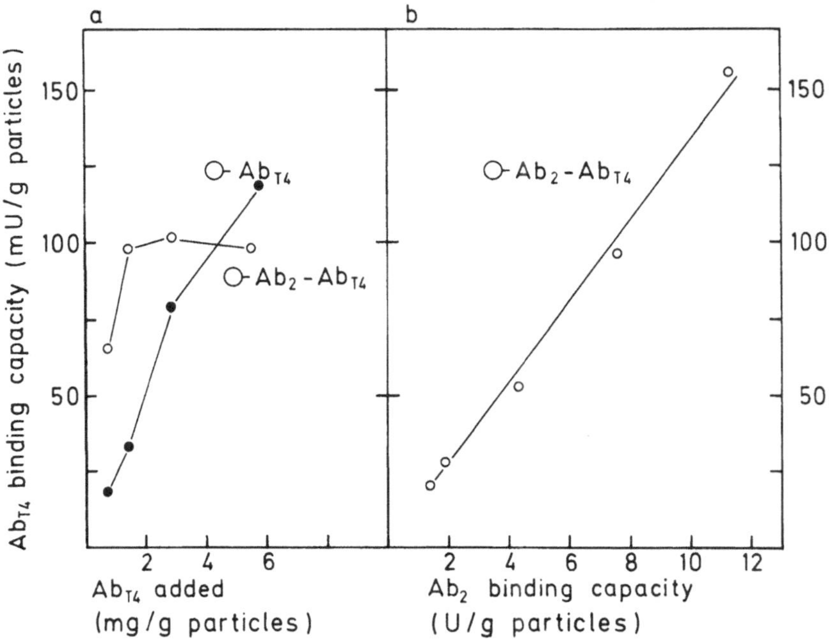

FIGURE 3. Rabbit anti-thyroxine (AbT4) antibodies immobilized on monodisperse polymer particles. (a) AbT4 was purified by protein A-Sepharose chromatography and coupled to tosylated particles (●), or adsorbed on to particles with coupled sheep anti-rabbit IgG (○). Chemical coupling of AbT4 was performed as described for Ab_2 in the legend to Figure 2. Immunoadsorption was performed by mixing AbT4 with particle-bound Ab_2 overnight at 20°C followed by three washings in T4 assay buffer (0.05 *M* Tris, 0.1% merthiolate, 0.01% anilino-naphthalene sulfonic acid, 0.01% BSA, 0.001% amidoblack, pH 8.0). The T4-binding capacity was determined by incubating 100 μℓ AbT4 with 10 μℓ serum devoid of T4 and 500 μℓ iodinated T4 in assay buffer for 2 hr at 20°C. Solid phase Ab_2 (6 mU) in 1-mℓ assay buffer was added, followed by 1-hr incubation at 20°C, centrifugation (3000 × *g* for 15 min), decantation and counting of the precipitate. Dilutions of AbT4 antiserum were used as standards and undiluted antiserum was defined as 1 unit per milliliter. Solid-phase Ab_2 was omitted when particle-bound AbT4 was assayed. (b) The amount of AbT4 immunoadsorbed on particles with Ab_2 was dependent on the rabbit-IgG binding capacity of different batches of solid-phase Ab_2. The plateau in Figure 3A was reached using 0.2 μg rabbit IgG per mU of Ab_2.

drophobic region in an antibody molecule is in the hinge region.[13] The hydrophobicity of this region can be further increased by acid treatment.[13] The same probably happens when preparing affinity-purified antibodies.[14]

Antibodies coupled to tosylated particles should have a similar orientation as adsorbed antibodies since the initial step is a rapid hydrophobic adsorption. The slow chemical coupling probably then takes place with available amino groups in the same area of the antibody molecule. In addition, by using anti-immunoglobulin antibodies on the particles which are Fc specific, we obtain an almost ideal orientation of the primary antibodies. The advantages of this approach are illustrated in Figure 3.[15] Direct coupling of rabbit antithyroxine (T4) antibodies on the particles gives a fair T4-binding capacity only when applying large amounts of antibody (Figure 3A). The same T4-binding capacity is obtained using far less antibody on particles with antirabbit immunoglobulin antibodies (Figure 3A). The plateau value in Figure 3A is determined by the rabbit IgG-binding capacity of the particle-bound antirabbit IgG antibody, and can be changed by using particles with a different amount of antirabbit IgG antibodies (Figure 3B). The primary antibody should therefore not be added using a constant amount of IgG per gram particles, but rather using a constant amount of IgG per

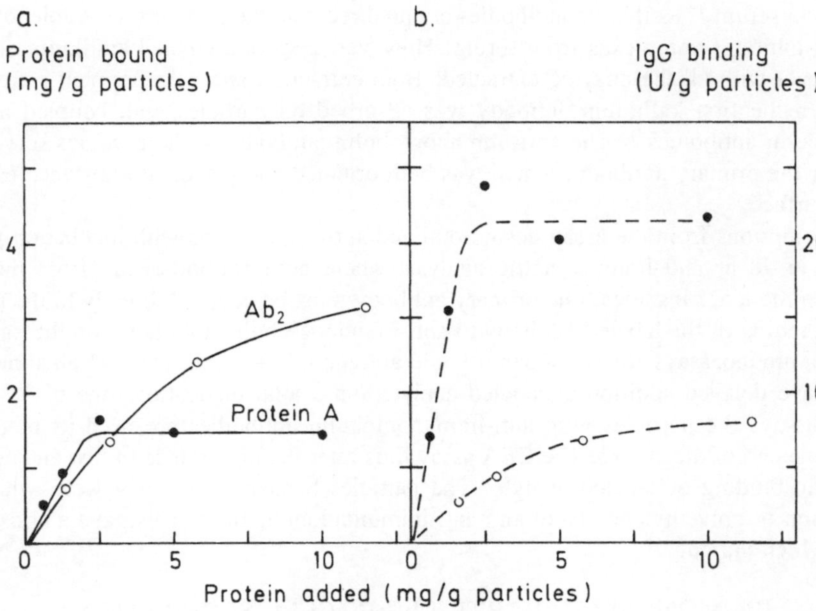

FIGURE 4. Affinity-purified sheep antirabbit IgG (Ab_2) and protein A coupled to tosylated particles. (a) Ab_2 and protein A (Pharmacia Ltd., Uppsala, Sweden) were coupled to tosylated particles as described for Ab_2 in the legend to Figure 2. (b) The rabbit-IgG binding capacity of both solid-phase preparations were tested as described in the legend to Figure 1B.

unit of anti-immunoglobulin antibody on the particles. We find that 0.2 to 0.3 μg rabbit or mouse IgG should be added per milliunit of binding capacity. Antimouse immunoglobulin antibodies are tested in an analogous assay and expressed in units using the same definition, as shown for antirabbit immunoglobulin antibodies in Figure 1B.

An alternative approach is the use of protein A from staphylococci of the Cowain strain on the particles, since this protein has five regions which bind to the Fc region of IgG molecules of certain types and species.[16] A comparison of protein A with our sheep antirabbit IgG antibody is shown in Figure 4. From Figure 4 we calculate that the antirabbit immunoglobulin antibodies (Ab_2) bound 0.5 molecules of rabbit IgG per molecule of Ab_2, whereas the corresponding figure for protein A was 0.6. The assumptions made were that saturated particles contained 2.5 mg Ab_2 or 1.5 mg protein A, and that the molecular weights were 160,000 and 42,000, respectively. Saturation of our sheep antirabbit IgG antibodies will increase the binding to about one molecule of IgG for each antibody molecule.[15]

D. Application of Monosized Particles in Immunoassays

The primary antibodies raised in rabbits and as mouse monoclonals are usually immunoadsorbed to particles with chemically coupled affinity-purified anti-immunoglobulin antibodies as discussed above. These preparations are stable for many months and give a fine orientation of the primary antibodies with well preserved antigen-binding capacities. This procedure also saves us the additional work of purifying and doing chemical coupling of all primary antibodies. Examples of such assays are: an immunoenzymometric assay for placental alkaline phosphatase and prostatic acid phosphatase.[5,17] In these assays the enzyme activity of the analytes are measured and the particles with their antibodies are used as a prior immunoextraction step. A particularly interesting analytical problem was the measurement of an enzyme present in complex with serum inhibitors. This was solved for rat glandular kallikrein, a trypsin-like serine proteinase, which is complexed with several trypsin inhibitors in serum.[18] Kallikrein antibodies bound directly to the particles were able to extract kallikrein-inhibitor complexes from serum. However, a second labeled kallikrein antibody did not react with all the enzyme extracted. Both extraction and measurement were easily done when the first kallikrein antibody was adsorbed on particles with coupled anti-immunoglobulin antibodies.[18] The anti-immunoglobulin antibody on the particles served as a spacer for the primary antibody, which was both oriented and placed at a distance from the particle surface.

The exemptions from the assay design outlined above are assays with incubation periods extended to 48 hr and immunometric analysis where both antibodies are from the same species. In the latter instance, one primary antibody must be coupled directly to the particle to avoid binding of the labeled antibody to anti-immunoglobulin antibodies on the particles. Our radioimmunoassays for carcinoembryonic antigen (CEA) and placental alkaline phosphatase have delayed addition of labeled analyte and a total incubation time of 48 hr.[19,20] In these assays the particles with anti-immunoglobulin antibodies are used as precipitant added at the end of the assays. The CEA assay illustrates the importance of low and constant nonspecific binding of labeled analyte. The particles behaved extremely well, whereas a combination of polyethylene glycol and anti-immunoglobulin antibodies gave a nonspecific binding which varied.[9,19]

III. MONOSIZED PARTICLES IN CELL SEPARATION

A. Why Use Monosized Particles for Cell Separation?

Cell separation based on the interaction between a cellular antigen and an antibody can be accomplished using soluble antibodies and flow cytometry with a cell sorter, or by antibodies on a stationary solid phase in a panning technique, or by antibodies on mobile particles. The cell sorter is mainly an analytical tool and can only cope with a limited number of cells within a reasonable time. We are therefore again faced with the choice of a mobile vs. a stationary solid-phase approach. A direct comparison of the two methods is not known to us, but the following facts are available: the binding of cells to a bottle coated with antibody is easily done and can be scaled up by use of multiple bottles. However, the binding is relatively weak, since collected cells can be recovered by simple manipulations with a Pasteur pipette.[21] This technique is therefore useful for both positive and negative selection of cells. The possibility of accidental loosening of bound cells indicates that the technique should not be used for binding of cells where absolutely all cells must be removed. The interaction between cells and particles with antibodies is extremely strong, as documented in Section III.E. An easy system for collecting complexes of cells and particles is obtained by using magnetic particles. Magnetic particles are probably therefore the method of choice for complete removal of a defined cell population from a mixture of cells.

B. Essential Properties for Particles to be Used in Cell Separation

The general characteristics of an ideal particle for immunoassays may also be applied for particles to be used for cell separation. In addition, the monodisperse particles should contain a constant amount of magnetizable material. Ugelstad and co-workers have prepared particles which besides being monodisperse have the same amount of maghemite inside each particle.[22] The method of preparation of these particles involved the use of modified monosized macroporous particles based upon styrene divinyl benzene. The monodispersity of these particles as shown in Figure 5 is of importance for the preparation of particles with a constant amount of magnetic material since iron was introduced as soluble Fe^{2+} and continuously oxidized inside the pores to give insoluble precipitates of maghemite, γFe_2O_3. The maghemite is evenly distributed throughout the particles as will be seen from the cross-sections pictured in Figures 6 to 13. The surface of the particles is 100 m^2/g. These 3-μm particles have been used for removal of neuroblastoma cells from bone marrow.[23]

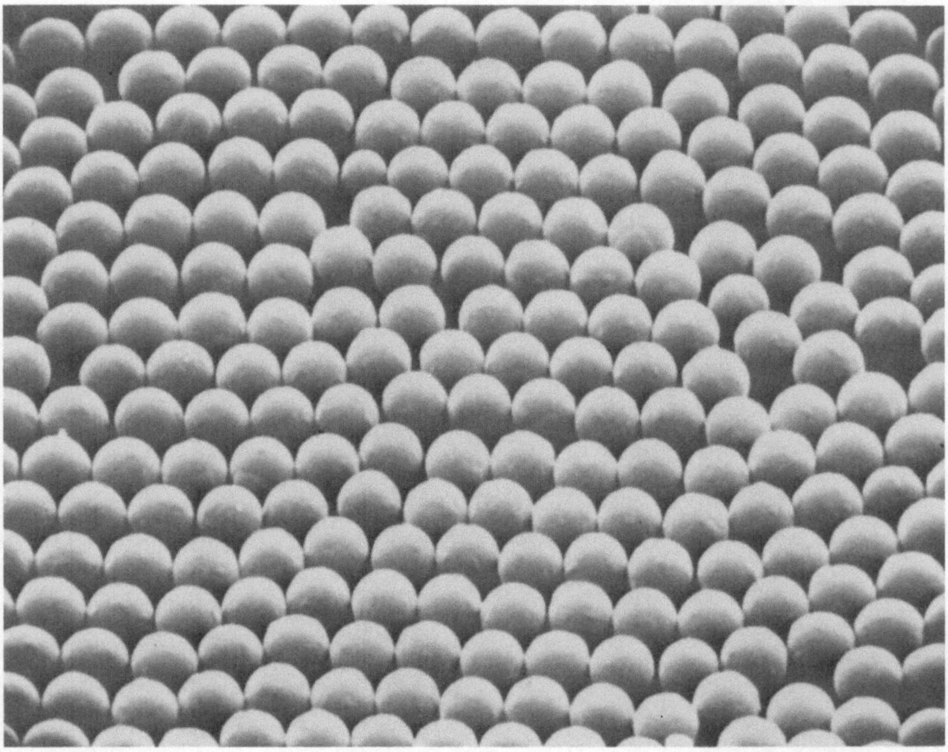

FIGURE 5. Survey view demonstrating the monomorph character of the magnetic sphere ME 228. (Magnification × 4000.)

The large surface caused by the porous structure is a disadvantage. First of all it necessitates the use of a large amount of antibody, 100 μg/mg particles, and obviously only a small fraction of this will be at the outer surface and thereby active in binding antigen. More recent work showed that the porous structure also led to a high nonspecific binding of cells. A number of particles were therefore prepared and different methods tried to fill and/or cover the pores with different oligomeric and polymeric materials. About 50 different particle preparations were prepared and tested using three different cell systems: mouse liver cells, permanent cell lines generated from B cell lymphoma, and T lymphocytes from peripheral blood. Two main conclusions concerning the particle properties emerged from these investigations. The particle surface should be small, with no open pores. It should be possible to make a dispersion of particles in water without any particle agglomeration before binding the antibody to the particles. Nonspecific binding of particles to the cells is observed if these two requirements are not fulfilled. The most efficient particles which so far have been tested, denoted M 450, have a diameter of 4.5 μm, a surface of about 3 to 5 m^2/g and contain about 20% Fe in the form of Fe_3O_4. The polymer used to fill and cover the pores is hydrophobic, so that it allows a binding of antibody by physical adsorption. However, the polymer also contains sufficient –OH groups on the surface to allow chemical coupling of antibody with the tosyl chloride method. The number of OH groups is much lower than for the core-and-shell particles discussed above, so that normally the tosyl chloride method will give a mixture of chemical and physical binding.

Particles with a more hydrophilic structure are under development as are particles with a thicker polymeric layer surrounding the magnetic particle. The latter is advantageous as it will diminish magnetic interaction between the particles.

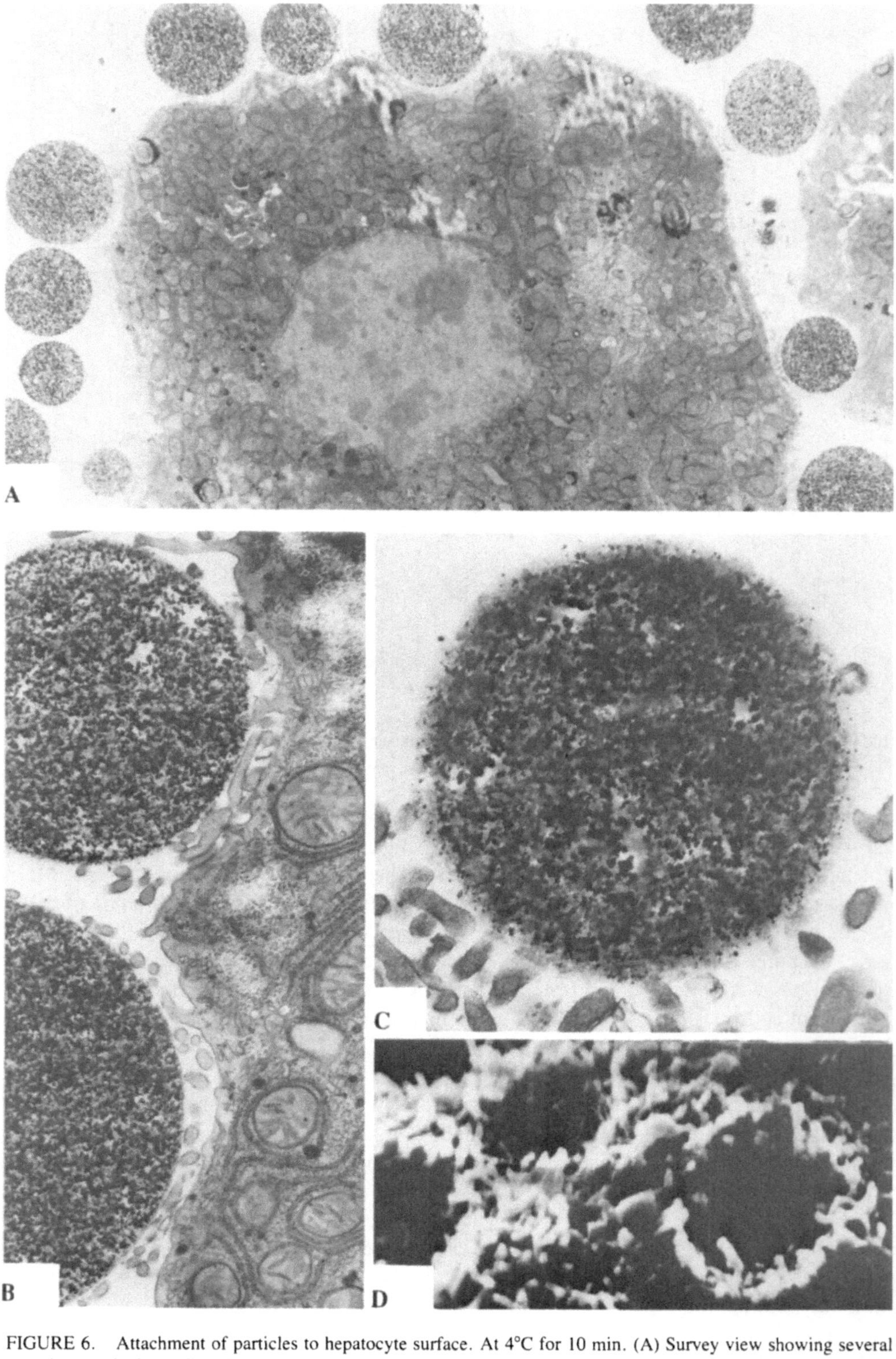

FIGURE 6. Attachment of particles to hepatocyte surface. At 4°C for 10 min. (A) Survey view showing several particles attached to cell surface inducing shallow depression (arrowheads). Microvilli extending from the surface are attached to the near side of the particles, mostly at the bottom. (Magnification × 2,000.); (B) attachment of particle bottom by microvilli. (Magnification × 20,000.); (C) a close contact between microvilli tip and particle surface. (Magnification × 40,000.); (D) SEM showing the nest of microvilli depression at cell surface. (Magnification × 5,000.)

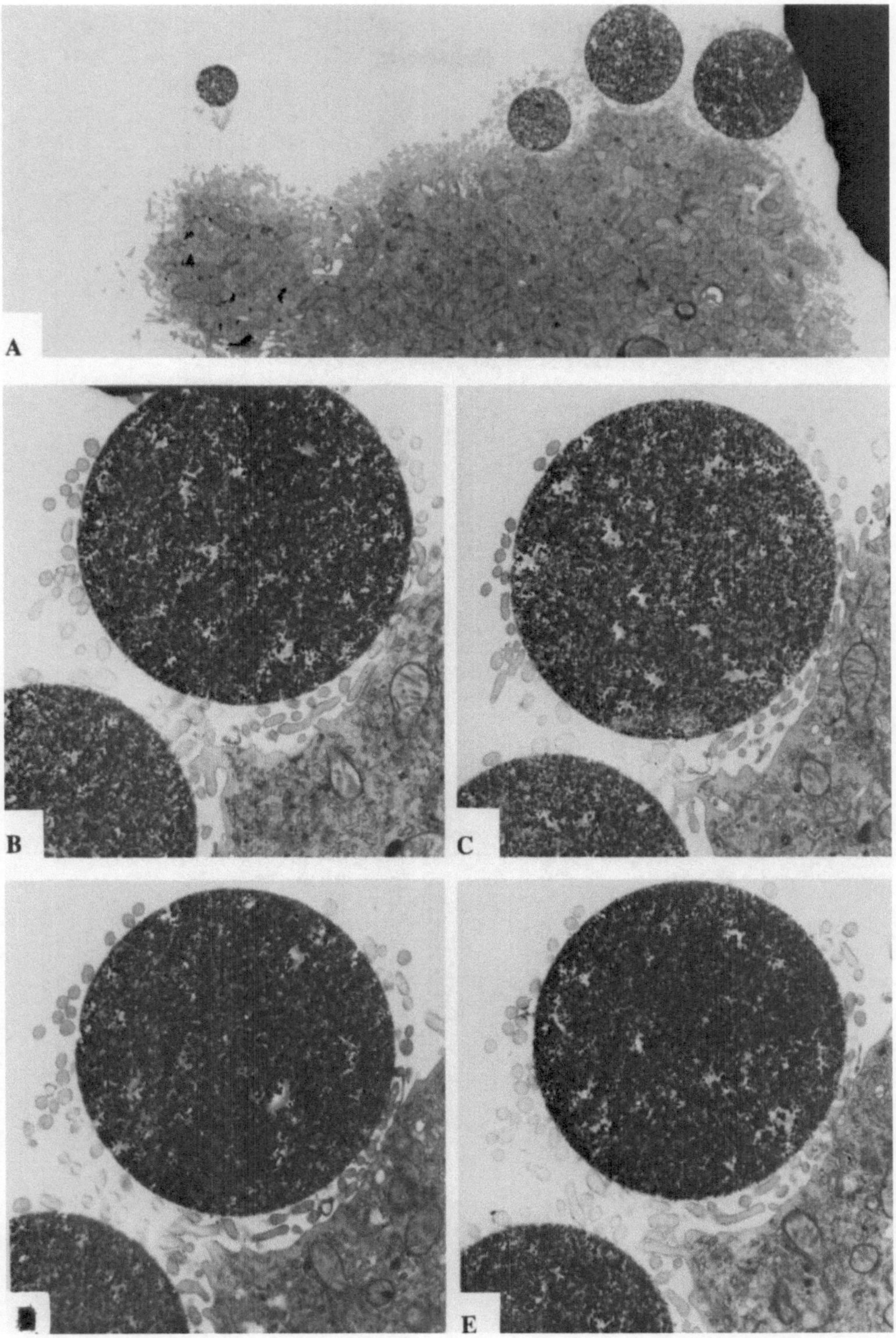

FIGURE 7. Attachment of particles to hepatocyte surface. At 4°C for 60 min. (A) Overview demonstrating particles covered totally by slender and long microvilli-like extensions. (Magnification × 2,000.) One particle (asterisk) in higher magnification; (B — E) serial sections of a particle with attached microvilli-like extensions. Note that the abundance of microvilli seen is not typical for cells incubated at the low temperature. (Magnification × 20,000.); (F) SEM demonstrates the slender microvilli-like extensions covering the particle surface. (Magnification × 5,000.)

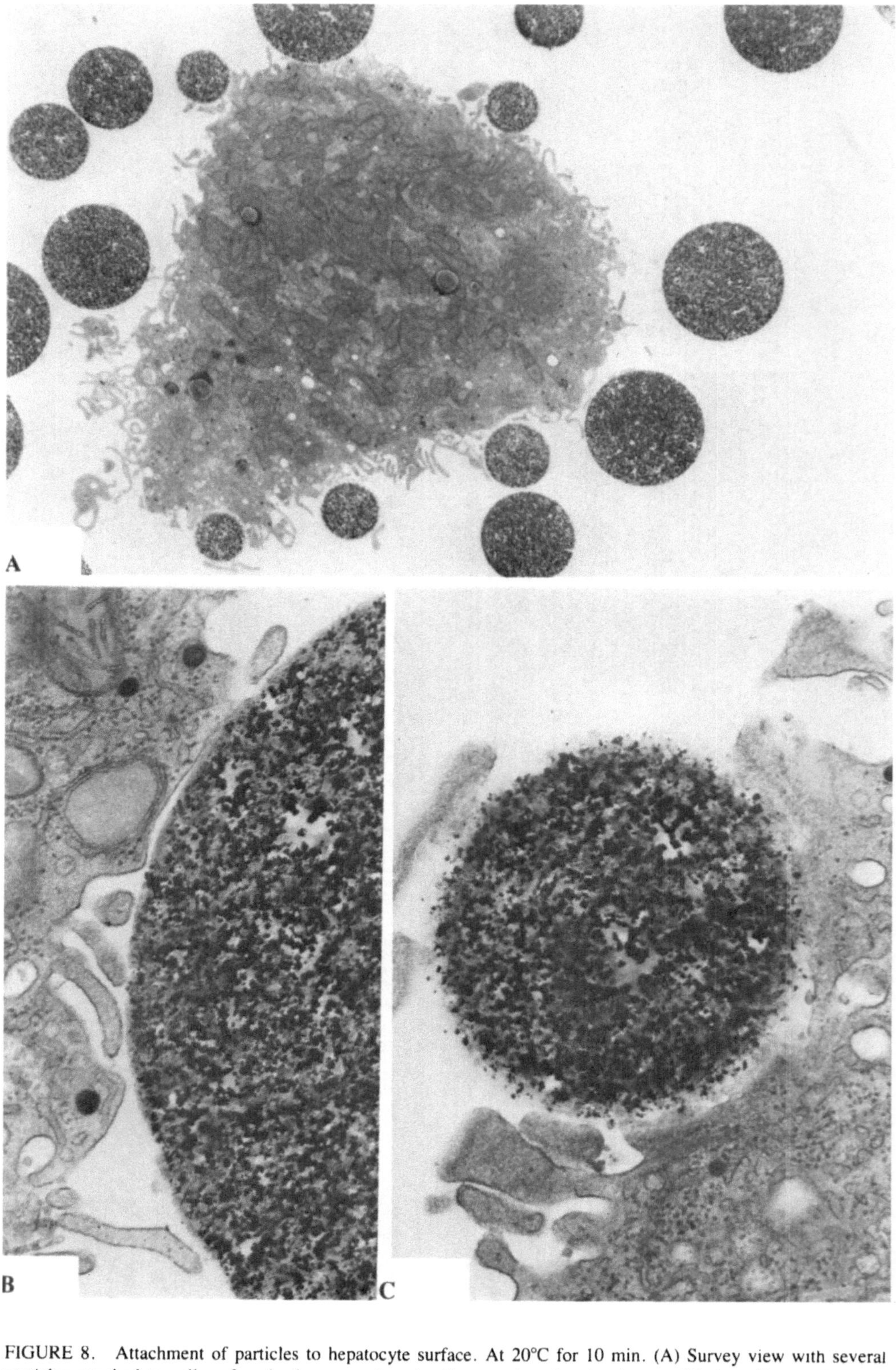

FIGURE 8. Attachment of particles to hepatocyte surface. At 20°C for 10 min. (A) Survey view with several particles attached to cell surface by long pseudopods. Compared to the situation at 4°C the cell extensions are generally much longer. (Magnification × 2,000.); (B) at higher magnification the close contact over larger distances (arrowheads) is seen. (Magnification × 40,000.); (C) adjacent to the particle the cytoplasm shows beginning of microfilament aggregations (arrows). (Magnification × 40,000.)

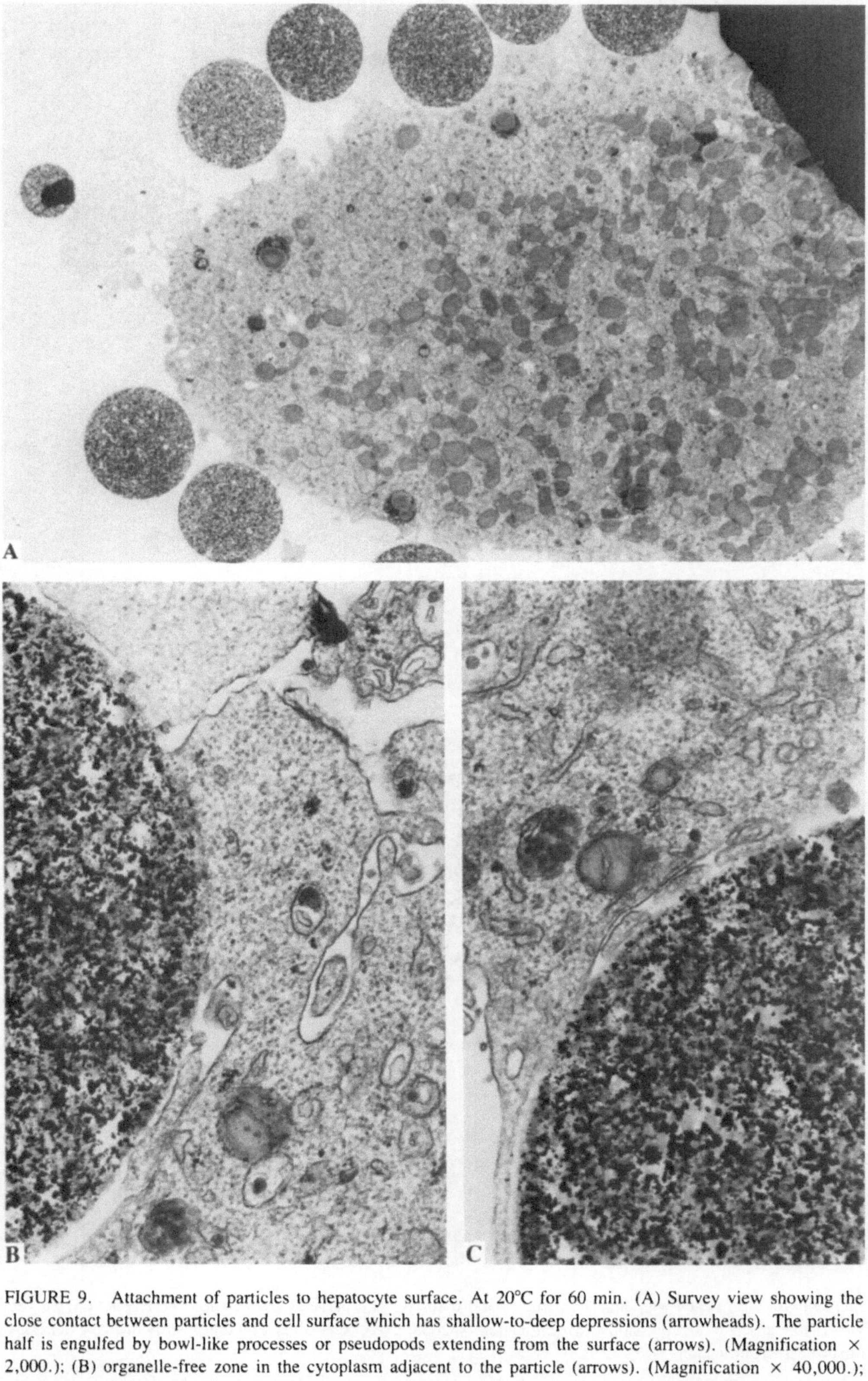

FIGURE 9. Attachment of particles to hepatocyte surface. At 20°C for 60 min. (A) Survey view showing the close contact between particles and cell surface which has shallow-to-deep depressions (arrowheads). The particle half is engulfed by bowl-like processes or pseudopods extending from the surface (arrows). (Magnification × 2,000.); (B) organelle-free zone in the cytoplasm adjacent to the particle (arrows). (Magnification × 40,000.); (C) the slender, bowl-like lamella enfolding the particle (arrowheads). (Magnification × 40,000.)

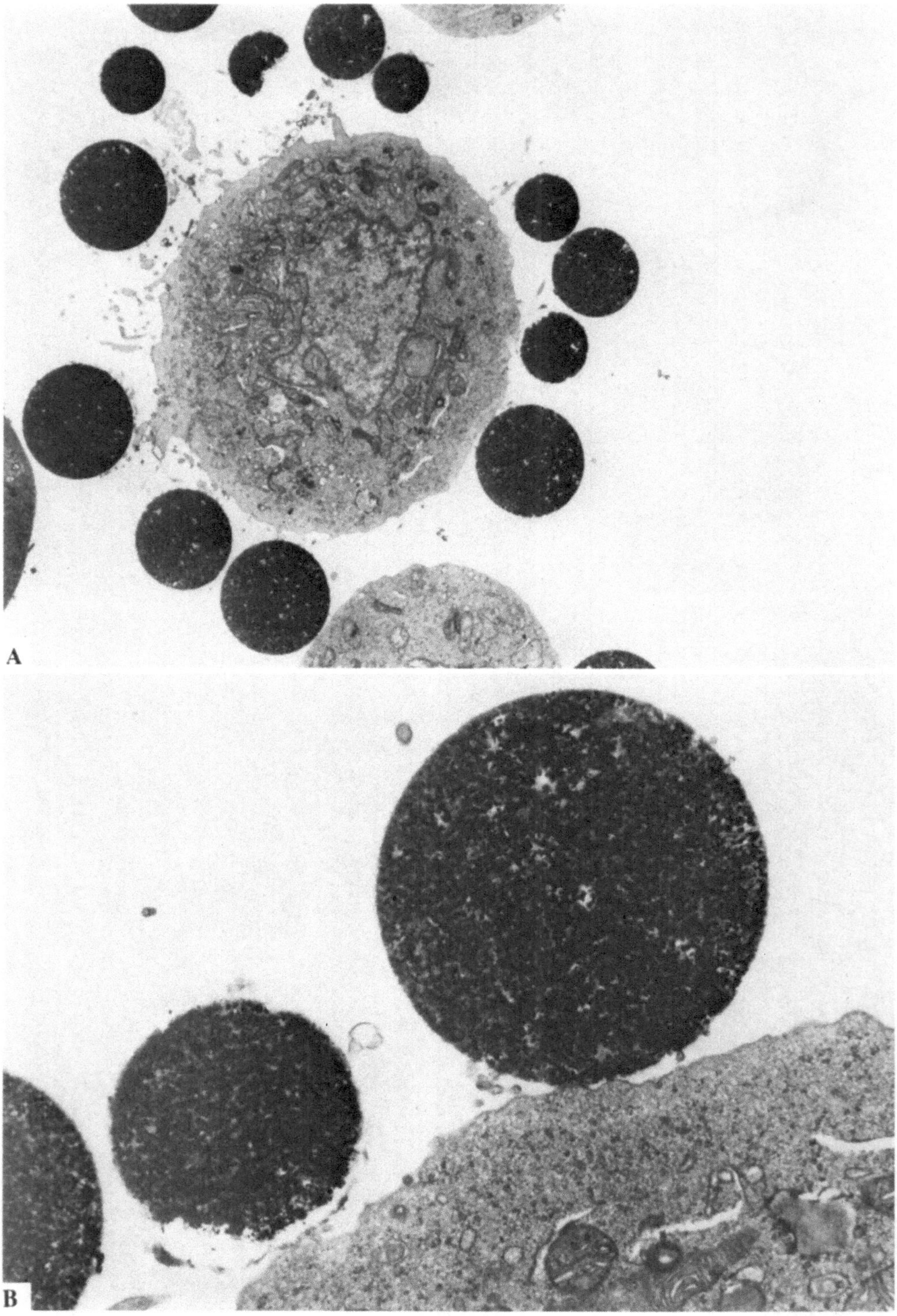

FIGURE 10. Raji cells with attached particles at 4°C for 10 min, particles are attached to the cells. (A) Survey view. Note absence of depressions. (Magnification × 2000.); (B) loose contact by thin microvilli-like extensions (arrows). (Magnification × 20,000.)

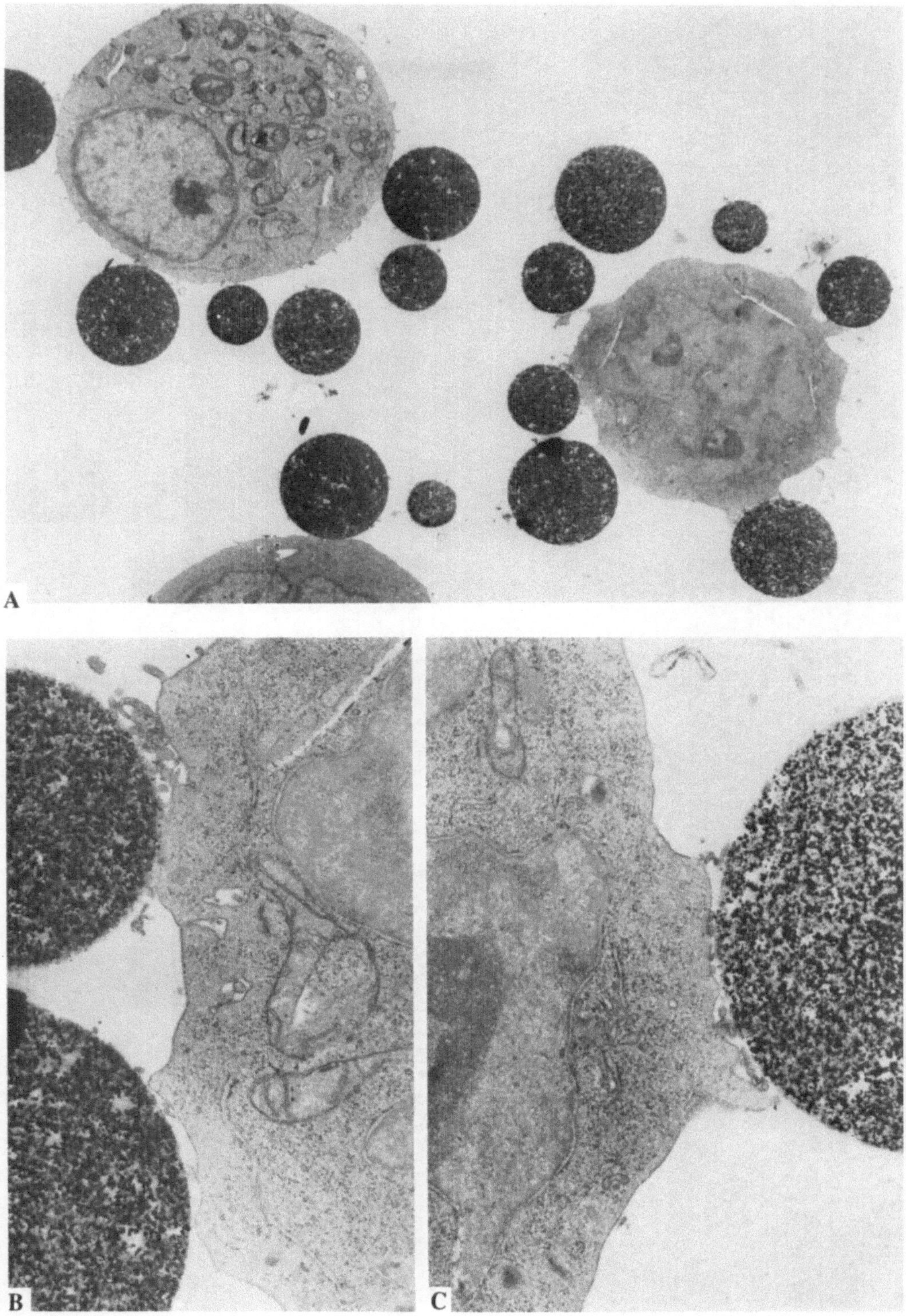

FIGURE 11. Raji cells with attached particles at 4°C for 60 min. (A) Survey shows the closer, compared to 10 min, attachment of particles to cell surface which has shallow depressions (arrowheads). (Magnification × 2000.); (B) the cell periphery is slightly elevated at the contact side with the particles. (Magnification × 20,000.); (C) between the elevated cytoplasm and the particle surface thin microvilli-like extensions are located (arrows). (Magnification × 20,000.)

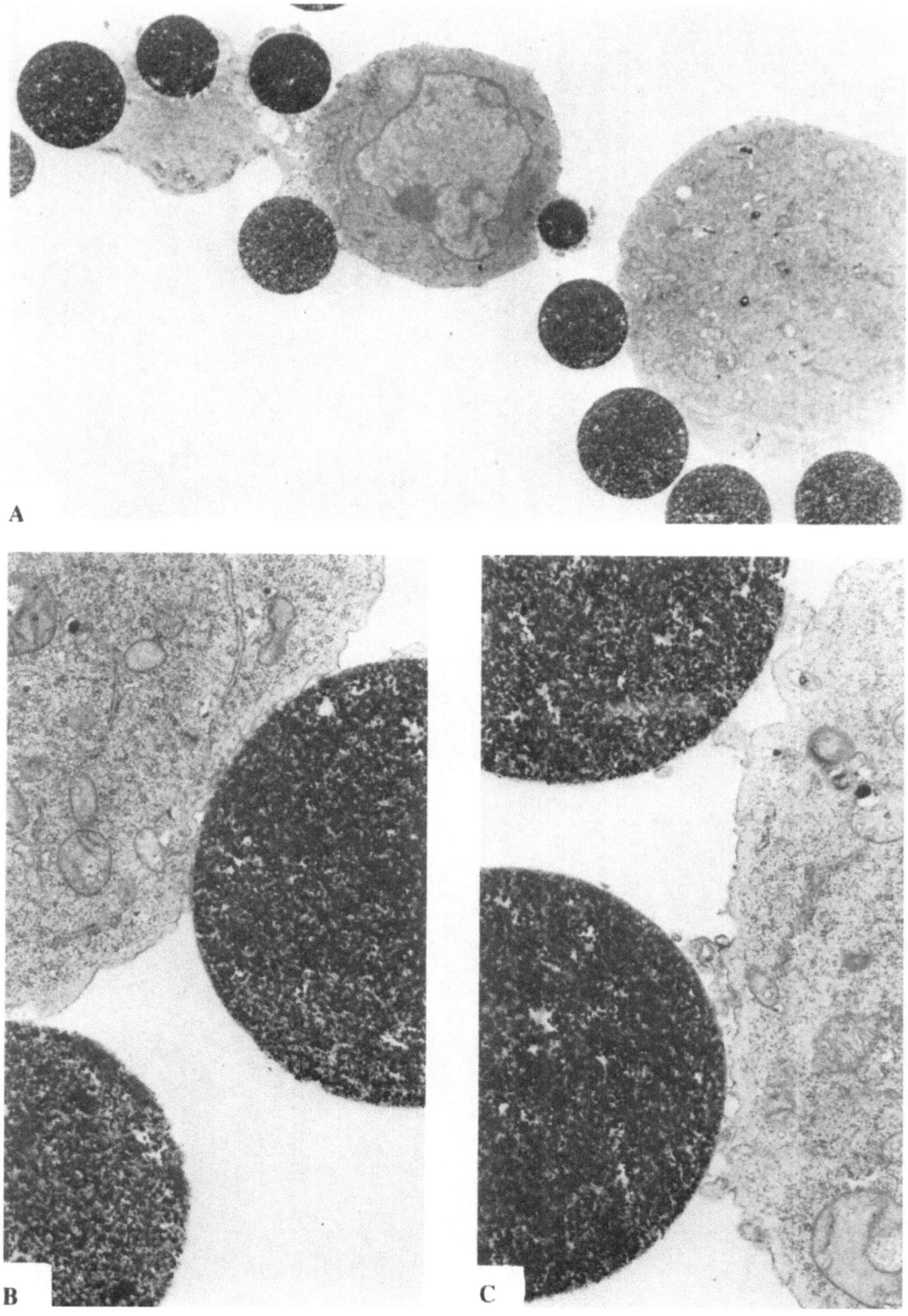

FIGURE 12. Raji cells with attached particles at 20°C for 10 min. (A) The survey view shows the close contact between particle surface and cell surface, which has deep depressions (arrowheads). (Magnification × 2000.); (B) Close contact between cell and particle surface over an extended area. (Magnification × 20,000.); (C) sometimes the contact between cell and particle surface is by means of small microvilli (arrow) or pseudopodlike extensions (arrowheads). (Magnification × 20,000.)

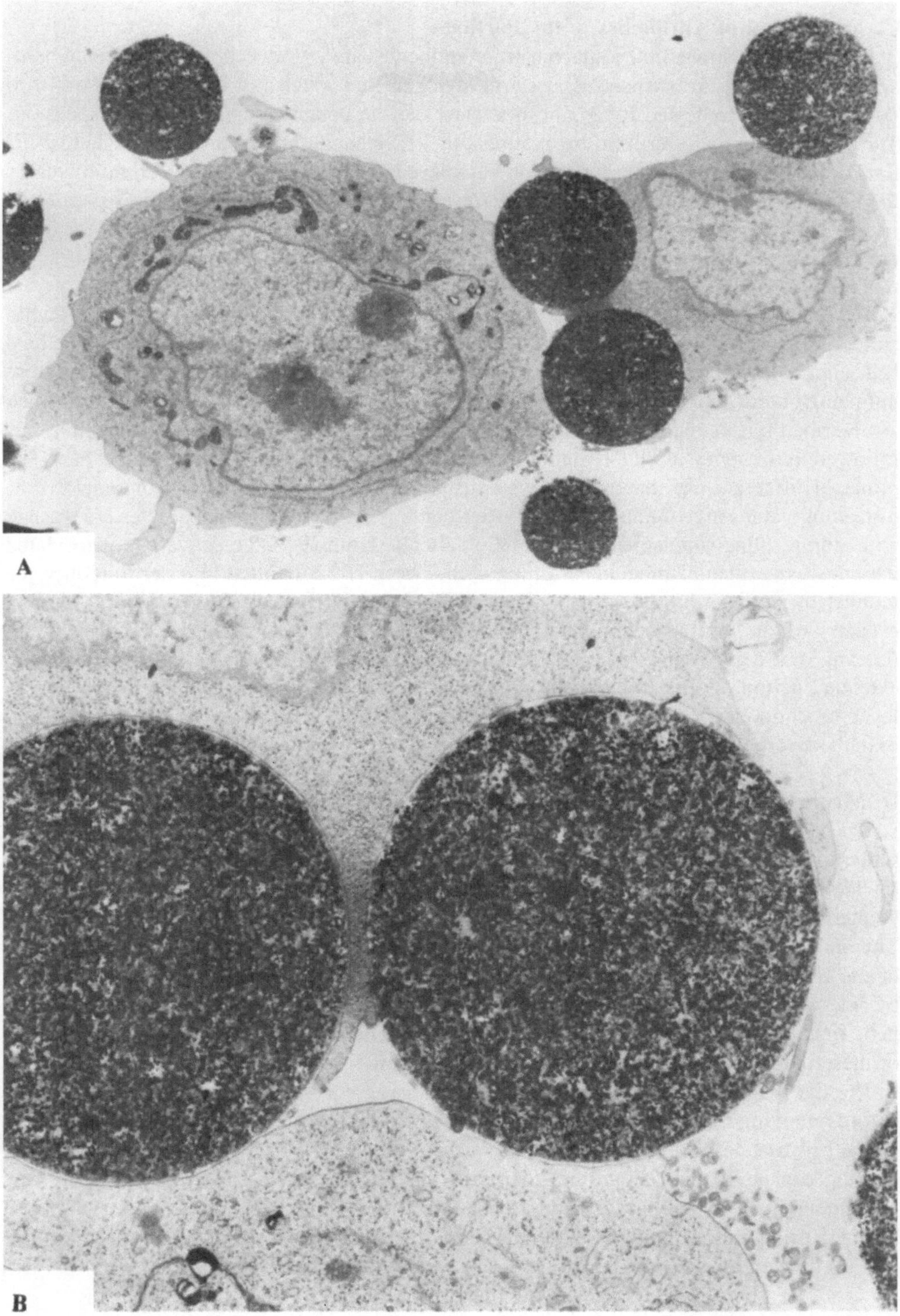

FIGURE 13. Raji cells with attached particles at 20°C for 60 min. (A) The particles are partly engulfed by thin bowl-like lamellae. Often two cells share common particles (arrowheads). (Magnification × 2000.); (B) these extensions show microfilament structure (arrows) at higher magnification. (Magnification × 20,000.)

C. Attachment of Antibodies to the Particles

Minor modifications in the attachment of antibodies are introduced to obtain sterile preparations. Particles are suspended in sterile pyrogen-free water at a concentration of 10 to 100 mg/mℓ and sonicated for 3 min in a water bath to obtain a single-particle suspension. The particles are then sterilized by treatment in 70% ethanol for 1 hr or autoclaved at 120°C for 20 min. Tosylation is performed on autoclaved particles since this procedure increases the number of hydroxyl groups available for activation. The entire procedure is performed in the bottle used for autoclaving. The tosylation procedure is similar to that used for nonmagnetic particles with 4 g tosyl chloride and 4 mℓ pyridine per gram particles.[8] A treatment with 70% ethanol is included after tosylation as an extra sterilization step.

The antibodies attached to the particles were affinity-purified antimouse immunoglobulin antibodies of sheep and rabbit origin, or monoclonal IgM antibodies, as stated. Monoclonal IgG antibodies were either added to the cells or adsorbed on particles with antimouse immunoglobulin antibodies. Adsorption and chemical coupling of antibodies were performed as described for nonmagnetic particles. However, a slightly different adsorption routine was followed in the removal of T lymphocytes (Section III.F.3), as described in Lea et al.[24] The principal difference was that particles were suspended in Hanks balanced salt solution (HBSS) instead of water as recommended here. After binding of antibodies, the particles are washed with sterile saline supplemented with 1% fetal calf serum (FCS). Inactivation of tosylated particles with ethanolamine and washing with Tween 20[8] is omitted in order to reduce the number of sterile solutions to a minimum. This has no effect on the mouse-IgG binding capacity of the antimouse immunoglobulins antibodies coupled. We are now including freezing or freeze drying in order to obtain preparations which can be kept sterile for an extended period of time. Freeze drying can also be performed on particles with a double layer of antibodies, i.e., chemically coupled antimouse immunoglobulin antibodies and immunoadsorbed monoclonal antibodies.[25]

D. Magnets for Collection of Particles

We have been through several stages in magnetic devices. We started up using cobalt-samarium magnets (10 × 10 × 3 mm) on the outside of a polycarbonate tube as described by Treleaven et al.[23] An increased magnetic force was obtained by placing these magnets at alternating sites along a V-shaped groove in a solid piece of iron (20 × 20 × 100 mm). This increased the magnetic field tremendously, but gave some trapping of cells at the site of the first magnet. At present we are using a "patchwork quilt" of magnets placed side by side with alternating north-south orientation. This gives a flat, fairly homogeneous magnetic field but with magnetic "holes" in the corners where four magnets meet. The thickness of these magnets may also be varied to obtain optimal magnetic strength.

The flat magnets are placed for 1 min. under tissue bottles with samples to be collected before removing free cells by decantation. The process is repeated once to reduce the number of trapped free cells. The number of cells can be at least 5 × 10^6 cells/mℓ and the height of fluid can be 10 mm above the surface of the magnet (see Sections III.F.1 and 2). A continuous flow system in the form of 3-mm thick tubing in conjunction with the V-shaped magnet, using a flow of 1 to 2 mℓ/min and a cell density of 5 × 10^6 cells per milliliter was used when a small number of cells was sorted (see Section III.F.3).

E. The Interaction between Particle-Bound Antibody and the Cellular Membrane[25]

Normal mouse liver cells were taken from the (C57BL/6J X C3H/HEJ)F_1 strain, and suspensions of vital hepatocytes were prepared by collagenase perfusion and differential centrifugation.[26] The hepatocytes were saturated with monoclonal antimouse H2[27] specific for the C57BL/6J strain using approximately 5 μg mouse IgG_{2a} per 10^6 cells and an incubation of 30 min at 4°C.[26] Excess antibodies were removed by three washings in HBSS supplemented

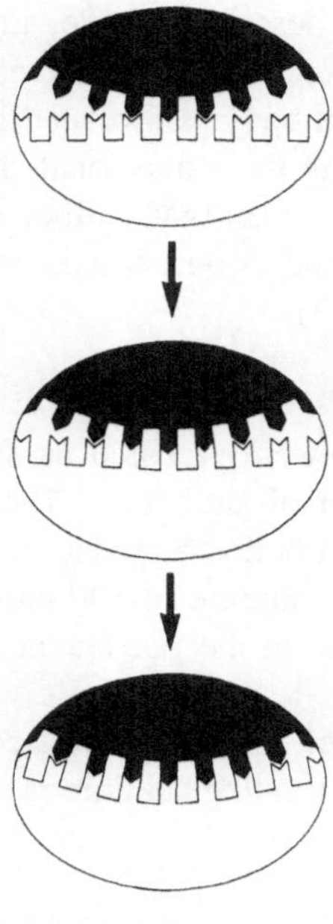

FIGURE 14. Schematic diagram of the interaction between particle with its antibody, i.e., affinity-purified sheep antimouse immunoglobulin antibody (▼) and the corresponding antigen on the cellular surface, i.e., monoclonal mouse IgG (M). The zipperlike binding is thought to create the depression in the cellular membrane and trigger the cellular response in the form of protruding villi and pseudopods.

with 5% FCS. The cells were incubated with end-over-end rotation (1 rpm) with particles coated with sheep antimouse immunoglobulin antibodies at 4 and 20°C for 10 to 60 min. The number of particles per cell was from 200 to 600. At 4°C, binding of particles to the cells was followed by a definite shallow depression in the cellular surface (Figures 6 to 9). This is probably due to multiple interactions between the antimouse immunoglobulin antibodies on the particles and mouse IgG on the cellular membrane by a zipper-like mechanism earlier described for macrophages.[28-30] When the temperature was increased to 20°C, a protrusion of fingerlike villi and pseudopods occurred on the cellular surface. These protrusions almost encircled some of the particles but fusion of the membrane on top of the particles was not observed (Figure 9). We surmise that this represents a kind of phagocytic process induced by the antigen-antibody interaction on the cellular membrane.

The general nature of this process was shown by repeating the experiments with another cell type. Raji, a cell line derived from a Burkitt lymphoma, were saturated with a mouse monoclonal antibody (AB-3) which is an IgG_{2a} reacting with class II antigen in the major histocompatibility complex.[31] The binding of particles with antimouse immunoglobulin antibodies to these cells was similar to that observed with the hepatocytes. The main difference was the enormous stretching of villi and pseudopods which occurred at 20°C in these cells with their sparse cytoplasm (Figures 10 to 13). The interaction of particles and cells is very strong. Mechanical manipulation, for instance with a Pasteur pipette, will not detach the particles. The zipper-like binding of multiple antibodies on a particle with the corresponding antigen on the cell membrane is shown schematically in Figure 14.

F. Cell Sorting Using Magnetic Particles

Mouse liver hepatocytes, Raji cells, and human T lymphocytes could be caught by magnetic particles using the procedure described above, i.e., saturating the cells with mouse monoclonal antibodies, removing excess monoclonal antibodies and finally adding particles with antimouse immunoglobulin antibody. However, the process could be shortened and more viable cells obtained by placing the monoclonal antibodies on the particles. This was done by direct coupling of the monoclonal IgM antibodies (Section III.F.2) and by immune adsorption of monoclonal IgG antibodies on particles with antimouse immunoglobulin antibodies (Sections III.F.1 and III.F.3).

1. Mouse Liver Parenchymal Cells[25]

Normal mouse liver hepatocytes were prepared as described above and incubated with particles M 450 with a double layer of antibodies. The bottom layer was affinity-purified sheep antimouse immunoglobulin antibodies bound to tosylated particles, and the upper layer was monoclonal antimouse H2.[27] A mixture of 300 particles per cell were rotated (1 rpm) for 30 min at 4°C before separation on the flat magnet. At least 99.9% of the liver cells were collected on the magnet after 1 min. No cells were captured on the magnet when control liver cells from the C3H/HeJ strain were treated in the same manner. These cells were not damaged, as judged by morphological criteria and the Trypan Blue exclusion test.

2. B Lymphocytes[25]

Raji cells were stained with the vital dye H 33342 from Hoechst as described by Reynolds et al.[32] Stained cells (5×10^6) were mixed with an equal number of normal human bone marrow cells in 2 mℓ RPMI medium with 5% FCS. The monoclonal antibody (AB-1), which is a mouse IgM antibody specific for human B lymphocytes, was coupled to tosylated M 450 particles.[31] Incubation with cells and collection on the flat magnet was performed as described for the direct system in III.F.1. After separation, more than 99.9% of the Raji cells were removed, whereas more than 90% of the marrow cells were present in the depleted cell suspension. Similar results were obtained with another pan B-cell monoclonal IgM antibody AB2.[31]

3. Removal of T Lymphocytes from Peripheral Blood[24]

Monodisperse polymer particles (M 450), previously sterilized in 70% ethanol, were suspended in HBSS at a concentration of 5 mg/mℓ and subsequently coated with affinity-purified, species-specific rabbit antimouse IgG antibodies (10 μg/mg particles) at 4°C for at least 24 hr on a rotation wheel. Unoccupied protein-binding sites on the particles were blocked by incubation of the particles with FCS (10 μℓ/mg particles) at 4°C for 24 hr. The particles were then washed twice for 30 min in HBSS supplemented with 1% FCS and subsequently incubated with a mixture of monoclonal anti-T3 and anti-T11 antibodies in saturating amounts for 1 hr at room temperature. Finally, the particles were washed twice for 30 min in HBSS-1% FCS. Particles coated in this way are in the following designated (A). For control experiments, particles coated with control ascitic fluid instead of T cell-specific antibodies and particles coated with FCS only were prepared and designated (B) and (C), respectively.

Peripheral blood mononuclear cells (PBM) were obtained by Isopaque Ficoll gradient centrifugation.[33] In these experiments 10^6 PBM were rosetted with 2 mg particles corresponding to a particle/cell ratio of approximately 20. Aliquots of PBM (5×10^6 cells/mℓ) were incubated in HBSS-1% FCS with particles (A), (B), and (C), respectively in plastic tubes for 30 min at room temperature. The tubes were carefully rotated for 1 min at 10-min intervals. Before separation, small samples were taken from the particle-cell suspensions and the number of rosette-forming cells was counted (Table 1). Magnetic separation of cells was carried out by letting the particle-cell mixture flow (1 to 2 mℓ/min) through a 3-mm plastic tube attached to a V-shaped iron block lined with cobalt-samarium magnets. Effluent

Table 1
E_{AET} (AMINOETHYLISOTHIOURONIUM-TREATED SHEEP ERYTHROCYTES) ROSETTE-FORMING CELLS AND ^{3}H-THYMIDINE INCORPORATION IN PBM (PERIPHERAL BLOOD MONONUCLEAR CELLS) DEPLETED WITH PARTICLES COATED IN THE SECOND LAYER WITH MONOCLONAL T LYMPHOCYTE-SPECIFIC ANTIBODIES

	Peripheral blood mononuclear cells (PBM) rosetted with[a]			
	Particle A	**Particle B**	**Particle C**	**Untreated PBM**
Before depletion				
Particle rosette-forming cells[b]	87/118 (73%)	3/108 (3%)	0/140 (0%)	
After depletion				
E_{AET}-rosette forming cells[b]	0/530 (<0.2%)	80/127 (63%)	75/128 (59%)	65/118 (55%)
Phytohemagglutinin (PHA)[c]	178	14,872	14,645	13,995
Pokeweed mitogen (PWM)[c]	90	4,645	3,862	2,634
Purified protein derivative (PPD)[c]	217	5,855	4,275	3,346
Allogeneic-irradiated PBM[c]	289	9,848	7,124	7,650
Medium[c]	234	226	231	257

[a] Particles coated with (A) specific antibodies, (B) control ascitic fluid, (C) with FCS (fetal calf serum) only.
[b] Results are given as the fraction of rosette-forming cells in the number of cells counted (percent in brackets).
[c] Results are given as counts per minute.

cells were centrifuged and the cell pellet was resuspended in RPMI 1640 and counted on a Coulter DN cell counter. Cells incubated with particles (A), (B), and (C), as well as untreated PBM, were assayed for rosette formation with 2-aminoethyl-isothiouronium bromide-treated sheep erythrocytes (E_{AET}) and for ^{3}H-thymidine incorporation after stimulation with phytohemagglutinin (PHA), pokeweed mitogen (PWM), purified protein derivative (PPD), and irradiated allogeneic PBM as described elsewhere.[34] The main results of these experiments are shown in Table 1.

The results indicate that no T cells are present after depletion, since no E_{AET} rosetting cells can be detected and no cell proliferation above background level is found after mitogen- or antigen-induced stimulation. Furthermore, experiments with particles not coated with T cell-specific antibodies showed mitogen- and antigen-induced stimulation comparable to that found in untreated PBM, indicating that the particles do not in themselves impair the functional reactivity of the cells to well known mitogens and antigens. We therefore conclude that magnetic monodisperse polymer particles coated with T cell-specific antibodies may be used for rapid, complete removal of T cells from peripheral blood. Similar experiments are now being carried out on normal bone marrow to see if these magnetic particles might be used for the removal of T cells in allogeneic bone marrow transplantation.

ACKNOWLEDGMENTS

The authors are indebted to SINTEF, Division of Applied Chemistry for the preparation of the magnet particles. The monoclonal mouse anti H2 was a generous gift from Professor Jan Klein, Department of Immunogenetics, Tübingen, West Germany.

We are grateful for the technical skills of Kari Thrane-Steen, Bärbel Schüler, Ruth Puntervold, and H. Vereide and the staff of the Photographic Department. We thank Iva Hirsch for carefully preparing the manuscript. This work was carried out with financial support from the Royal Norwegian Council for Scientific and Industrial Research and the Norwegian Cancer Society to whom we are deeply indebted.

REFERENCES

1. **Ugelstad, J., Mørk, P. C., Kaggerud, K. H., Ellingsen, T., and Berge, A.,** Swelling of oligomer-polymer particles. New methods of preparation of emulsions and polymer dispersions, *Adv. Colloid Interface Sci.*, 13, 101, 1980.
2. **Ugelstad, J., Mfutakamba, H. R., Mork, P. C., Ellingsen, T., Berge, A., Schmid, R., Holm, L., Jørgedal, A., Hansen, F. K., and Nustad, K.,** Preparation and application of monodisperse polymer particles, *J. Polym. Sci.*, 72, 225, 1985.
3. **Hamilton, R. G., Hussain, R., Alexander, E., and Adkinson, N. F.,** Limitations of the radioimmunoprecipitation polyethylene glycol assay (RIPEGA) for detection of filarial antigens in serum, *J. Immunol. Methods*, 68, 349, 1984.
4. **Chen, I.-W., Heminger, L., Maxon, H. R., and Tsay, J. Y.,** Nonspecific binding as a source of error in thyrotropin radioimmunoassay with polyethylene glycol as a separating agent, *Clin. Chem.*, 26, 487, 1980.
5. **Millán, J. L., Nustad, K., and Nørgaard-Pedersen, B.,** Highly sensitive solid-phase immunoenzymometric assay for placental and placental-like alkaline phosphatases with a monoclonal antibody and monodisperse polymer particles, *Clin. Chem.*, 31, 54, 1985.
6. **Masson, P. L., Cambiaso, C. L., Collet-Cassart, D., Magnusson, C. G. M., Richards, C. B., and Sindic, C. J. M.,** Particle counting immunoassay (PACIA), *Methods Enzymol.*, 74, 106, 1981.
7. **Hunter, W. M. and Budd, P. S.,** Immunoradiometric versus radioimmunoassay: a comparison using alpha-fetoprotein as the model analyte, *J. Immunol. Methods*, 45, 255, 1981.
8. **Nustad, K., Johansen, L., Ugelstad, J., Ellingsen, T., and Berge, A.,** Hydrophilic monodisperse particles as solid-phase material immunoassays: comparison of shell-and-core particles with compact particles, *Eur. Surg. Res.* 16 (Suppl. 2), 80, 1984.
9. **Nustad, K., Ugelstad, J., Berge, A., Ellingsen, T., Schmid, R., Johansen, L., and Børmer, O.,** Use of monosized polymer particles in immunoassays. Evaluation of particles with hydroxyl groups, in *Radioimmunoassay and Related Procedures in Medicine 1982*, International Atomic Energy Agency, Vienna, 1982, 45.
10. **Johansen, L., Nustad, K., Ørstavik, T. B., Ugelstad, J., Berge, A., and Ellingsen, T.,** Excess antibody immunoassay for rat glandular kallikrein. Monosized polymer particles as the preferred solid phase material, *J. Immunol. Methods*, 59, 255, 1983.
11. **Nilsson, K. and Mosbach, K.,** Immobilization of ligands with organic sulfonyl chlorides, *Methods Enzymol.*, 104, 56, 1984.
12. **Zaborsky, O. R.,** Immobilized enzymes — miscellaneous methods and general classification, *Methods Enzymol.*, 44, 317, 1976.
13. **Vandenbranden, M., Kayser, G., Banerjee, S., and Ruysschaert, J. M.,** Immunoglobulin-lipid interaction. A model membrane study, *Biochim. Biophys. Acta*, 685, 177, 1982.
14. **Conradie, J. D., Govender, M., and Visser, L.,** ELISA solid phase: partial denaturation of coating antibody yields a more efficient solid phase, *J. Immunol. Methods*, 59, 289, 1983.
15. **Nustad, K., Closs, O., and Ugelstad, J.,** Mouse monoclonal anti rabbit IgG coupled to monodisperse polymer particles: comparison with polyclonal antibodies in radioimmunoassays for thyroid hormones, in *Developments in Biological Standardization*, Vol. 57, Barme, M. and Hennessen, W., Eds., S. Karger, Basel, 1984, 321.
16. **Langone, J. J.,** Protein A of *Staphylococcus aureus* and related immunoglobulin receptors produced by streptococci and pneumonococci, *Advan. Immunol.*, 32, 157, 1982.
17. **Theodorsen, L., Nustad, K., and Skinningsrud, A.,** unpublished data, 1985.
18. **Johansen, L., Nustad, K., Berg, T., and Pierce, J. V.,** Excess antibody immunoassay for rat glandular kallikrein. Measurement of kallikrein complexed with inhibitors and in plasma, *J. Immunol. Methods*, 69, 253, 1984.
19. **Børmer, O.,** A direct assay for carcinoembryonic antigen in serum and its diagnostic value in metastatic breast cancer, *Clin. Biochem.*, 15, 128, 1982.

20. **Nustad, K., Monrad-Hansen, H. P., Paus, E., Millán, J. L., Nørgaard-Pedersen, B., and the DATECA study group,** Evaluation of a new, sensitive radioimmunoassay for placental alkaline phosphatase in pre- and post-operative sera from the Danish testicular cancer material, in *Human Alkaline Phosphatases*, Stigbrand, T. and Fishman, W. H., Eds., Alan R. Liss, New York, 1984, 337.
21. **Smeland, E., Funderud, S., Ruud, E., Blomhoff, H. K., and Godal, T.,** Characterization of two murine monoclonal antibodies reactive with human B cells. Their use in a high-yield, high-purity method for isolation of B cells and utilization of such cells in an assay for B-cell stimulating factor, *Scand. J. Immunol.*, 21, 205, 1985.
22. **Ugelstad, J., Ellingsen, T., Berge, A., and Helgé, B.,** Norwegian Patent No. 155316.
23. **Treleavan, J. G., Ugelstad, J., Philip, T., Gibson, F. M., Rembaum, A., Caine, G. D., and Kemshead, J. T.,** Removal of neuroblastoma cells from bone marrow with monoclonal antibodies conjugated to magnetic microspheres, *Lancet*, 1, 70, 1984.
24. **Lea, T., Vartdal, F., Davis, C., and Ugelstad, J.,** Magnetic monosized macroporous polymer particles for fast and specific fractionation of human mononuclear cells, *Scand. J. Immunol.*, 22, 207, 1985.
25. **Danielsen, H., Funderud, S., Nustad, K., Reith, A., and Ugelstad, J.,** The interaction between cell surface antigens and antibodies bound to monodisperse polymer particles in normal and malignant cells, *Scand. J. Immunol.*, 24, 179, 1986.
26. **Seglen, P. O.,** Preparation of isolated rat liver cells, *Methods Cell Biol.*, 13, 29, 1976.
27. **Klein, J.,** personal communication.
28. **Griffin, F. M., Jr., Griffin, J. A., Leider, J. E., and Silverstein, S. C.,** Studies on the mechanism of phagocytosis. I. Requirements for circumferential attachment of particle-bound ligands to specific receptors on the macrophage plasma membrane, *J. Exp. Med.*, 142, 1263, 1975.
29. **Griffin, F. M., Jr., Griffin, J. A., and Silverstein, S. C.,** Studies on the mechanism of phagocytosis. II. The interaction of macrophages with anti-immunoglobulin IgG-coated bone marrow-derived lymphocytes, *J. Exp. Med.*, 144, 788, 1976.
30. **Silverstein, S. C., Steinman, R. M., and Cohn, Z. A.,** Endocytosis, *Ann. Rev. Biochem.*, 46, 669, 1977.
31. **Melsom, H., Funderud, S., Lie, S. O., and Godal, T.,** The great majority of childhood lymphoblastic leukaemias are identified by monoclonal antibodies as neoplasias of the B-cell progenitor compartment, *Scand. J. Haematol.*, 33, 27, 1984.
32. **Reynolds, C. P., Black, A. T., and Woody, J. N.,** Sensitive method for detecting viable cells seeded into bone marrow, *Cancer Res.*, 46, 5878, 1986.
33. **Bøyum, A.,** Separation of blood leukocytes, granulocytes and lymphocytes, *Tissue Antigens*, 4, 269, 1974.
34. **Lea, T., Rasmussen, A. M., and Michaelsen, T. E.,** Differentiation antigens in human T cell activation, *Cell. Immunol.*, 81, 209, 1983.

Chapter 5

MICROSPHERES AND CELL SEPARATION

Jerome A. Streifel

TABLE OF CONTENTS

I. INTRODUCTION AND BACKGROUND

Separation of cells relies on cells having either functional or physical differences that can be exploited in the separation process. Phenotypic differences which have been exploited include phagocytosis, adherence to surfaces, resistance to various antimetabolites, repairing damage or taking advantage of unique nutritional requirements. Additionally, physical characteristics useful in cell separations include differences in cellular density, size, cell surface charge, optical properties, and ligand binding. Separation methods have been recently reviewed.[1] Physical characteristics often do not differ greatly among cells such that there is incomplete separation. Sometimes the separations can be improved by altering the physical characteristics of either the desired or the contaminating cells. Improvement may also be achieved by taking only part of the fractions produced in the separation process and/or by repeating the separation process; however, these last two techniques can greatly reduce the yield of desired cells.

Separation techniques which offer the greatest selectivity with the greatest variability are those which are based on the ability of cells to differentially bind molecules or ligands. The binding sites on the cell can be antigens, antibodies, receptors or receptor sites, or moieties bound by effectors. Binding of a ligand to a cell may alter its surface properties such that its adherent behavior is changed. The ligand may also alter the charge distribution on the surface of the cell, thus changing its electrophoretic mobility. Another method is based on binding an optically detectable ligand, such as a fluorescent antibody with separations performed with a fluorescent-activated cell sorter. Alternatively, the binding ligand may be attached to the surface of a solid support such as a vessel surface or a column matrix.

Microspheres have been used as a method of cell labeling as well as a method to facilitate cell separation. When coupled to the surface of a cell, they have been used to enhance separation techniques based on differences in density or cell charge. Microspheres have also been used to impart an entirely new characteristic to the target cell population, such as fluorescence or magnetic susceptibility.

For the purpose of this review, microspheres are defined as particles which are in the size range of cells or smaller, which possess a ligand for binding cells, and which are used in a monodispersed state. The matrix of microspheres can be inorganic, organic, proteinaceous, or any combination of these. The density, hydrophilicity, surface charge, or chemical reactivity can be varied through selection of the matrix material or through chemical modification. They may be solid or semipermeable and have a ligand or optically active molecule attached to the surface, or distributed homogeneously throughout the microsphere. In addition, they may be made magnetic by including magnetic material in the formation process.

Microspheres for cell separation have certain theoretical advantages over methods using column chromatography or surface adherence. Due to their relatively small size, microspheres have a greater surface area per weight in comparison to either the surfaces of vessels or the surface of beads used in columns. The lower limit of the size of beads used in columns is determined by the necessity of maintaining spaces around the beads through which the cells can pass. In addition, because the microspheres are used in a monidispersed state rather than having a cell mixture flow through them, the problem of obstructing the column with debris is eliminated. Both the completeness of separation and the recovery of cells from a column is limited by nonspecific adherence and physical entrapment of cells of nontarget populations. This problem is greatly reduced by the use of microspheres in a monodispersed state. The major limitations of separation techniques using selective adherence to surfaces is nonspecific adherence and dislodgment of cells by sheer forces. The advantage of microspheres over surfaces for separating cells is that several microspheres can be bound per cell such that sheer forces alone will not easily remove all of the microspheres. In addition, the sheer forces between an attached microsphere and a cell are considerably less than that between an immobile surface and a cell.

While it is one of the most precise and controllable separation techniques available, the use of a flow cytometer to sort cells has three major limitations. These limitations are the complexity, the cost, and the time required to separate sufficient quantities of cells. In some cases insufficient signal strength is a problem when there is a low titer of surface antigens to which fluorescent antibodies may bind. The lower limit is approximately 300 fluorescent molecules per cell.[2] However, a weak fluorescent signal can be increased by using antibody-coated fluorescent microspheres. The use of fluorescent microspheres is advantageous when the signal strength of each microsphere times the number of microspheres bound is greater than the signal strength of each antibody times the number of antibodies bound. The greatest advantage magnetic microspheres have over fluorescence-activated cell sorting is in the speed of separation. Instead of processing the cells one at a time, the cells are removed in bulk. Speed of separation becomes particularly important when isolating a cell type that comprises a very small percentage of the total cell population.

Adherent or phagocytic cells may attach to or ingest microspheres. If they are not the target cells, they must be removed from the mixed population of cells before the microspheres are added. This can be done by incubating them with magnetic compounds, such as magnetite, followed with removal of magnetically responsive cells with a magnet,[3] or by letting them adhere to the surface of a vessel.

II. SUMMARY OF MICROSPHERE SEPARATIONS

Cell separations using microspheres are summarized in Table 1. The data listed are reported values, calculation from data, or interpretations from graphs. Complete or nearly complete separations have been achieved in most cases. The table gives the method of separation, characteristics of the microsphere-ligand and cell-ligand complexes, and the results of the separation in terms of purity of the fractions.

The effectiveness of any separation is judged by the type of separation performed. Separations can be either positive, where a cell type is isolated for further use, or negative, where a cell type is removed and the remaining cells are used. The relevant data for positive selections is the percent of target cells in the target fraction. Where reported, the purity of the target fraction varied from 34 to 100% (see Table 1). Conversely, the relevant data for a negative selection is the percent of target cells in the nontarget fraction. When reported, the target cell comprised from 0 to 10% of the nontarget fraction with one exception of 23%.[13] In terms of total target cells removed the worst reported separation was for a series of neuroblastoma cell lines using a mixture of six antineuroblastoma monoclonal antibodies incubated with the cells. The values for the percent of cells removed for four cell lines were 82, 3.0, 2.6, and 4.5% (data calculated from graphs). When one antibody was dropped from the mixture the cell line that had resulted in only 3.0% removal subsequently allowed for 99.89% removal.[29] These results reflect cell-type variability possibly due to antibody-mediated changes.

III. SEPARATION METHODS

Separation of cells after they are labeled with one or more microspheres has been done using four different methods which were density equilibrium centrifugation, cell electrophoresis, fluorescent-activated cell sorting, and magnetic separation. The first three methods are done in the same manner as with unlabeled cells. The theory and technique of magnetic separations with microspheres has been reviewed recently.[12,30,31] Flow systems have been devised for the magnetic separation of large amounts of cells bound to microspheres. The important factors with flow systems are the strength and gradient of the magnetic field and the flow rate.[12,26,30,31] However, all that is necessary for processing 100 million cells in less

Table 1
MICROSPHERE CELL SEPARATION SYSTEMS

No.	Separation type[a]	Microsphere matrix[b]	Attachment by[c]	Ligand on microsphere: Primary	Ligand on microsphere: Secondary	Target cell	Nontarget cell, culture or tissue	Ligand on	Percent of target cells: Initial	Percent of target cells: Target fraction	Percent of target cells: Nontarget fraction	Analysis by[d]	Comment	Ref.
1	Mag.	Polyhydroxyethyl methacrylate	Ald.	g × rIg	—	m Thymocytes prelabeled with ms	hRBC	r × thymocyte	82.3	98.0	0.4	LM, FS		4
						m Ig^+ cells	m Splenocytes	—	30.9 36.5 39.1	91.5 81.6 76.4	0.4 1.4 3.0			
2	Mag.	nr	nr	g × rig	—	m Oligodendrocytes	m Brain glial cell fraction	r × bovine oligodendrocytes	nr	>90	nr	SEM, TEM	Target cells active after separation	5
3	Mag.	Hydrogel	Ald.	Choleragen	—	High-titre GM1 C-1300 neuroblastoma	C-1300	—	12.5 15.0	99.0 99.5	1.0 1.5	HRP-, FITC-, ms-choleragen binding	Target cells grew after separation	6
4	Den.	Polyacrylamide	Entr.	horse × h lymphocyte	—	T cells	hPBL	—	46 47	nr nr	10 8	ms binding E-rosetting	Retreated with ms for analysis	7
5	Mag.	Polyglutaraldehyde	Ald.	g × rIgG	—	hRBC prelabeled with ms	hRBC	r × hRBC	10—50	95	5	ms binding		8
6	FACS	Polystyrene	CDI	Myeloma protein GPC-8	—	Ig(1.6)2.4	11—5.2	—	0.2	61.1	nr	RIA	Separated cells grew in culture	9
						Ig(1.6)2.4 fluorescent	11—5.2	—	0.2	62.1	nr	FS	Ms removable with pronase	
				Myeloma protein C.BPC-101	—	20—6 fluorescent	20.9	—	0.2	34.1	nr	FS		
7	Flow Eltr.	Polyglutaraldehyde	Ald.	g × rIgG	—	hRBC prelabeled with ms hRBC	Turkey RBC Turkey RBC	r × hRBC r × hRBC	7.2 55.7	100 100	0 0	SEM	Data from graph Data from graph	10
		Polyuynyl-pyridine	Ald.	g × rIgG	—	hRBC prelabeled with ms	hRBC	r × hRBC	nr	100	0		—	
8	Mag.	Human serum albumin	Entr.	SpA	r × cIgG r × rat Ig	cRBC Ig^+ cells Ig^+ cells	sRBC Rat splenocytes Rat thymocytes	— — —	50.0 44—51 4—6	>90 nr nr	1—3 0.5 0	LM FS FS	Nonspecific removal of nontarget cells was 5—10%	11
9	Mag.	Hydrogel	Ald.	Ricin	—	HELA wild type	RRII (low ricin binding HELA)	—	50	nr	<10	Ricin resistance	RRII has 20% of wild type HELA ricin receptors	12
10	Mag.	Polyglutaraldehyde	Ald.	Swine × rIg	—	Ig^+ cells	m spleen lymphocytes	r × Ig	50.0 ± 2.2	79.2 ± 2.8	22.8 ± 2.0	FS		13
						Ig^+ cells	m peripheral lymphocytes	r × Ig	23.8 ± 3.9	53.6 ± 3.5	10.0 ± 1.1			
				r × hIg	—	Ig^+ cells	m peripheral lymphocytes	—	23.8 ± 3.9	51.2 ± 2.7	10.5 ± 1.3			
11	Mag.	Human serum albumin	Entr.	SpA	HLA-BW6 MAB	HLA-$BW6^+$ cells	HLA-$BW4^+$ cells	—	50	nr	0	Cytotoxicity		14
12	Mag.	Polyacrolein	Ald.	g × rIgG	—	hRBC	Turkey RBC	r × hRBC	50	95	nr	LM, FS		15

13	Mag.	Human serum albumin	Entr.	SpA	r × cRBC	cRBC	sRBC	—	50	91.1	nr	Cr-51	0.3% of target cells remaining in sup.	16
						Ig + cells	Rat splenocytes	—	43—51	nr	nr	FS	0.5% of target cells remaining in sup.	
					r × ratIg	Ig + cells	Rat thymocytes	—	4—6	nr	0			
					r × hIg	Ig + cells	hPBL	—	15—13	nr	0			
14	FACS	Polystyrene	Ald.	r × m myosin	—	Dead m myocytes	m myocytes in O_2 high glucose	—	16	nr	nr	LM, FM, growth	No growth in dead cell fraction	17
							m myocytes in N_2 high glucose	—	16	nr	nr		Init. percent from FACS analysis	
							m myocytes in glucose free	—	15	nr	nr			
							m myocytes in N_2 glucose free	—	40	nr	nr			
15	Mag.	Polyacrolein	Ald.	g × rIgG	—	hRBC	Turkey RBC	r × hRBC	50	95	nr	LM	Separation repeated twice	18
16	Mag.	Polyacrylamide agarose hybrid	Ald.	r × mIg	04 MAB	04 antigen +	m cerebellum	—	1.5 ± 0.2	91 ± 4	nr	FS	Cells put out processes in 1 hr	19
17	Mag.	Dextran	Ald.	SpA	—	hRBC	SP2 m myeloma	r × hRBC	50	92	2	LM	Data from graph	20
18	Mag.	Polystyrene polyacrolein hybrid	Ald.	r × sRBC	—	sRBC	cRBC	—	1.0	99.85	nr	LM	Reseparated eight times, vortex to remove ms	21
19	Mag.	Polystyrene	NSA	g × mIg	—	h neuroblastoma	h bone marrow	6 Antineuroblastoma MABs	nr	nr	nr	nr	0.1% of target cells remaining in supply	22
20	Mag.	Polystyrene	NSA	g × mIg	—	h neuroblastoma C-1300	h leukemic cell Nalm-6	Mixed antineuroblastoma MABs	9—50	100	nr	nr	1—3% of target cells remaining in supply	23
21	Mag.	Distearoyl- or dipalmitoyl-phosphatidyl choline liposome	Entr.	r × h plasma fibronectin	—	m embryo fibroblasts prelabeled with ms and carmine dye	m embryo fibroblasts	—	47	84	nr	LM for carmine dye	Target cells were prelabeled with carmine dye	24
									44	78	nr			
									41	80	nr			
									23	58	nr			
22	Mag.	Metallic cobalt colloid	NSA	g × mIg	—	CALL antigen +	h leukemic bone marrow	J5 MAB	3	nr	0	FS	No loss of CFUs Marrow re-engrafted	25
23	Mag.	Polystyrene	NSA	s × mIg	—	h neuroblastoma CHP 100	h bone marrow	6 Antineuroblastoma MABs	1.0	nr	nr	FS, growth	<0.1% of target cells remaining in supply	26
									0.1—3.0	nr	0	FS	Four patients, CFUs not lost, marrow re-engrafted	
24	Den.	Colloidal gold	NSA	g × mIgG	—	OKT3 antigen +	h bone marrow	OKT3 MAB	nr	nr	nr	PHA response	14.7% of target cells remaining in supply	27
	Mag.	Dextran	NSA	g × mIgG	—	OKT3 antigen +	h bone marrow	OKT3 MAB	nr	nr	nr		32.9% of target cells remaining in supply	
25	Mag.	Albumin	SPDP	g × rIg	—	cRBC	sRBC	r × cRBC	50	94	nr	LM	1—7% of nontarget cells were in pellet	28
									59	95	nr			
									5	41	nr			
									5	50	nr			
						sRBC	cRBC	r × sRBC	55	91	nr			

Table 1 (continued)
MICROSPHERE CELL SEPARATION SYSTEMS

No.	Separation type[a]	Microsphere matrix[b]	Attachment by[c]	Ligand on microsphere: Primary	Ligand on microsphere: Secondary	Target cell	Nontarget cell, culture or tissue	Ligand on	Percent of target cells: Initial	Percent of target cells: Target fraction	Percent of target cells: Nontarget fraction	Analysis by[d]	Comment	Ref.
									9	54	nr			
									1	37	nr			
26	Mag.	Polystyrene poly-acrolein hybrid	Ald.	g × mIg		h neuroblastoma KCNR	h bone marrow h bone marrow	6 antineuroblastoma MABs	5.0	nr	nr	Immuno-peroxidase staining	18% of target cells remaining in supply	29
						h neuroblastoma LAN5	h bone marrow h bone marrow		5.0	nr	nr		97% of target cells remaining in supply	
						h neuroblastoma LAN1	h bone marrow h bone marrow		5.0	nr	nr		97.4% of target cells remaining in supply	
						h neuroblastoma KAN	h bone marrow		.50	nr	nr		95.5% of target cells remaining in supply	
						h neuroblastoma CHP 100		5 antineuroblastoma MABs	4.8	nr	.0003		Above data from graph 0.04% of target cells remaining in supply	
						h neuroblastoma LA-N-5			1.0	nr	.002		0.11% of target cells remaining in supply CFUs increased	
						h neuroblastoma patient			14.9	nr	0		0% of target cells remaining in supply CFUs not affected; All separations were done twice with antibody and ms	

Note: Abbreviations are: (c) chicken; (h) human; (m) mouse; (MAB) monoclonal antibody; (ms) microspheres; (nr) not recorded; (s) sheep; and (SpA) staphylococcal protein A.

[a] Separation types are: (Den.) density centrifugation; (Flow Eltr.) flow electrophoresis; (FACS) fluorescent-activated cell sorter; and (Mag.) magnetic pelleting.

[b] All microspheres separated by a magnet were made magnetic by including magnetic material in the microsphere formation process and all microspheres used in separations with a fluorescent-activated cell sorter were made fluorescent by including a fluorescent dye.

[c] Attachment was by: (Ald.) reaction with aldehyde groups; (CDI) coupling with carbodiimide; (Entr.) entrapment; (NSA) nonspecific adsosrption; and (SPDP) coupling with succinimidyl pyridyldithio proprionate.

[d] Analysis abbreviations are: (LM) light microscopy; (EM) electron microscopy; (FS) fluorescent staining; and (RIA) radioimmunoassay.

than 6 min is a strong permanent magnet.[6] There are additional benefits to magnetic separation in that washing the cells is easy and any entrapment of nontarget cells can be easily overcome by resuspension and reseparation.[2,21]

In one report, density sedimentation was used to separate T cells from human peripheral blood lymphocytes mediated by antibody-labeled polyacrylamide microspheres. The T cells remaining in the supernatent comprised 8% of that fraction.[7] Density sedimentation was also used to remove T cells from human bone marrow with specific antibody adsorbed onto colloidal gold. In this case, 14% of the T cells were left in solution.[27] Both of these values are higher than those obtained with other methods and may reflect incomplete separation due to an insufficient difference in density. Since both of these systems used nonspecific adsorption of antibody onto the particles the results may also be due in part to problems discussed below inherent with this method of attaching antibodies to microspheres. There was only one example of cell separation done by altering the electrophoretic mobility of cells. Cells that had nearly identical mobilities were separated such that upon elution there were only a few fractions between the cell populations that contained a mixture of target and nontarget cells. Proper choice of fractions could produce pure populations.[10]

There are numerous examples in the literature of labeling cells with fluorescent microspheres; however, there have been only two reported cases where the cells were actually physically separated and not just analyzed for fluorescent distribution. There is insufficient data to evaluate the utility of this technique. No purity data is presented in one study[17] and in the other the purity of the target fractions was 46 to 64%, although, in this study a high purity was achieved by growing out single separated cells.[9]

The majority of the microsphere cell separation systems used magnetic separation (see Table 1). In these systems the initial distribution of cells affected the final purity of the target fraction. The smaller the initial percent of target cells the lower final purity.[24,28] This contamination of the target fraction is due to a constant percentage of the nontarget cells being isolated with the target cells. Values for this removal were approximately 5 to 10%,[11] 17 to 22%,[24] and 1 to 6%[28] for three different systems.

A comparison of cell separations using magnetic microspheres vs. columns gave similar results for separating surface Ig-positive cells. The major differences were the method of attachment of the antibody and the recovery of cells. For the microsphere and column separations the percent of target cells in the nontarget fraction were 22.8 ± 3.0% and 38.8 ± 5.4%, respectively while the purity of the target fraction was 79.2 ± 2.8% and 90.8 ± 5.1%. In the same report, when separating human B and T cells, microspheres gave similar separation results to those obtained by using rosetting followed by density fractionation.[13] This study shows that microspheres are capable of producing a purer target fraction than does separation of the cells on a column.

IV. MATRIX OF MICROSPHERES

There are four basic types of microspheres used for cell separation — colloidal metals, proteins, liposomes and various organic polymers (see Table 1). Reported sizes ranged from 0.03 to 10 μm. In several cases the microspheres were made fluorescent by including a dye in the polymerization reaction or attaching the dye after formation of the microsphere.[9,17] Most of the separations used microspheres which were made magnetic by including magnetic materials, such as magnetite, in the formation process. The one exception was a magnetic microsphere composed of magnetic colloidal cobalt.[25] There is insufficient data to evaluate the effect of the different matrices or sizes on the separation process in terms of purity of target and nontarget cells. The major difference between the matrices was in the method of attachment of the primary ligand which is discussed below.

V. COUPLING OF LIGANDS TO MICROSPHERES

Ligands have been attached to microspheres by nonspecific adsorption, entrapment, and covalent attachment. It is know that when a protein is exposed to a hydrophobic surface it denatures onto the surface.[32] Nonspecific adsorption of antibodies to microspheres occurs through a similar mechanism.[33] With this method of attachment there is a lower limit to the concentration or amount of the antibody incubated with the microspheres necessary to prevent complete denaturation of the antibody.[34] The surface of the microsphere most probably consists of a mixture of inactivated antibodies and antibodies with one or both of the Fab arms oriented in an active configuration. This method of coupling antibodies to microspheres may not be very efficient when using rare or expensive antibodies because a large portion of the antibodies may not contribute to the attachment of the cell. Also, the strength of the nonspecific adsorption may not be great enough to prevent dissolution or displacement of the antibody from the surface of the microsphere. With this type of binding, serum was found to interfere with microsphere-to-cell binding for some systems.[35,36] An additional theoretical problem is that redissolved antibodies may mask target sites on the cell. For cell separations, nonspecific adsorption has been used to attach antibodies to microspheres made of metallic cobalt colloid,[25] polystyrene,[22,23,25,27] colloidal gold, and dextran.[26] Despite the theoretical disadvantages of this method of attachment several systems resulted in a removal of 97 to 100% of the target cells.

Where there is a possibility of ligand loss or exchange it is preferable to have the ligand bound to the microsphere by entrapment or by covalent bonds rather than by adsorption. Entrapment is done by mixing the ligand with the monomers from which a microsphere is formed. This results in the ligand being held by physical entanglement and possibly covalent coupling reactive groups formed during polymerization. For cell separation purposes this was used to couple antibody to polyacrylamide[7] and liposomal microspheres.[24] It was also used to couple staphylococcal protein A (SpA) to human serum albumin microspheres.[11,14,16] This mode of coupling for the SpA microspheres yielded 99.7 to 100% removal of target cells from suspension[11,14] and could give greater than 90% target cell purity.[11,16] However, antibodies coupled by entrapment also gave poorer than average results in some cases. For the polyacrylamide microspheres, target cells comprised 8 to 10% of the nontarget fraction.[7] For liposomes the target fraction was only 58 to 84% pure.[24] These lower performances may reflect inactivation of the antibody in the polyacrylamide polymerization process and nonspecific adhesion of the nontarget cells to liposomal microspheres.

Methods for attachment are based on coupling ligands to aldehyde, carboxyl, or sulfhydryl groups already on the surface of the microsphere or placed there by chemical modification. The number and activity of ligands attached to microspheres is determined by the number of functional groups available on the microsphere and by the orientation of the ligand. For cell separation purposes the predominant method for covalent attachment has been via aldehyde groups which were inherent to polyglutaraldehyde and polyacrolein microspheres or added by chemical modification for other types of matrices. There was one report each for coupling with carbodiimide[9] and succinimidyl pyrodyldithioproprionate.[25] In addition to attached ligands, the toxins choleragen and ricin,[6,12] two myeloma antigen proteins[9] and the antibody-binding protein SpA[20] have also been used.

Covalent attachment yielded good results except for three systems. One of these was a separation method based on a myeloma protein attached to a fluorescent polystyrene microsphere via carbodiimide which yielded 34.1 to 64.1% target-cell purity. This poor purity was due to the fact that the number of microspheres bound did not correlate with the rate of antibody secretion.[9] A second example was a separation of antibody-positive lymphocytes using antibody coupled to a polyglutaraldehyde microsphere which yielded only 51.2 ± 2.7% to 79.2 ± 2.8% target-cell fraction purity and 10.0 ± 1.1% to 22.8 ± 3.0% target-cell contamination of the nontarget fraction.[13] The third was the previously mentioned removal of different mouse monoclonal antibody-labeled neuroblastoma cell lines using a polystyrene, polyacrolein-hybrid microsphere coupled to an antimouse antibody. With this system, removal of target cells ranged between 2.6 to 100% and was dependent on cell type and mixture of monoclonal antibody.[29]

As with the method of nonspecific adsorption, covalent attachment of antibodies to microspheres most likely results in a portion of them being bound in a nonfunctional manner. One way to overcome this is to attach first a less expensive antibody to the microsphere. This first antibody is directed against a more expensive second antibody. In this manner, the less expensive antibody can be attached to the microsphere either covalently or by nonspecific adsorption. The use of a first less expensive antibody bound to the microsphere can significantly reduce the cost but would not necessarily overcome the problem of the loss of antibody by dissolution if bound by nonspecific adsorption. This method was used to couple microspheres with attached antimouse immunoglobulin to human or T cells via monoclonal antibodies.[22,23,25-27,29] Another method to increase the efficiency is to use SpA as the primary ligand on the microsphere. SpA binds the Fc region of various antibodies[37] which results in orienting the Fab regions which possess the antigen binding capacity toward the solution. This technique gave good separation with SpA covalently attached to a microsphere[20] or when attached by entrapment.[11,14,16] Methods used to attach antibodies to microspheres have recently been reviewed[38] and will not be covered in further detail here.

VI. COUPLING OF MICROSPHERES TO CELLS

During the microsphere cell-coupling phase, there are several factors in addition to the buffer conditions which affect the rate and the extent of the reaction. These are the temperature, the concentration, and the relative ratio of microspheres to cells. The temperature is limited to a range which allows the coupling reaction to proceed without the cells losing function. The range of temperatures for optimum viability of cells is usually between 0 and 37°C. Many antibody-antigen bonds can be formed throughout this temperature range; although some systems including those involving receptor sites as the primary target on the cell may show a temperature dependence on the equilibrium of binding or on the availability of target. When comparing temperature dependence of microsphere binding to cells there was only a slight increase observed at 23°C over that at 4°C.[7]

The stability of the microsphere cell complex is determined by the number of interactions between the microsphere and cell, by the stability of those interactions, and by the presence of moities capable of displacing one of the ligands or of cleaving apart the complex.

Cell separations have been successful with various coupling sequences. Although the effect of capping separation was not reported cells that may be induced to cap should be kept at 4°C. In one comparison where the cell antigen-recognizing antibody was placed either first on the microsphere or first on the cell, both systems yielding the same result for removal of Ig-positive cells.[13]

In cases where the target cell consists of different antigenic subpopulations, more than one set of targets can be used to bind the microspheres to the cell. This was done for removing neuroblastoma cells from human bone marrows. In three studies different cultured-neuroblastoma cells were added to human bone marrow samples followed by the addition of mixtures of mouse antineuroblastoma antibodies. Subsequent removal of cells with microspheres which were coated with an antimouse antibody resulted in bone marrows with a contamination of 1 to 3%,[23] less than 0.1%,[26] and 0.04 to 97.4%[29] of the original tumor cell population. The results were much better when these systems were used to remove neuroblastoma cells from the marrow of patients with neuroblastoma tumors. In these cases the contamination of the bone marrow samples was 0.1[23] and 0%[26,29] of the original tumor cell population.

The optimal concentration of microspheres and cells is one that allows good initial mixing to achieve a homogeneous distribution of reactants. As a first approximation the concentration should be kept as high as possible to minimize loss of reactants to vessel surfaces and to maximize the rate of microsphere-to-cell collisions. Mixing should be at speeds that do not shear the microspheres from the cell. A gentle procedure which minimizes microsphere-cell disruption is to mix the microspheres and cells together and allow them to settle under unit gravity or at low centrifugal forces. This ensures a homogeneous distribution while achieving a high localized concentration of microspheres and cells. The time for coupling microspheres to cells is typically between 0.5 and 2 hr.

The probability of a microsphere coupling to a cell is in part determined by the ratio of microspheres to cells. The higher the ratio the greater the probability of coupling to every target cell. Since one cell can usually bind more than one microsphere this ratio should generally be greater than one. Since there is an upper limit to the number of microspheres that any one cell can bind, increasing the ratio of microspheres to cells above a certain value should theoretically have no effect on the number of cells labeled with microspheres.

Since one microsphere can also bind to more than one cell, there is a range in the ratio of microspheres to cells at which there is the possibility of agglutination. This may interfere with the fractionation step and may also result in the entrapment of nontarget cells. If agglutination occurs it can be remedied by either lowering or increasing the ratio of microspheres to cells. Reducing the number of microspheres bound per cell may result not only in reducing the yield but might also result in selecting for a subpopulation of cells that possess a high density or number of surface targets. Reduction of yield may not be important for systems using positive selection, i.e., where the goal is to isolate and use a certain cell type. Missing some cells is a more serious problem for systems using negative selection, i.e., where the goal is to completely remove a certain cell type and use the remaining cells. Entrapment of nontarget cells can also occur in microagglutinates which form with increasing concentration of microspheres incubated with a high initial percentage of target cells.[39] The best strategy for minimizing this effect is to use microspheres in excess of the maximum amount that the cells are capable of binding. This means using a ratio of microspheres to cells well out on the plateau of the binding curve. There is however, a theoretical upper limit to the concentration of free microspheres or microsphere-labeled cells where unlabeled cells are swept out of solution during the magnetic pelleting step. If this is a problem it can be reduced by removing excess microspheres on a density gradient[6,9,12,13,20] or by velocity sedimentation.[8,15,18] It can also be reduced by repeated separations. In an attempt to reduce nonspecific contamination of the target-cell fraction, one separation was repeated eight times to wash out nontarget cells. This was also done with a low ratio of 0.4 microspheres to cells. These conditions resulted in a target cell purity of 99.85% when isolating sheep RBCs which were mixed with chicken RBCs.[21]

VII. FUNCTION OF SEPARATED CELLS

Of the studies reviewed, separations were analyzed for purity by morphological characteristics, staining, radioimmunoassay, microsphere binding, or E-rosetting. Several were tested for functional response, which consist of ricin resistance,[12] cytotoxicity,[14] and phytohemaglutinin response.[27] In addition, there were several reported cases of positively selected cells that were active in culture. These were either subsets of tumor cells[6,12] or hybridomas[9] that grew in culture, or subsets of oligodendrytes that put out processes in culture.[5,19] In these cases the cells were used with the microspheres attached. Although not tested for functional response, microspheres were removed after separation by vortexing[21] or by pronase.[9] Microspheres have also been removed by treatment with trypsin.[40]

In negative separations where the cells with attached microspheres are discarded and the remaining cells kept for use, there has been growth observed in culture of mouse myocytes after microsphere removal of dead myocytes.[17] Also, in several studies there was retention of colony-forming units in human bone marrow samples after removal of either cultured or endogenous tumor cells.[25,26,29] A similar result was observed for the removal of T cells from human bone marrow.[27] In another study the bone marrow once cleared of a leukemic tumor was successfully reimplanted in a patient treated with extensive chemotherapy. There was, however, reappearance of tumor antigen-positive cells at 21 days.[25]

The data presented show that cell separations with microspheres can be a very selective technique especially in negative separations. In addition, with magnetic microspheres the separations are relatively easily performed. The limited results also show that microsphere-mediated separations may be a very effective technique for therapeutic purposes.

REFERENCES

1. **Kumar, R. K. and Lykke, A. W. J.,** Cell separation: a review, *Pathology,* 16, 53, 1984.
2. **Rembaum, A. and Dreyer, W. J.,** Immunomicrospheres: reagents for cell labeling and separation, *Science,* 208, 364, 1980.
3. **St. George, S., Friedman, M., and Byers, S. O.,** Mass separation of reticuloendothelial and parenchymal cells of rat's liver, *Science,* 120, 463, 1954.
4. **Molday, R. S., Yen, S. P. S., and Rembaum, A.,** Application of magnetic microspheres in labeling and separation of cells, *Nature (London),* 268, 437, 1977.
5. **Campbell, G., Abraomsky, O., and Silberberg, A. D.,** Isolation of oligodeandrocytes from mouse cerebellum using magnetic microspheres, *Soc. Neurosci. Abstr.,* 4, 64, 1978.
6. **Kronick, P. L., Campbell, G. L., and Joseph, K.,** Magnetic microspheres prepared by redox polymerization used in a cell separation based on gangliosides, *Science,* 200, 1074, 1978.
7. **Ljungstedt, I., Ekman, B., and Sjoholm, I.,** Detection and separation of lymphocytes with specific surface receptors, by using microparticles, *Biochem. J.,* 170, 161, 1981.
8. **Margel, S., Zisblatt, S., and Rembaum, A.,** Polyglutaraldehyde: a new reagent for coupling proteins to microspheres and for labeling cell-surface receptors. II. Simplified method labeling method by means of non-magnetic and magnetic polyglutaraldehyde microspheres, *J. Immunol. Methods,* 28, 341, 1979.
9. **Parks, D. R., Bryan, V. M., Oi, V. T., and Herzenberg, L. A.,** Antigen-specific identification and cloning of hybridomas with a fluorescent activated cell sorter, *Proc. Natl. Acad. Sci. U.S.A.,* 76, 1962, 1979.
10. **Smolka, A. J. K., Margel, S., Nerren, B. H., and Rembaum, A.,** Electrophoretic cell separation by means of microspheres, *Biochem. Biophys. Acta,* 588, 246, 1979.
11. **Widder, K. J., Senyei, A. E., Ovadia, H., and Paterson, P. Y.,** Protein A magnetic microspheres: a rapid method for cell separation, *Clin. Immunol. Immunopathol.,* 14, 395, 1979.
12. **Kronick, P. L.,** Magnetic microspheres in cell separation, in *Methods of Cell Separation,* Vol. 3, Catsimpoulas, N., Ed., Plenum Press, New York, 1980, chap. 3.
13. **Merivuori, H., de la Chapelle, A., and Schroder, J.,** Cell labeling and separation with polyglutaraldehyde microspheres, *Exp. Cell Res.,* 130, 464, 1980.
14. **Kandzia, J., Anderson, M. J. D., and Muller-Ruchholtz, W.,** Cell separation by antibody-coupled magnetic microspheres and their application in conjunction with monoclonal HLA antibodies, *J. Cancer Res. Clin. Oncol.,* 101, 165, 1981.
15. **Margel, S., Beitler, U., and Ofarim, M.,** A novel synthesis of polyacrolein microspheres and their application for cell labeling and separation, *Immunol. Commun.,* 10, 567, 1981.
16. **Widder, K. J., Senyei, A. E., Ovadia, H., and Paterson, P. Y.,** Specific cell binding using staphylococcal protein A magnetic microspheres, *J. Pharm. Sci.,* 70, 387, 1981.
17. **Khaw, B. A., Scott, J., Fallon, J. T., Cahill, S. L., Haber, E., and Homcy, G.,** Myocardial injury: quantitation by cell sorting initiated with antimyosin fluorescent spheres, *Science,* 217, 1040, 1982.
18. **Margel, S., Beitler, U., and Ofarim, M.,** Polyacrolein microspheres as a new tool in cell biology, *J. Cell. Sci.,* 56, 157, 1982.
19. **Meier, D. H., Lagenaur, C., and Schacher, M.,** Immunoslection of oligodendrocytes by magnetic beads. I. Determination of antibody coupling parameters and cell binding conditions, *J. Neurosci. Res.,* 7, 119, 1982.

20. **Molday, R. S. and MacKenzie, D.,** Immunospecific ferromagnetic non-dextran reagents for the labeling and magnetic separation of cells, *J. Immunol. Methods,* 52, 353, 1982.
21. **Rembaum, A., Yen, R. C. K., Kempner, D. H., and Ugelstad, J.,** Cell labeling and magnetic separation by means of immunoreagents based on polyacrolein microspheres, *J. Immunol. Methods,* 52, 341, 1982.
22. **Kemshead, J. T., Gibson, F. J., Ugelstad, J., and Rembaum, A.,** A flow system for the *in vitro* separation of tumor cells from bone marrow using monoclonal antibodies and magnetic microspheres, *Proceedings of the American Association for Cancer Research,* Magee, P. N., Ed., Waverly Press, Baltimore, Md., 1983, 217.
23. **Kemshead, J. T., Rembaum, A., and Ugelstadt, J.,** The potential use of monoclonal antibodies and microspheres containing magnetic compounds to remove neuroblastoma cells from bone marrow to be used in autologous transplantation programmes, *Proceedings of the American Society of Clinical Oncology,* Magee, P. N., Ed., Waverly Press, Baltimore, Maryland, 1983, 36.
24. **Margolis, L. B., Namiot, V. A., and Kljukin, L. M.,** Magnetoliposomes: another principle of cell sorting, *Biochem. Biophys. Acta,* 735, 193, 1983.
25. **Poynton, C. H., Dicke, K. A., Culbert, S., Frankel, L. S., Jagannath, S., and Reading, C. L.,** Immunomagnetic removal of CALLA positive cells from human bone marrow, *Lancet,* 1, 524, 1983.
26. **Treleaven, J. G., Ugelstad, J., Philip, T., Gibson, F. M., Rembaum, A., Caine, G. D., and Kemshead, J. T.,** Removal of neuroblastoma cells from bone marrow with monoclonal antibodies conjugated to magnetic microspheres, *Lancet,* 1, 70, 1984.
27. **Vellekoop, L., Reading, C. L., Poynton, C. H., Chandran, M., Hickey, C. M., and Dicke, K. A.,** Cell separation on the basis of monoclonal antibody coated colloidal gold particles and magnetic dextran particles, in *Recent Advances in Bone Marrow Transplantation,* Gale, R. P., Ed., Alan R. Liss, New York, 1983, 77.
28. **Owen, C. S. and Sykes, N. L.,** Magnetic labeling and cell sorting, *J. Immunol. Methods,* 73, 41, 1984.
29. **Seeger, R. C., Reynolds, C. P., Vo, D. D., Ugelstad, J., and Wells, J.,** Depletion of neuroblastoma cells from bone marrow with monoclonal antibodies and magnetic immunobeads, in *Advances in Neuroblastoma Research,* Evans, A. E., D'Angio, G. H. and Seeger, R. C., Eds., Alan R. Liss, New York, 1984, 443.
30. **Owen, C. S.,** Magnetic cell sorting, in *Cell Separation: Methods and Selected Applications, Vol. 2,* Pretlow, T. G. and Pretlow, T. P., Eds., Academic Press, New York, 1983, chap. 8.
31. **Whitesides, G. M., Kazlauskas, R. J., and Josephson, L.,** Magnetic separations in biotechnology, *Trends Biochem.,* 1, 144, 1983.
32. **Shaw, D. J.,** *Introduction to Colloid and Surface Chemistry,* Buttersworth, London, 1966, 67.
33. **Bagchi, P. and Birnbaum, S. M.,** Effect of pH on the adsorption of immunoglobulin G on anionic poly(vinyltoluene) model latex particles, *J. Colloid Interface Sci.,* 83, 460, 1981.
34. **Morrissey, B. W. and Han, C. C.,** The conformation of γ-globulin adsorped on polystyrene lattices determined by quasielastic light scattering, *J. Colloid Interface Sci.,* 65, 423, 1977.
35. **Illum, L., Jones, P. D. E., Kreuter, J., Baldwin, R. W., and Davis, S. S.,** Adsorption of monoclonal antibodies to poly(hexyl cyanoacrylate) nanoparticles and subsequent immunospecific binding to tumor cells *in vitro, Int. J. Pharm.,* 17, 65, 1983.
36. **Brash, J. L. and Davidson, V. J.,** Adsorption on glass and polyethylene from solutions of fibrinogen and albumin, *Thromb. Res.,* 9, 249, 1976.
37. **Lindmark, R., Thoren-Tolling, K., and Sjoquist, J.,** Binding of immunoglobulins to protein A and immunoglobulin levels in mammalian sera, *J. Immunol. Methods,* 62, 1, 1983.
38. **Illum, L. and Jones, P. D. E.,** Attachment of monoclonal antibodies to microspheres, in *Methods in Enzymology,* Widder, K. J. and Green, R., Eds., Academic Press, Orlando, Fla., 1985, chap. 6.
39. **Mirro, J. Jr., Schwartz, J. F., and Civin, C. I.,** Simultaneous analysis of cell surface antigens and cell morphology using monoclonal antibodies conjugated to fluorescent microspheres, *J. Immunol. Methods,* 47, 39, 1981.
40. **Ghetie, V., Mota, G., and Sjoquist, J.,** Separation of cells by affinity chromatography on SpA-Sepharose 6MB, *J. Immunol. Methods,* 21, 133, 1978.

Chapter 6

THE USE OF MAGNETIC MONOSIZED POLYMER PARTICLES FOR THE REMOVAL OF T CELLS FROM HUMAN BONE MARROW CELL SUSPENSIONS

C. D. Platsoucas, F. H. Chae, N. Collins, N. Kernan, J. Laver, T. Ellingsen, P. Stenstad, J. Bjørgum, A. Rembaum, R. O'Reilly, and J. Ugelstad

TABLE OF CONTENTS

I. INTRODUCTION

Bone marrow transplantation is the method of choice for the treatment of a number of fatal diseases including severe combined immunodeficiency, leukemia, aplastic anemia, and others. However, the applicability of the method has been limited to those patients for whom matched donors (siblings or others) are available because of the development of graft-vs.-host disease (GvHD).[1] Graft-vs.-host disease (GvHD) develops in more than 50% of the recipients of allogeneic-matched bone marrow and is fatal in up to 25% of those affected.[2-5] Immunosuppressive therapy has been used to decrease or eliminate the symptoms of the GvHD in certain patients; however, these patients continue to be at high risk of fatal complications due to infection.[5-9]

It has been clearly demonstrated in experimental animals[10-15] that mature T cells present in the graft are responsible for the generation of both acute and chronic GvHD. Similar results have been obtained in man.[16-25] Removal of T cells from the graft resulted in elimination or partial reduction of the incidence and severity of GvHD in man. Various methods have been used to remove human T cells from bone marrow cell suspensions. These include soybean lectin agglutination followed by rosetting with sheep erythrocytes,[17-19] and treatment of the bone marrow before infusion with antibody and complement in vitro[20-24] or with anti-T cell immunotoxins.[25] We investigated and report here the use of magnetic monosized particles and anti-T cell monoclonal antibodies for the removal of T cells from human bone marrow cell suspensions. Depletion of T cells of the order of 2 to 3 logs was achieved by this method in a single step.

II. MAGNETIC MONOSIZED POLYMER PARTICLES

A. Characteristics of the Particles

The following monosized particles were used:

1. Highly porous magnetic monosized particles of 3 μm in diameter. These polystyrene particles have surface area of approximately 100 m^2/g. These particles were made by Ugelstad and collaborators and are described elsewhere.[26,27]
2. M-450 particles. These particles have been recently made by Ugelstad and collaborators (unpublished results) from preformed highly porous magnetic particles by filling and covering the pores with a polymeric material. The surface of these particles is thereby reduced from about 100 m^2/g to 3 to 5 m^2/g. Their size is 4.5 μm. The final particles have a hydrophobic character which allows physical adsorption of protein (noncovalent binding). Antibody adsorbed to these particles is located on the outer surface and probably is more accessible for binding to the cells. Also, the M-450 particles are more easily dispersed than the 3-μm highly porous particles. Microconjugates formed of two or three particles are not observed with the M-450 particles, whereas these microconjugates are commonly observed with the highly porous 3-μm particles.

Furthermore, sufficient hydroxyl groups remain on the surface of the M-450 particles to allow chemical (covalent) binding of proteins via the tosyl chloride method of Nilsson and Mosbach.[28] These particles were also employed in this study and permit both covalent and noncovalent binding of antibody.

B. Coating of the Particles with Antibody

The highly porous 3-μm particles or the M-450 particles were resuspended in sterile water or phosphate-buffered saline solution (PBS) at pH 7.2 and counted on a hemacytometer. If microaggregates (defined as clustering together of three or more spheres) were observed

$$\text{Microsphere-OH} + \underset{\text{Tosyl Chloride}}{\text{Cl-SO}_2\text{-C}_6\text{H}_4\text{-CH}_3} \longrightarrow$$

$$\text{Microsphere-O-SO}_2\text{-C}_6\text{H}_4\text{-CH}_3 \xrightarrow{\text{Antibody-NH}_2}$$

$$\text{CH}_3\text{-C}_6\text{H}_4\text{-SO}_3^- + \text{Microsphere-NH-Antibody}$$

FIGURE 1. The tosyl-chloride method for covalent conjugation of antibody to magnetic microspheres.

under the microscope, as it was usually the case with the highly porous 3-μm particles, the mixture was sonicated at 4°C for 10 sec. Sonication was rarely required with the M-450 microspheres which were not forming microconjugates. However, the monodispersion of the particles should be checked always under the microscope. The particles were sterilized by radiation (5000 rad) and washed once with PBS.

The M-450 particles were coated with affinity-purified goat antimouse IgG or rabbit antimouse IgG (Cappel, Cochranville, Pa.). One milligram of antibody was used per 8×10^9 M-450 particles in total volume of 1 mℓ of PBS without calcium and magnesium. The mixture was incubated at 4°C for 14 hr with continuous mixing on a rotator. The particles were washed twice with PBS (without calcium and magnesium) supplemented with 1% fetal calf serum (FCS). If the microspheres were not used within 30 min, they were washed two more times with 1% FCS in PBS before use.

The presence of hydroxyl groups in sufficient numbers on the surface of the M-450 particles permitted covalent binding of antibodies by the method of tosyl chloride as described by Nilsson and Mosbach.[28] This method is diagramatically shown in Figure 1.

The binding of ^{125}I-labeled purified sheep antimouse immunoglobulin (Ab_2) to native M-450 particles by physical adsorption (noncovalent binding) and to tosylated M-450 particles by physical adsorption and chemical binding (covalent) was investigated systematically by Ugelstad and colleagues. Representative results are shown in Figure 2. In these experiments, various amounts of antibody (5 to 50 mg) were incubated with native M-450 or with Tosylated M-450 particles and washed several times to determine the amount of immunoglobulin retained and the amount that was removed by washing. In these experiments, various amounts of particles (2 to 20 mg/mℓ) were added to 100 μg of antibody in a total volume of 1 mℓ of PBS and incubated with rotation for 16 hr at room temperature. Native M-450 particles were washed in buffer consisting of 0.05 *M* Tris, 0.1 *M* NaCl, 0.01% BSA, 0.01% merthiolate, pH 7.7. Initially the particles were washed by resuspension in 1 mℓ of buffer and centrifugation. This short wash is marked with the letter S in Figure 2. The particles were washed ten additional times. They were resuspended in 1 mℓ of washing buffer, rotated for 24 hr at room temperature and the amount of ^{125}I-labeled immunoglobulin that remained bound to the particles was determined using a gamma counter (Figure 2). Tosylated M-450 particles were coupled with antibody in borate buffer (0.1 *M* pH 9.4), and were incubated with rotation for 16 hr at room temperature. The tosylated M-450 particles were washed the

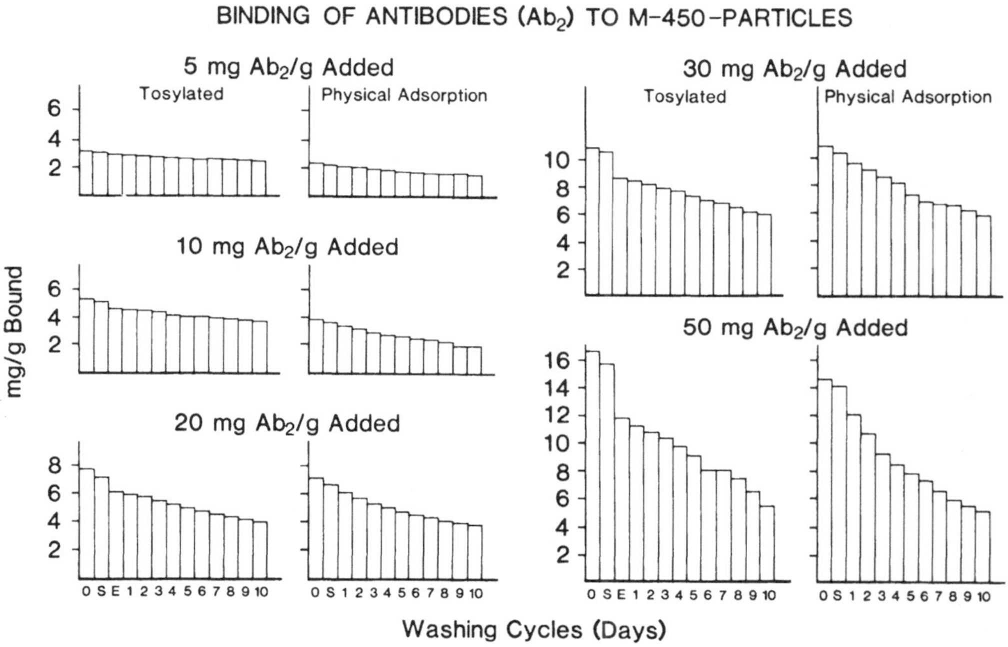

FIGURE 2. Studies of the binding and retention of ^{125}I-labeled sheep antimouse immunoglobulin (Ab_2) to native M-450 particles by physical adsorption and tosylated M-450 particles by physical adsorption and chemical binding. The particles were coated with ^{125}I-labeled sheep antimouse immunoglobulin by incubation in PBS for 16 hr at room temperature, washed 11 times, and the amount of immunoglobulin attached to the particles were determined. O designates the stage immediately after the coating of the particles with antibody, and S the stage immediately after an initial wash of the particles by centrifugation in the Tris buffer mentioned previously. The particles were washed ten additional times (steps 1—10), with Tris buffer for 24 hr at room temperature, and the ^{125}I immunoglobulin that remained attached was determined. The tosylated M-450 particles were washed by the same procedure. In addition they were treated with 1 *M* ethanolamine-HCl, pH 9.5, with rotation for 4 hr between the initial and the first of the ten washings, to mask remaining tosyl groups.

same way except that an additional step was added between the initial and the first of the ten washings to mask any remaining unreactive tosyl groups (step E in Figure 2). This was accomplished by treating the particles with 1 mℓ of 1 *M* ethanolamine-HCl, pH 9.5 for 4 hr with rotation. Comparison of the amount of immunoglobulin bound per gram of particles revealed a rather consistent pattern, with the tosylated particles retaining modestly higher amounts of immunoglobulin.

III. IMMUNOMAGNETIC SEPARATIONS

A. Labeling of Cells with Magnetic Particles

Bone marrow was obtained from healthy adult normal donors of both sexes by multiple aspirations from the posterior iliac crest. The aspirates were diluted 1:10 with Hanks' balanced salt solution (HBSS) and separated on Ficoll/Hypaque density cushion as previously described.[29] In certain experiments, phagocytic cells were removed by allowing them to ingest iron carbonyl particles followed by centrifugation on a Ficoll/Hypaque density cushion as described elsewhere.[30] the cells were washed three times with PBS supplemented with 1% FCS and incubated for 2 hr at 4°C with rotation with mixtures of two (anti-Leu 5 and anti-Leu 2a) or four (anti-Leu 1, anti-Leu 2a, anti-Leu 3a, and anti-Leu 5) monoclonal antibodies (2.5 μg of each monoclonal antibody was added to 10 × 10^6 cells). The cells were washed three times with PBS containing 1% FCS by centrifugation and resuspended in 1 mℓ of this medium.

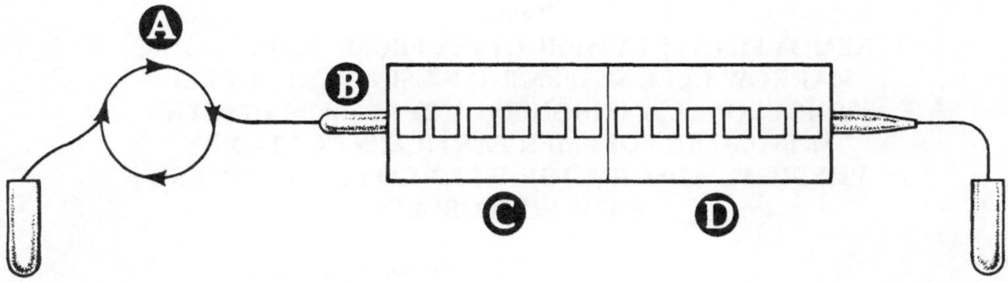

FIGURE 3. The apparatus for immunomagnetic separation employed in these studies consists of two polycarbonate chambers with samarium-cobalt magnets attached to the bottom (C and D) and a peristaltic pump (A). These chambers are described in Section III.B. The actual separation took place in a sterile disposable pipet (B), which was placed in the chambers and was connected to the peristaltic pump and to the collection tube with plastic tubing.

Monoclonal antibody-coated bone marrow cells (up to 20 × 10^6 in 1 mℓ) and antimouse polyclonal antibody-coated magnetic particles were mixed at the ratio of 100:1 in a 5-mℓ round-bottom test tube at a total volume of 2 mℓ, and were incubated for 2 hr at 4°C with gentle agitation every 15 min. The cells were separated in the immunomagnetic apparatus as described in following paragraphs.

B. The Apparatus

A sterile continuous flow system (Figure 3) similar to that described by Treleaven et al.[27] was employed in these experiments. This system consists of two polycarbonate chambers with samarium-cobalt permanent magnets attached to the bottom and a peristaltic pump. Six samarium-cobalt magnets (1 × 1 × 0.4 cm; Magnet Sales and Manufacturing Company, Culver City, Calif.) were attached to the bottom of each chamber. In the first chamber the magnets were separated by the interior by 4 mm of polycarbonate and the distance between two successive magnets was 1 cm. In the second chamber the magnets were separated by the interior by 2 mm of polycarbonate and the distance between two successive magnets was 6 mm. The objective of this design is to minimize nonspecific trapping of the cells that may occur because of the formation of aggregates at the beginning of the chamber due to strong magnetic fields. A sterile disposable 2-mℓ pipette was placed in the chambers and was connected to the peristaltic pump and to the collection tube with plastic tubing (Figure 3). The separation system was sterilized by washing with 70% alcohol and UV irradiation and washed with PBS supplemented with 1% FCS.

C. The Separation Procedure

Up to 20 × 10^6 bone marrow cells labeled with magnetic microspheres as previously described, in 2 mℓ of PBS supplemented with 1% FCS, were pumped through the separation chambers using the peristaltic pump at a rate of 6 mℓ/min. Eluted cells were collected in a test tube and were washed three times with PBS supplemented with 1% FCS and resuspended in RPMI-1640 supplemented with 10% FCS, 25 m*M* Hepes buffer, 2 m*M* L-glutamine, and antibiotics. The separation procedure was carried out in a sterile hood. Eluted cells were characterized by functional and phenotypic tests. It was not possible to dissociate the cell-microsphere complexes and recover the cells from the retained fraction.

D. Characterization of Eluted Cells

Magnetic monosized M-450 polymer particles (4.5 μm) coated by physical adsorption with goat antimouse immunoglobulin were effective in removing T lymphocytes from human bone marrow cell suspensions, incubated with the anti-Leu 5 and anti-Leu 2a monoclonal antibodies. Representative results are shown in Table 1. Proliferative response to PHA-P

Table 1
REMOVAL OF T LYMPHOCYTES FROM HUMAN BONE MARROW CELL SUSPENSIONS USING ANTI T-CELL MONOCLONAL ANTIBODIES[a] AND 4.5-μm MAGNETIC-MONOSIZED POLYMER PARTICLES COATED BY PHYSICAL ADSORPTION WITH GOAT ANTIMOUSE IMMUNOGLOBULIN

	E-rosettes (%)	Proliferative responses (counts per min)		
		Medium	PHA[b]	MLC[b]
Initial sample[c]	15	2,451 ± 542	26,109 ± 1,673	17,288 ± 459
Eluted cells[d]	1	120 ± 55	280 ± 98	952 ± 190

[a] Monoclonal antibodies used: anti-Leu 5 and anti-Leu 2a.
[b] Proliferative responses to PHA and to allogeneic cells MLC were determined as previously described.[30] Cultures of unfractionated human bone marrow cells and cells depleted to T lymphocytes were incubated for 6 days at 37°C, pulsed with tritiated thymidine during the last 18 hr of the culture, and harvested using an automated cell harvester.
[c] 6×10^6 human bone marrow cells purified by centrifugation on Ficoll/hypaque density cushion.
[d] 2.4×10^6 cells were recovered (40%).

and to allogeneic mononuclear leukocytes in mixed lymphocyte culture were absent in the eluted fraction. In contrast, unseparated cells responded vigorously to these stimuli (Table 1). These responses were determined as previously described.[30] E-rosette forming cells were determined by rosetting with sheep erythrocytes by a standard method[30] and were 0.25 ± 0.42% in the eluted fraction vs. 22.5 ± 10.4% in the unseparated cells (mean ± 1 SD of six experiments). Recovery of eluted cells was 38.8 ± 8.7% (mean ± 1 SD of six experiments). The mixture of the anti-Leu 5 and anti-Leu 2a monoclonal antibodies appeared to be sufficient for the removal of T cells. M-450 magnetic particles coated with rabbit antimouse immunoglobulin or sheep antimouse immunoglobulin were equally effective in removing T cells from bone marrow cell suspensions coated with anti-T cell monoclonal antibodies. In contrast, highly porous 3-μm magnetic monosized particles coated with goat antimouse immunoglobulin were not effective in removing T cells from human bone marrow cell suspensions (data not shown).

Tosylated M-450 particles coated with sheep antimouse immunoglobulin[31] by physical adsorption and chemical binding were equally effective to the M-450 particles coated with antimouse Ig in removing T cells from human bone marrow cell suspensions, previously incubated with anti-T cell monoclonal antibodies. In these experiments tosylated M-450 particles coated with affinity-purified sheep antimouse immunoglobulin antibody were used and were kindly provided by Dr. K. Nustad of The Norwegian Radium Hospital, Oslo, Norway. Approximately 3.4 μm of sheep antimouse immunoglobulin per milligram of particles were attached by adsorption and covalent binding. Representative results are shown in Table 2. Removal of T cells from the bone marrow resulted in elimination of the proliferative responses to allogeneic cells in mixed lymphocyte culture, whereas unseparated bone marrow cells responded vigorously. Quantitation of T-cell removal from the bone marrow using the tosylated M-450 particles was achieved by a limiting dilution microculture assay.[32] Various numbers of bone marrow cells were cultured in Terasaki microculture plates with

Table 2
REMOVAL OF T LYMPHOCYTES FROM HUMAN BONE MARROW CELL SUSPENSIONS USING ANTI-T CELL MONOCLONAL ANTIBODIES[a] AND 4.5-μm TOSYLATED-MAGNETIC MONOSIZED POLYMER PARTICLES COATED BY PHYSICAL ADSORPTION AND COVALENT BINDING WITH SHEEP ANTIMOUSE IMMUNOGLOBULIN

	Proliferative responses (counts per min)	
	Medium	**MLC[b]**
Initial sample[c]	297 ± 300	39,888 ± 95
Eluted cells[d]	161 ± 97	718 ± 492

[a] Monoclonal antibodies used: anti-Leu 1, anti-Leu 2a, anti-Leu 3a, and anti-Leu 5.
[b] Proliferative responses to allogeneic leukocytes in mixed lymphocyte culture were determined as previously described.[30] Cultures of unfractionated human bone marrow cells and cells depleted of T lymphocytes were incubated for 6 days at 37°C, pulsed with tritiated thymidine during the last 18 hr of the culture, and harvested using an automated cell harvester.
[c] 6×10^6 human bone marrow cells purified by centrifugation on Ficoll/hypaque density cushion.
[d] 3.2×10^6 cells were recovered (32%).

irradiated (4000 rad) autologous or allogeneic peripheral blood mononuclear cells (feeders), PHA-P (20 μm/mℓ) and an optimal concentration of IL-2 (supernatant from a gibbon lymphoma cell line). After 16 days the wells were scored as positive or negative by microscopic examination, on the basis of the presence or absence of cell growth. Statistical analysis was carried out as described by Taswell.[33] A representative experiment is shown in Figure 4. The frequency of clonable T lymphocytes in the T-cell depleted marrow is 1 in 1659 cells and the total number of clonable T cells 1.65×10^3. In this experiment, 7.5×10^6 bone marrow cells were subject to separation and 2.74×10^6 cells (36.5%) were recovered in the eluted fraction. The frequency of clonable lymphocytes in the unseparated bone marrow is approximately 1 in 15 cells (data not shown) and the total number of clonable T cells present in this sample (of 7.5×10^6 cells) is 5×10^5. Therefore, a 2.5 log reduction in T cells was achieved in a single step using these magnetic particles. Results on the recovery of committed hemopoietic progenitor cells of the granulocyte/macrophage lineage is shown in Table 3. Selective loss of CFU-c (determined as described by Pike and Robinson[34]) was not observed during the immunomagnetic separation.

IV. CONCLUSIONS

M-450 magnetic monosized polymer particles coated with goat antimouse immunoglobulin were successfully used to achieve 2 to 3 log depletion in a single step of T cells from human bone marrow cell suspensions, previously incubated with anti-T cell monoclonal antibodies.

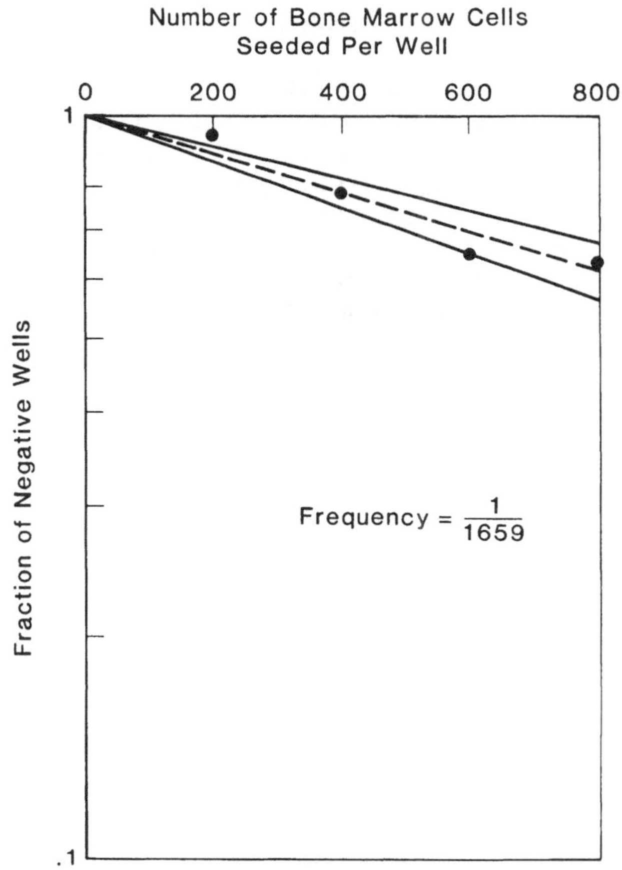

FIGURE 4. Quantitation of T-cell removal from human bone marrow cell suspensions by limiting dilution analysis. T cell-depleted bone marrow cells seeded per well are plotted against the fraction of negative wells; (---), line for best fit as determined by the minimum chi-square method; (—), 95% confidence limits.

Table 3
COLONY FORMING UNITS IN VITRO AFTER IMMUNOMAGNETIC DEPLETION OF T CELLS FROM HUMAN BONE MARROW CELL SUSPENSIONS USING TOSYLATED M-450 PARTICLES

	Total cells	CFU-C/1 × 10^5 cells	Total CFU-C
Initial sample	7.50 × 10^6	140	1.05 × 10^4
Eluted cells	1.65 × 10^6 (yield 22%)	138	2.30 × 10^3

The separation procedure is carried out in a sterile continuous-flow system, using chambers with permanent samarium-cobalt magnets and a peristaltic pump. Although in the experiments presented here, small numbers (up to 20 × 10^6) of human bone marrow cells were depleted of T cells at a time, it is anticipated that the method can be scaled up to accommodate large numbers of cells (up to 1 × 10^{11}) without major difficulties.

The advantages of the immunomagnetic method for removal of T cells from human bone marrow cell suspensions are

1. The time needed for the separation is significantly reduced. Cumbersome cell separation procedures are not employed.
2. The method is simple and the separation takes place in a closed sterile continuous flow system.
3. Agents such as sheep erythrocytes, whose viral or other contamination cannot be effectively controlled, are not employed.
4. Lysed or inactivated T cells are not infused into the patient as it happens with methods based on cell elimination by complement or immunotoxin treatment in vitro.
5. The end point of the procedure is well defined.

Depletion of T cells higher than that of 2 to 3 logs achieved by one cycle of immunomagnetic separation can be presumably accomplished by repeating the procedure for a second time or by adding to the procedure an agglutination step with soybean lectin, preceding the immunomagnetic separation. It is anticipated that the introduction of magnetic monosized particles and closed sterile continuous flow systems will further facilitate the use of T-cell depletion methods for bone marrow transplantation across histocompatibility barriers.

ACKNOWLEDGMENTS

This work has been supported by grant PCM-8119178 from the National Science Foundation, program project P01 CA 23766 from The National Institute of Health and a grant from the Whitaker Foundation. We thank SINTEF for providing the magnetic particles used in this study and Dr. K. Nustad for coating certain M-450 particles used in this study with sheep antimouse immunoglobulin.

REFERENCES

1. **Simonsen, M.,** The impact on the developing embryo and newborn animal of adult homologous cells, *Acta Pathol. Microbiol. Scand.*, 40, 480, 1957.
2. **Weiden, P. L. and the Seattle Marrow Transplant Team,** Graft-vs-host disease in allogeneic marrow transplantation, in *Biology of Bone Marrow Transplantation,* Gale, R. P., Fox, C. F., Eds., Academic Press, New York, 1980.
3. **Storb, R., Prentice, R. L., Buckner, C. D., Clift, R. A., Appelbaum, F., Deeg, J., Doney, K., Hansen, J. A., Mason, M., Sanders, J. E., Singer, J., Sullivan, K. M., Witherspoon, R. P., and Thomas, E. D.,** Graft versus host disease and survival in patients with aplastic anemia treated by marrow grafts from HLA-identical siblings. Beneficial effect of a protective environment, *N. Engl. J. Med.*, 308, 302, 1983.
4. **Gale, R. P., Kersey, J. H., Bortin, M. M., Dicke, K. A., Good, R. A., and Zwaan, F. E.,** Bone marrow transplantation for acute lymphoblastic leukaemia, *Lancet,* 2, 639, 1983.
5. **Thomas, E., Storb, R., Clift, R., et al.,** Bone marrow transplantation, *N. Engl. J. Med.*, 292, 895, 1975.
6. **Winston, D., Gale, R., Meyer, D., and Young, L.,** Infectious complications of human bone-marrow transplantation, *Medicine,* 58, 1, 1979.
7. **Ramsay, N. K. C., Kersey, J. H., Robison, L. L., et al.,** Prevention of acute graft-versus-host disease: a randomized study demonstrating the influence of treatment regimen and age, *N. Engl. J. Med.*, 306, 392, 1982.
8. **Powles, R. L., Clink, H., Sloane, J., Barrett, A. J., Kay, H. E., and McElwain, T. J.,** Cyclosporin A for the treatment of graft versus host disease in man, *Lancet,* 2, 1327, 1978.
9. **Deeg, H. J., Storb, R., Thomas, E. D., et al.,** Marrow transplantation for acute, nonlymphoblastic leukaemia in first remission: Preliminary results of a randomised trial comparing cyclosporin and methotrexate for the prophylaxis of graft-versus-host disease, *Transplant. Proc.*, 15, 1385, 1983.

10. **Dicke, K. A., Van Hooft, J., and Van Bekkum, D. W.,** The selective elimination of immunologically competent cells from bone marrow and lymphatic cell mixture, *Transplantation,* 6, 562, 1968.
11. **Rodt, H., Thierfelder, S., and Eulitz, M.,** Anti-lymphocytic antibodies and marrow transplantation. III. Effect of heterologous anti-brain antibodies on acute secondary disease in mice, *Eur. J. Immunol.,* 4, 25, 1974.
12. **Tyan, M.,** Modification of severe graft-versus-host disease with antisera to the theta antigen or to whole serum, *Transplantation,* 15, 601, 1973.
13. **Onoe, K., Fernandes, G., and Good, R. A.,** Humoral and cell-mediated immune responses in fully allogeneic bone marrow chimera in mice, *J. Exp. Med.,* 151, 115, 1980.
14. **Korngold, R. and Sprent, J.,** Surface markers of T cells causing lethal graft-versus-host disease in mice, in *Recent Advances in Bone Marrow Transplantation,* Gale, R. P., Ed., Alan R. Liss, New York, 1983, 199.
15. **Vallera, D. A., Youle, R. J., Neville, D. M., and Kersey, J. H.,** Bone marrow transplantation across major histocompatibility barriers. V. Protection of mice from lethal graft-vs-host disease by pretreatment of donor cells with monoclonal anti-Thy-1-2 coupled to the toxin ricin, *J. Exp. Med.,* 155, 949, 1982.
16. **Reisner, Y., Itzicovitch, L., Meshorer, A., and Sharon, N.,** Hemopoietic stem cell transplantation using mouse bone marrow and spleen cells fractionated by lectins, *Proc. Natl. Acad. Sci.,* 75, 2933, 1978.
17. **Reisner, Y., Kapoor, N., O'Reilly, R. J., and Good, R. A.,** Allogeneic bone marrow transplantation using stem cells fractionated by lectins. VI. In vitro analysis of human and monkey bone marrow cells fractionated by sheep red blood cells and soybean agglutinin, *Lancet,* 2, 1320, 1980.
18. **Reisner, Y., Kapoor, N., Kirkpatrick, D., Pollack, M., Dupont, B., Good, R. A., and O'Reilly, R. J.,** Transplantation for acute leukaemia with HLA-A and B nonidentical parental marrow cells fractionated with soybean agglutinin and sheep red blood cells, *Lancet,* 2, 327, 1981.
19. **Reisner, Y., Kapoor, N., Kirkpatrick, D., Cunningham-Rundles, S., Dupont, B., Hodes, M. Z., Good, R. A., and O'Reilly, R. J.,** Transplantation for severe combined immunodeficiency with HLA-A, B, D, DR incompatible parental marrow cells fractionated by soybean agglutinin and sheep red blood cells, *Blood,* 61, 341, 1983.
20. **Rodt, H., Kolb, H. J., Netzel, B., et al.,** Effect of anti-T cell globulin on GvHD in leukaemic patients treated with BMT, *Transplant. Proc.,* 13, 257, 1981.
21. **Blacklock, H. A., Prentice, H. G., and Gilmore, M.,** Attempts at T cell depletion using OKT3 and rabbit complement to prevent a GvHD in allogeneic BMT, *Exp. Hematol.,* 11 (Suppl. 13), 37, 1983.
22. **Sharp, T. G., Sachs, D. H., Fauci, A. S., Messerschmidt, G. L., and Rosenberg, S. A.,** T cell depletion of human bone marrow using monoclonal antibody and complement-mediated lysis, *Transplantation,* 35, 112, 1983.
23. **Prentice, H. G., Janossy, G., Price-Jones, L., Trejdosiewicz, L. K., Panjwani, D., Graphakos, S., Ivory, K., Blacklock, H. A., Gilmore, M. J. M. L., Tidman, N., Skeggs, D. B. L., Ball, S., Patterson, J., and Hoffbrand, A. V.,** Depletion of T lymphocytes in donor marrow prevents significant graft-versus-host disease in matched allogeneic leukaemic marrow transplant recipients, *Lancet,* 1, 472, 1984.
24. **Ash, R. C., Serwint, M., Doukas, M., Romond, E., Bradley, P., Metcalfe, M., Marshall, E., Geil, J., Greenwood, M., MacDonald, J. S., and Thompson, J. S.,** Marrow T cell depletion with antihuman T cell antibodies is effective for graft versus host disease (GvHD) prophylaxis in human allogeneic marrow transplantation, *Blood,* 64 (Suppl. 1), 744, 1984.
25. **Filipovich, A. H., Youle, R. J., Neville, D. M., Jr., Vallera, D. A., Quinones, R. R., and Kersey, J. H.,** Ex-vivo treatment of donor bone marrow with anti-T-cell immunotoxins for prevention of graft-versus-host disease, *Lancet,* 1, 469, 1984.
26. **Ugelstad, J., Soderberg, L., Berge, A., and Bergstrom, J.,** Monodisperse polymer particles — a step forward for chromatography, *Nature (London),* 303, 96, 1983.
27. **Treleaven, J. G., Ugelstad, J., Philip, T., Gibson, F. M., Rembaum, A., Caine, G. D., and Kemshead, J. T.,** Removal of neuroblastoma cells from bone marrow with monoclonal antibodies conjugated to magnetic microspheres, *Lancet,* 1, 70, 1984.
28. **Nilsson, K. and Mosbach, K.,** *p*-Toluenesulfonyl chloride as an activating agent of agarose for the preparation of immobilized affinity ligands and proteins, *Eur. J. Biochem.,* 112, 397, 1980.
29. **Platsoucas, C. D., Beck, J. D., Kapoor, N., Good, R. A., and Gupta, S.,** Separation of human bone marrow cell populations by density gradient electrophoresis: differential mobilities of myeloid (CFU-C), monocytoid, and lymphoid cells, *Cell. Immunol.,* 59, 345, 1981.
30. **Platsoucas, C. D. and Good, R. A.,** Inhibition of specific cell mediated cytotoxicity by monoclonal antibodies to human T-cell antigens, *Proc. Natl. Acad. Sci. U.S.A.,* 78, 4500, 1981.
31. **Nustad, K., Danielsen, H., Reith, A., Funderud, S., Lea, T., Vartdal, F., and Ugelstad, J.,** Monodisperse polymer particles in immunoassays and cell separation, in *Microspheres: Medical and Biological Applications,* Rembaum, A., Ed., CRC Press, Boca Raton, Fla., 1988, in press.
32. **Kernan, N. A., Burns, M. J., Collins, N. H., O'Reilly, R. J., and Dupont, B.,** Limiting dilution microculture assay for quantitation of T lymphocytes in bone marrow, *Transplant. Proc.,* 17, 437, 1985.

33. **Taswell, C.,** Limiting dilution assays for the determination of immunocompetent cell frequencies, *J. Immunol.*, 126, 1614, 1981.
34. **Pike, B. L. and Robinson, W. A.,** Human bone marrow colony growth in agar-gel, *J. Cell. Physiol.*, 76, 77, 1970.

Chapter 7

POLYMER COLLOIDS AS CONTROLLED RELEASE DEVICES

R. M. Fitch and K. M. Scholsky

TABLE OF CONTENTS

I. INTRODUCTION

Controlled release technology is playing an increasingly important role in modern industry. This is evidenced by the wide variety of new products introduced during the last two decades which utilize this technology.

The purpose of a controlled-release device is to provide a more efficient means of releasing a bioactive agent into its surrounding environment. For example, Figure 1 shows the blood plasma drug levels obtained by conventional, periodic administration of a drug in the form of a pill or tablet. This method of administration may result in periods of time when the desired upper and lower limits have been exceeded. By substituting an appropriate controlled-release device, the drug release can be maintained an an optimal level for long periods of time. This reduces toxic side effects and also decreases the number of administrations required, thereby improving patient compliance. Equally important benefits can also be realized in agriculture and pest control where it is desirable to maintain a critical level of a bioactive agent over long time periods.

A variety of colloidal delivery systems in the form of microspheres,[1-5] microcapsules,[6-10] nanoparticles,[11-17] and liposomes[18-21] have been developed for controlled release. Their colloidal nature lends itself to oral, topical, and parenteral administration.

This paper describes the preparation of polyacrylate ester latex particles, 40 to 400 nm in diameter, capable of sustained drug release. A latex is a dispersion of colloidal polymer particles in a continuous, aqueous phase. These particles are prepared by emulsion polymerization and possess pendant drug moieties attached by ester linkages. Types of drugs which can be incorporated by this method include analgesics, antidepressants, tranquilizers, respiratory and cerebral stimulants, estrogens, antibiotics, antitumor drugs, and others.

Pendant drug delivery systems based on synthetic and naturally occurring macromolecules are well known.[22-28] Advantages of these systems include higher specificity towards the target, reduced drug toxicity, and modification in the pharmacokinetics and cellular uptake of the therapeutic agent. By linking to a suitable carrier, a slow release of the drug can also be achieved by hydrolysis or enzyme-catalyzed cleavage of the pendant moieties.

In an earlier paper Fitch et al.[29] prepared model polyacrylate ester colloids which were found to release various pendant phenols and alcohols via acid-catalyzed hydrolysis. The redox system used to initiate emulsion polymerization produced latex particles with sulfate and sulfonate end groups at the polymer-water interface. Conversion of these groups to their strong acid form by ion exchange caused the particles to "self-catalyze" hydrolysis of the pendant ester linkages and the subsequent release of the active agent. Hydrolysis of these systems followed pseudo zero-order kinetics and resulted in steady rates of release over long periods of time.

This paper presents the kinetics of drug release from latex particles. The drugs released were salicylic acid and chloramphenicol. Since such systems have implications for controlled release, it is desirable to determine the release mechanisms involved. The observed hydrolysis kinetics have been explained by a reaction zone model in which a steady state concentration of polymeric ester is maintained. Acid-catalyzed hydrolysis is believed to occur within a thin interfacial reaction volume which slowly recedes inward toward the particle center. Base-catalyzed hydrolysis is believed to occur simultaneously in both an interfacial and a gel reaction zone. Hydrolysis causes the particles to ultimately assume a core/shell morphology with a central core of unhydrolyzed polymeric ester surrounded by a diffuse shell consisting primarily of polyacrylic acid. Mathematical models have been proposed to explain the kinetics of drug release observed. They are based on principles similar to those described by Hopfenberg for the controlled release of drugs from erodible spheres.[30]

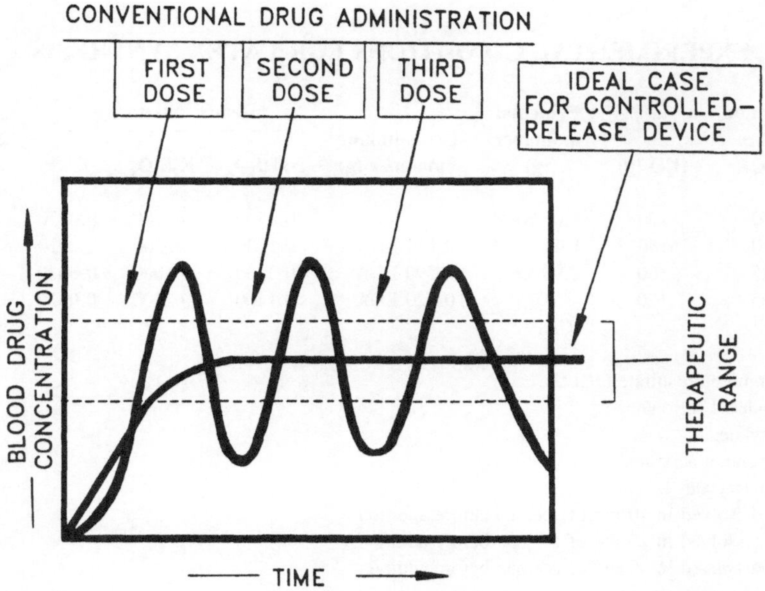

FIGURE 1. Example of drug delivery obtained by conventional administration and from a controlled-release device.

II. EXPERIMENTAL

A. Monomer and Latex Preparation

The monomers used were 2-carboxyphenyl acrylate (salicyl acrylate), 2-(dichloroacetamido)-3-hydroxy-3-(4-nitrophenyl) propyl acrylate (chloramphenicol acrylate), and ethylene diacrylate. Salicyl and chloramphenicol acrylate monomers were prepared using reagent grades of acryloyl chloride, chloramphenicol, and salicylic acid using procedures similar to those described by Banks et al.[31] Batz et al.[32] have synthesized pharmacologically active, water-soluble polymers using monomers prepared by analogous methods. Chloramphenicol acrylate and salicyl acrylate monomers were purified by recrystallization or other means prior to use.[33] Ethylene diacrylate was purified as described previously.[29]

Latexes were prepared in magnetically stirred, round-bottom flasks which were freed of oxygen and maintained under nitrogen atmospheres. Polymerizations were initiated using the persulfate/bisulfite/iron (PBI) redox system.[34] Reagent grades of potassium persulfate, sodium bisulfite, and ferrous ammonium sulfate ($\cdot 9\ H_2O$) were used as received. Sodium dodecyl sulfonate was prepared by a modified Strecker reaction.[35] After charging the reaction vessel with water, surfactant, and initiator, monomers dissolved in acetone were added continuously from a pressure-equalized dropping funnel at a rate of 1 to 2 drops per second. Polymerizations were allowed to run overnight. Table 1 lists the recipes used to prepare the various latexes.

B. Purification and Ion Exchange of Latexes

Latexes were purified of emulsifier, unreacted monomers, and initiator byproducts by either continuous, liquid-liquid extraction, or hollow-fiber dialysis followed by treatment with ion-exchange resins. Continuous, liquid-liquid extraction with ether was used for the initial purification of latexes 1, 2, and 3. A Bio-fiber model 50 ''miniplant'' hollow fiber dialysis unit and 60°C distilled water were used for the initial purification of latex 4. Final purification and ion exchange of each latex was achieved by two consecutive treatments with mixed beds of Dowex® 50W-X4 and Dowex® 1-X4 ion-exchange resins in their strong

Table 1
EXPERIMENTAL CONDITIONS FOR LATEX SYNTHESIS

Latex no.	Reaction temp. (°C)	H_2O (g)	Principal monomer (g)	Cross-linking monomer (g)	Redox initiator (g) $NaHSO_3$	$K_2S_2O_8$	FAS[a]	Surfactant[b] (g)
1	30	150	4.00 SA[c,f]	—	0.2500	0.5000	0.0030	0.2000
2	30	150	1.45 SA[c,g]	0.145 EDA[e,g]	0.1200	0.2500	0.0020	0.2000
3	45	300	2.90 SA[c,f]	0.290 EDA[e,f]	0.2400	0.5000	0.0040	0.4000
4	65	120	0.75 CAPA[d,h]	0.120 EDA[e,h]	0.1200	0.2500	0.0020	0.5000

[a] Ferrous ammonium sulfate ($\cdot 9H_2O$).
[b] Sodium dodecyl sulfonate.
[c] Salicyl acrylate.
[d] Chloramphenicol acrylate.
[e] Ethylene diacrylate.
[f] Monomer dissolved in 40 mℓ of acetone before addition.
[g] Monomer dissolved in 20 mℓ of acetone before addition.
[h] Monomer dissolved in 30 mℓ of acetone before addition.

acid and strong base forms, respectively. These resins were conditioned prior to use according to the methods described by Vanderhoff et al.[36] After ion exchange the percentage of polymer present in each latex was determined gravimetrically by drying in an oven at 70°C. To retard hydrolysis, purified latexes were stored in a refrigerator until kinetic studies were undertaken.

C. Determination of Particle Size

The average particle sizes of latexes 1, 2, and 3 were determined using a Hitachi® model HU-200F transmission electron microscope and a phosphotungstic acid (PTA) negative staining technique.[37,38] PTA is an electron-opaque material which darkens the area around the particles producing ''negative'' type images.

The average particle size of latex 4 was determined by angular dissymmetry light scattering using an Oriel Optics® model 6697 He-Ne laser light source and a modified Brice-Phoenix Model 200 DM light scattering photometer. The dissymmetry in intensity of light scattered by the particles at 45 and 135° was measured for a wavelength of 632 nm. The average particle diameter was then calculated using tabulated dissymmetry values for spheres.[39]

III. SURFACE ANALYSIS

The number of strong acid end groups present per gram of polymer was determined by conductometric titration. After purging with nitrogen for about 10 min to remove dissolved CO_2, each latex was titrated with standardized $Ba(OH)_2$ solution added by means of a motor-driven Manostat ultramicropipette.

Conductance was measured using a Beckman dip-type conductance cell connected to a conductivity amplifier. The surface charge density derived from strong acid end groups was then calculated using values obtained for the percent solids polymer, average particle size, and number of millequivalents of strong acid per gram of polymer.

IV. MEASUREMENT OF DRUG RELEASE RATES

The kinetics of drug release from latex particles into the continuous phase were measured using Beckman Model 25 and Acta V UV-Visible spectrophotometers. Kinetic experiments were performed on latex particles dispersed in distilled water, phosphate buffer, Hanks'

Table 2
EXPERIMENTAL RESULTS

Latex no.	Average particle size (nm)	Surface strong acid conc. (μC/cm²)	pH	Medium	Temp. (°C)	Initial[a] rate (R_1)	Second[a] rate (R_2)	Fig. no.
1[d]	400	345	5.0[b]	Water	37	6.4×10^{-7}	4.5×10^{-7}	1
			5.0[b]	Water	50	2.7×10^{-6}	7.5×10^{-7}	1
			7.4	Phosphate Buffer (PB)	37	(1.1×10^{-3})[c]	—	2
			7.4	PB	50	(1.6×10^{-3})[c]	—	2
2[e]	40	11.5	7.4	HBSS	4	1.1×10^{-5}	5.3×10^{-6}	3
			7.4	Blood Serum	4	1.4×10^{-5}	5.1×10^{-6}	3
			2.0	PB	37	1.1×10^{-7}	—	1
			5.0[b]	Water	37	2.4×10^{-7}	1.1×10^{-10}	1
3[e]	100	54	7.4	PB	37	(2.3×10^{-4})[c]	—	2
			7.4	PB	50	(4.0×10^{-4})[c]	—	2
			7.8	PB	37	(2.3×10^{-4})[c]	—	2
4[e]	126	64	5.0[b]	Water	37	4.1×10^{-8}	1.1×10^{-8}	4
			7.4	PB	37	(2.4×10^{-7})[c]	(4×10^{-8})[c]	4

[a] Moles per gram of polymer per hour.
[b] Self-catalyzed hydrolysis experiments.
[c] Approximate rate of release.
[d] Non-cross-linked particles.
[e] Cross-linked particles.

Balanced Salt Solution (HBSS), and human blood serum. PBSs of pH 2.0, 7.4, and 7.8 were prepared using pHydrion® capsules obtained from Grand Island Biological Co. Human blood serum was obtained by centrifuging whole blood at 3000 to 4000 rpm for 10 min to remove red blood cells. Whole blood was obtained from a volunteer. The serum was decanted and filtered through a 0.45-μm Millipore® membrane prior to use.

The various latex mixtures were kept in 50-mℓ volumetric flasks, and maintained at the desired temperature by immersing in a thermostatically controlled glycerin bath. Samples were removed periodically and the dispersed phase was flocculated by mixing with equal volumes of a high ionic-strength solution. Flocculated polymer was removed by filtration through a 0.45 μm membrane. The absorbance of the filtered continuous phase was then measured at either 540 nm or 278 nm relative to a 50/50 mixture of distilled water and the flocculating agent. Phenol reagent was used as the flocculating agent for latexes 1, 2, and 3. It was prepared by dissolving 40.0 g of $Fe(NO_3)_3 \cdot 9H_2O$ and 120 mℓ of 1 *N* HCl into 850 mℓ of distilled water. Fe(III) ions in phenol reagent react with salicylic acid to form purple-colored complexes exhibiting a λ_{max} at 540 nm. Schols reagent, prepared by dissolving 40.0 g of $Al(NO_3)_3 \cdot 9H_2O$ in 500 mℓ of 0.1 *N* HCl was used as the flocculant for latex 4. The absorbance of the filtered continuous phase from latex 4 was measured at the λ_{max} for chloramphenicol which occurs at 278 nm. Appropriate calibration curves following Beer's law were used to convert absorbance readings to actual concentration values. From these, Q′ values, the quantities of drug released per gram of polymer were calculated and plotted as a function of time.

V. RESULTS

Table 2 is a summary of the experimental results. The average particle size of the various latexes ranged between 40 and 400 nm. The surface charge density of strong acid end groups, as determined by titration, varied from 11.5 to 345 $\mu C/cm^2$. However, the value of 345 $\mu C/cm^2$ obtained for latex 1 is probably not a true surface charge density. The strongly hydrophilic nature of the polymers comprising these uncross-linked latex particles probably exposes strong acid end groups normally buried within the particle interior to hydroxyl ion during titration. Thus the apparent value for the surface charge density is increased.

After completion of emulsion polymerization, latexes 1, 2, and 3 exhibited a light purple tint. This was attributed to a reaction of salicylic acid with Fe(III) ions present in the continuous phase. These free salicylic acid molecules were probably produced by hydrolysis of monomer and polymer ester moieties during polymerization. Continuous extraction (1 to 2 hr) of these latexes using ether was found to remove this color.

Latex 1, which was not cross-linked, yielded water-soluble polymer at pHs greater than 7.0. This was attributed to ionization of carboxyl groups on the pendant salicyl-ester moieties and the consequent solvation and swelling by water. However, these water-soluble polymers could be converted back to latex particles simply by lowering the pH sufficiently. Latexes 2 and 3, which were cross-linked, did not yield water-soluble polymers at basic pH but rather exhibited a translucent, opalescent appearance. Latex 4, which lacked the carboxyl functionality of latexes 1, 2, and 3, exhibited no noticeable change in appearance upon raising the pH from acidic to basic. All four latexes were found to remain stable under the various experimental conditions employed.

Plots of the various experimental data can be seen in Figures 2 to 5, where Q' represents the number of moles of drug released per gram of polymer. In all cases the rates of release were found to decrease steadily with time and also varied as a function of temperature, pH, and the chemical composition of the particles.

Latexes 1, 2, and 3 exhibited rates of hydrolysis which were strongly pH dependent. For latex 1 at 37°C, the initial rate of release (R_1) of salicylic acid at pH 7.4 was about 1700 times faster than the R_1 observed for the same latex at pH 5.0 (curves B in Figures 2 and 3). For latex 3 the R_1 value at pH 7.4 (curve C in Figure 3) was 1000 times higher than at pH 5.0 (curve C in Figure 2) or at pH 2.0 (curve D in Figure 2). While the initial rates of self-catalyzed acid hydrolysis of latexes 3 and 4 differ by only a factor of 6, the initial rates of base-catalyzed hydrolysis of the same latexes differ by a factor of about 1000 (Figures 2, 3, and 5).

Increasing the temperature from 37 to 50°C during self-catalyzed acid hydrolysis of latex 1 (curves A and B of Figure 2) resulted in a doubling of the R_1 and a more abrupt transition between initial and secondary slopes. However, at pH 7.4 the same change in temperature resulted in only a slight increase in R_1 (curves A and B in Figure 3).

Latex 3, when buffered at pH 7.4, exhibited a doubling of R_1 (curves C and D in Figure 3) upon raising the temperature from 37 to 50°C. At 37°C latex 3 exhibited the same R_1 when buffered at pH values of 7.4 and 7.8 (curves D and E in Figure 3). Figure 4 shows the drug release profiles obtained for latex 2 when dispersed in HBSS and human blood serum buffered at pH 7.4 and at a temperature of 4°C.

VI. DISCUSSION

A. Self-Catalyzed Acid Hydrolysis

The kinetics of drug release observed for self-catalyzed acid hydrolysis of latexes 1, 3, and 4 are consistent with results previously published for other polyacrylate ester latexes.[29] The drug release kinetics from these latexes appears to decrease slowly with time, exhibiting a more or less abrupt change in rate at approximately 150 to 200 hr. This results in two essentially pseudo zero-order kinetic regions, indicated by R_1 and R_2 in Figures 2 and 5. The R_2 values comprise steady rates of drug release over hundreds of hours in some instances.

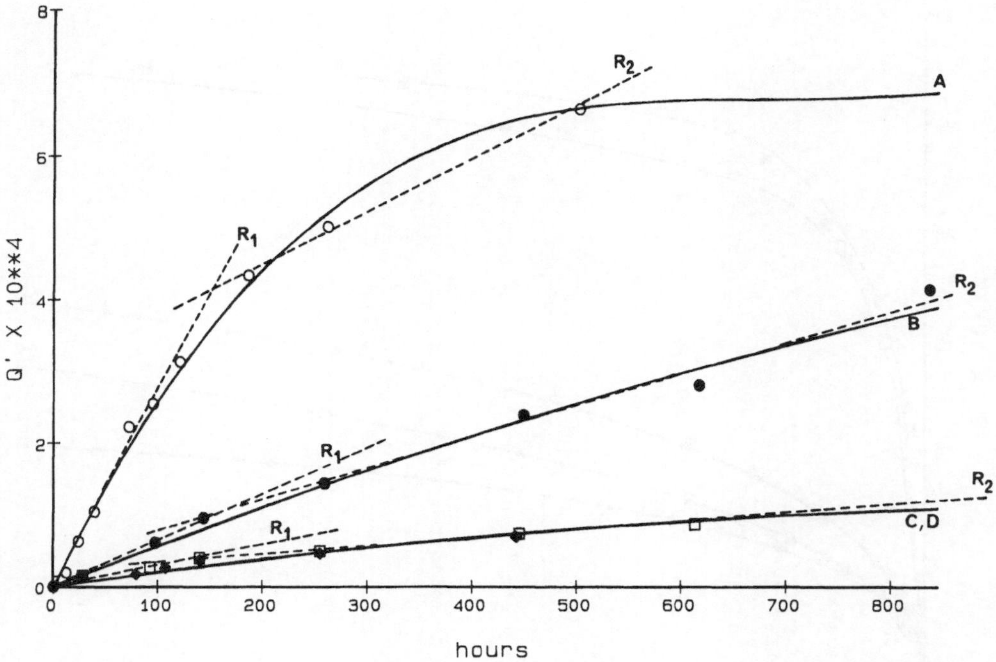

FIGURE 2. Self-catalyzed acid hydrolysis. Experimental data: (○) latex 1 at 50°C, pH 5.0; (●) latex 1 at 37°C, pH 5.0; (□) latex 2 at 37°C, pH 5.0; (◇) latex 2 at 37°C, pH 2.0.

Theoretical data:

Curve	$K'' \times 10^4$	$E_o \times 10^4$	n	Correlation coefficient
A	1.34	6.69	1.20	0.999
B	0.198	0.869	1.50	0.993
C	0.323	1.21	6.03	0.989

The pendant drug molecules are released continuously from these systems via hydrolysis, self-catalyzed by strong acid end groups attached at the latex particle interface. By varying the surface charge density of end groups and latex particle size, the rate of hydrolysis can be modified. Previous experiments have suggested that the overall rate of hydrolysis is pseudo first order with respect to acid concentration.[29]

Figure 6 is a plot of ln [Q'(t)/Q'(inf)], the fraction of drug released, vs. ln [t] for the self-catalyzed hydrolysis of latexes 1 and 3. Curve B in Figure 7 shows a similar plot for latex 4.

The data in Table 3 show that with the exception of latex 4, all the slopes are between 0.527 and 0.675. These slopes have been reported along with their corresponding correlation coefficients and 95% confidence intervals. Values between 0.5 and 1.0 suggest that a non-Fickian (anomalous) mechanism of drug transport is operative.[40,41] The value of 0.334 obtained for latex 4, however, would not fit this mechanism.

A mathematical model has been proposed to explain the drug release kinetics observed. It is based on the assumption that ester hydrolysis occurs within a thin ''interfacial reaction zone'' of thickness Δr with a volume equal to:

$$V_i = 4\pi r^2 \Delta r \quad (1)$$

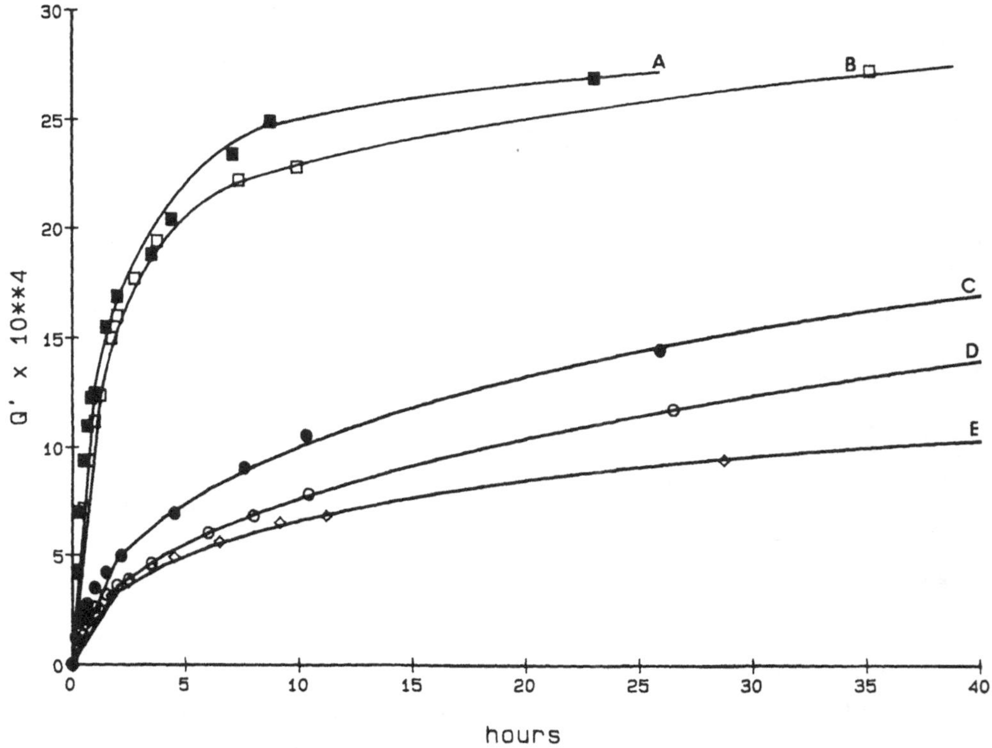

FIGURE 3. Base-catalyzed hydrolysis. Experimental data: (■) latex 1 at 50°C, pH 7.4; (□) latex 1 at 37°C, pH 7.4; (●) latex 3 at 50°C, pH 7.4; (○) latex 3 at 37°C, pH 7.4; (◇) latex 3 at 37°C, pH 7.8. Theoretical data:

Curve	k_5	k_3	n	Correlation coefficient
C	1.3×10^{-7}	-1.6×10^{-5}	0.83	0.999
D	6.8×10^{-8}	-6.3×10^{-6}	0.67	0.999
E	6.4×10^{-8}	-1.4×10^{-5}	1.33	0.999

where Δr remains constant throughout the course of the hydrolysis reaction. This reaction zone recedes as hydrolysis proceeds, presumably producing a core/shell type morphology. Earlier studies on polymethyl acrylate latices which had undergone partial self-catalyzed hydrolysis, utilizing carbon-13 NMR and differential scanning calorimetry have supported this model.[42]

The central core is composed of unhydrolyzed polymeric ester while the outer shell is believed to consist mostly of hydrolyzed polymer in the form of polyacrylic acid. This process can be visualized by reference to Figure 8. Since the rate of hydrolysis is proportional to the reaction-zone volume it decreases slowly as the radius of the core diminishes.

Hydrolysis is assumed to be pseudo first order with respect to ester. The concentration of ester in the interfacial reaction zone summed over all particles, E_i, can then be given by:

$$E_i = N \cdot e_i 4\pi r^2 \Delta r \tag{2}$$

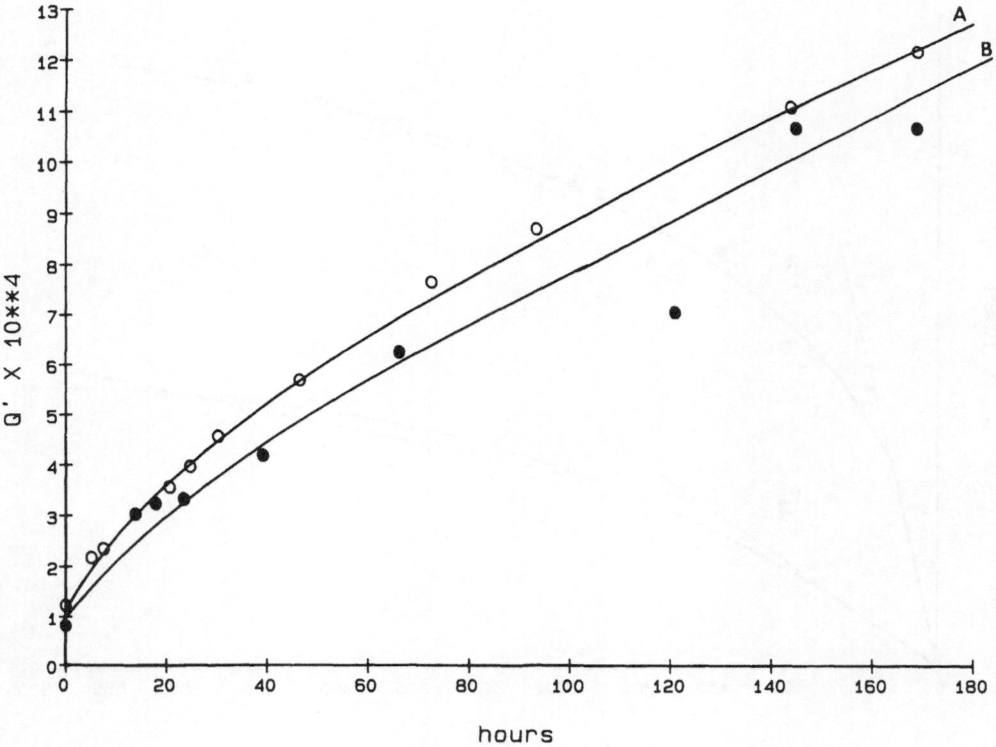

FIGURE 4. Based-catalyzed hydrolysis: Curve A, for latex 2 in HBSS buffered at pH 7.4 and at 4°C; curve B, for latex in human blood serum buffered at pH 7.4 and at 4°C.

where e_i is the steady-state concentration of ester within the reaction zone, and N is the total number of particles per liter of latex.

The overall decrease in ester is then given by:

$$-dE/dt = k'E_i(H+)^o = kE_i \tag{3}$$

where E is the total concentration of ester per liter of latex. [H+], the concentration of H+ in the interfacial zone, is considered to remain constant throughout the course of particle hydrolysis.

Next it is assumed that the total volume of polymer present, V, is proportional to E;

$$V \propto E; \quad V = cE \tag{4}$$

Therefore,

$$-dV/dt = -cdE/dt = kcE_i \tag{5}$$

and since $V = 4/3\pi r^3$,

$$dV/dt = 4\pi r^2 dr/dt = c \cdot dE/dt \tag{6}$$

Ultimately (see Appendix), an equation is obtained which describes the quantity of drug released per gram of polymer, Q′, as a function of time:

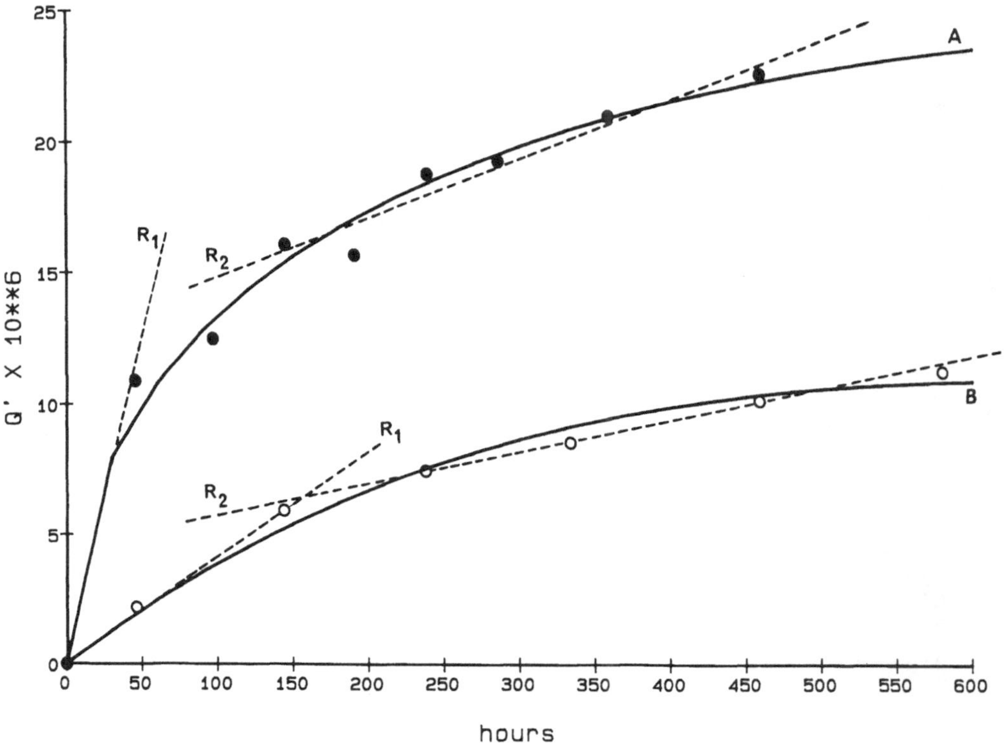

FIGURE 5. Based-catalyzed hydrolysis. Experimental data: (●) latex 4 at 37°C, pH 7.4; (○) latex 4 at 37°C, pH 5.0. Theoretical data:

Curve	K″	E_o	n	k_5	k_3	Correlation coefficient
A	—	—	1.10	2.6×10^{-12}	-2.6×10^{-8}	0.998
B	3.01×10^{-5}	1.10×10^{-5}	2.40	—	—	0.997

$$Q' = K^*t(3E_0^{2/3} - 3E_0^{1/3} K''t + K''^2 t^2)$$

where $K^* = K''/n$ and n is the number of grams of polymer present per liter of latex.

After inserting values for the constants E_o, K″, and n, the previous equation was used to plot Q′ vs. t. Theoretical curves obtained by using a fit-function routine in RS1 software on a Digital VAX 785 computer have been superposed on the experimental data in Figures 2 and 5. The experimental and theoretical curves appear to match well as evidenced by the correlation coefficients obtained for each fit.

The development of this model was based on the simplifying assumptions that the surface strong-acid concentration and the interfacial reaction-zone thickness remained constant during particle hydrolysis. Since the sulfate ester end groups present at the particle interface are themselves known to be self-hydrolyzed,[34] the surface strong-acid concentration actually decreases slowly with time. This lowers the drug release rate. Eventually after all the surface end groups have been removed by hydrolysis, the surface strong-acid concentration would be expected to reach a constant value. This would have the effect of stabilizing the drug release rate.

The surface reaction-zone thickness may also increase as the particles hydrolyze. The

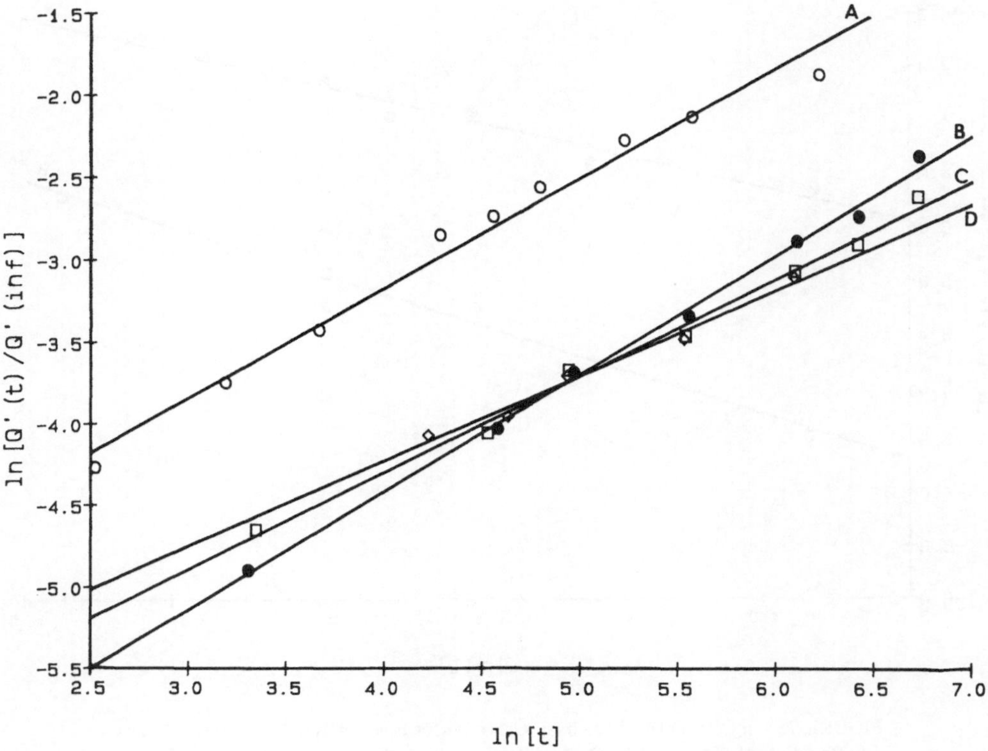

FIGURE 6. ln[Q'(t)/Q'(inf)] vs. ln[t]: Curves A and B, for self-catalyzed acid hydrolysis of latex 1 at 50 and 37°C, respectively; curves C and D, for self-catalyzed acid hydrolysis of latex 3 at 37°C and pH 5.0 and 2.0, respectively.

formation of polyacrylic acid at the interface after drug release would be expected to cause increased plasticization of the reaction zone with water producing a swollen hydrogen network. An increased reaction zone would in turn cause a corresponding increase in the rate of ester hydrolysis. This could partially offset the decrease in [H+] and might explain the relatively steady release rates observed over long periods of time in these systems.

To produce a more accurate model, considerations for changes in the interfacial [H+] and reaction-zone thickness will have to be taken into account. By preparing latices stabilized solely by sulfonate end groups which are known for their hydrolytic stability, this problem could be avoided. For example, during the initial stages of hydrolysis, an all-sulfonate-stabilized polymethyl acrylate latex was shown to yield a more constant hydrolysis rate than similar mixed sulfonate/sulfate systems.[29] Alternatively, in addition to determining drug release kinetics, simultaneous studies could be performed on latexes to quantify the rate of sulfate end group hydrolysis in mixed end group systems. Appropriate changes could then be incorporated into Equation 3. Quasi-elastic light scattering might also prove useful for determination of changes in particle size occuring as a function of hydrolysis. This might help to further elucidate the type of core/shell morphology mechanism involved and the extent to which the surface reaction-zone thickness changes with time.

B. Base-Catalyzed Hydrolysis of Latexes

Latexes were placed in environments at pH 7.4 to estimate drug-release kinetics which might be expected in the human bloodstream or small intestine.[43] However, it should be noted that the pH of the latter may actually vary between 4.8 and 8.2.[43]

The first base-catalyzed hydrolysis experiments performed by placing latex 2 in human

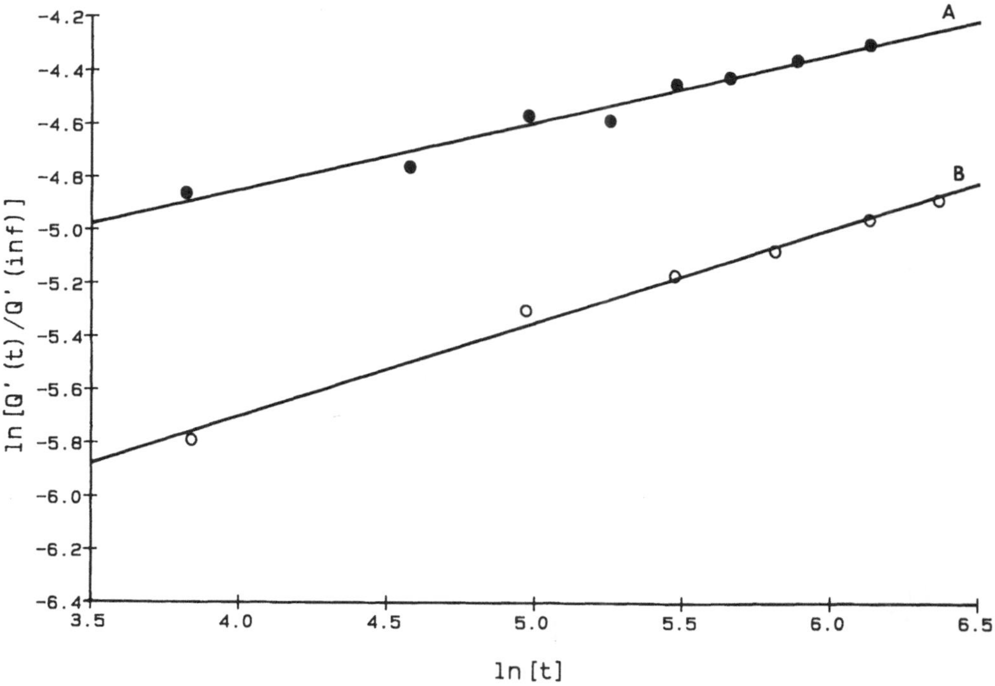

FIGURE 7. ln[Q'(t)/Q'(inf)] vs. ln[t]: Curve A, for base-catalyzed hydrolysis of latex 4 buffered at pH 7.4 and at 37°C; curve B, for self-catalyzed acid hydrolysis of latex 4 at 37°C.

blood serum and HBSS buffered at pH 7.4 and at 4°C demonstrated the stability of these particles in bloodlike environments. Latexes 1, 3, and 4 buffered at pH 7.4 and 7.8, and maintained at 37°C, yielded quantitative estimates of the drug delivery which might be expected in the bloodstream or small intestine at body temperature. The very large difference in R_1 values obtained for latexes 1 and 3 upon changing the pH from acidic to basic contrasts sharply with the much smaller change observed for latex 4 exposed to similar conditions. Base-catalyzed hydrolysis is believed to follow Ingold's B_{AC^2} mechanism:[44]

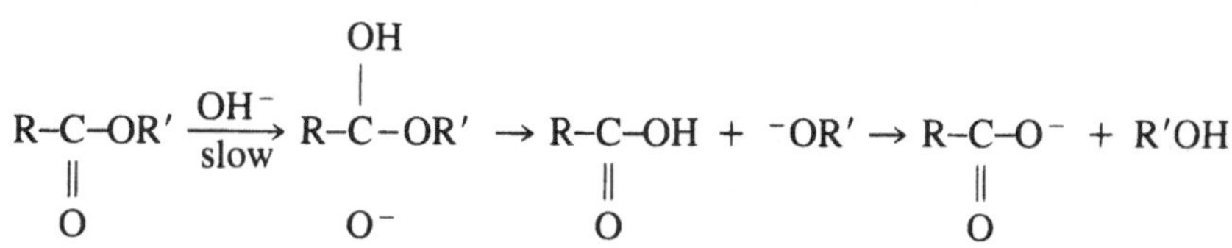

However, ionized, ortho-carboxylate groups on the pendant salicyl ester moieties in latexes 1, 2, and 3 are capable of lending anchimeric assistance[45,46] to the hydrolysis reaction. These groups polarize the ester carbonyl carbon atoms increasing their susceptibility to nucleophilic attack by hydroxyl ions. This results in rate enhancement. This process can be visualized by reference to Figure 9.

Anchimeric assistance cannot occur during hydrolysis of latex 4 which lacks the appropriate neighboring groups. This would explain the comparatively small sixfold increase in rate observed for latex 4 upon increasing the pH.

The fact that latex 1 hydrolyzes faster than latex 3 at basic pH is attributed to its water solubility which exposes a greater number of ester linkages to hydroxyl ion. The relatively large increase in R_1 for latex 3 with increased temperature relative to that for latex 1 is

Table 3
HYDROLYSIS KINETIC DATA

Self-catalyzed acid hydrolysis

Latex no.	pH	Temp. (°C)	Slope	+/−	r^2	Drug released
1	5.0	37	.675	.066	.841	Salicylic acid
1	5.0	50	.727	.019	.997	Salicylic acid
3	2.0	37	.527	.053	.984	Salicylic acid
3	5.0	37	.597	.024	.992	Salicylic acid
4	5.0	37	.334	.020	.991	Chloramphenicol
Base-catalyzed hydrolysis						
1	7.4	37	.295	.018	.902	Salicylic acid
1	7.4	50	.318	.016	.956	Salicylic acid
3	7.4	37	.321	.0034	.998	Salicylic acid
3	7.4	50	.373	.0030	.999	Salicylic acid
3	7.8	37	.404	.0075	.990	Salicyclic acid
4	7.4	37	.256	.018	.963	Chloramphenicol

Note:
$$\text{Slope} = \frac{d\ \ln[Q'(t)/Q'(\text{inf})]}{d\ \ln[t]}$$
r^2 = correlation coefficient for slope

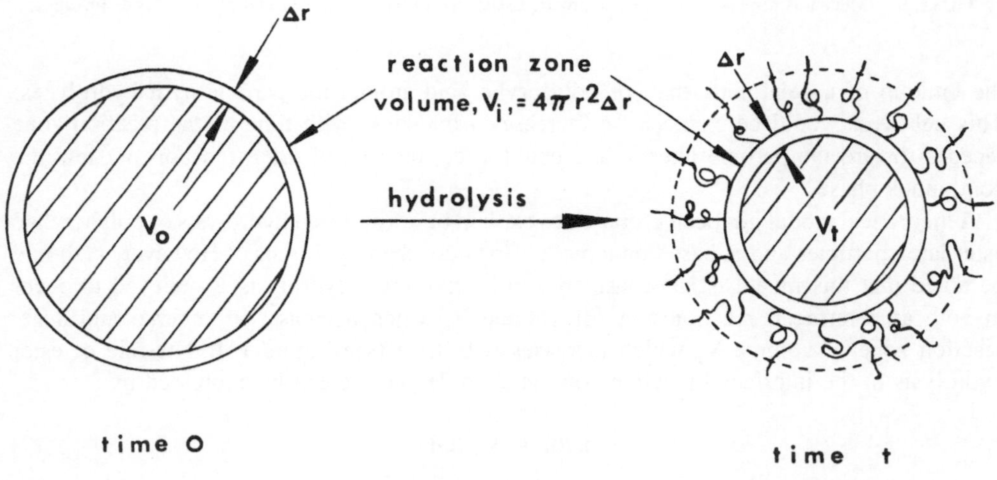

FIGURE 8. Schematic representation of self-catalyzed acid hydrolysis of latex particles.

attributed to greater swelling of the particles by water at the higher temperature. This would increase the activity of hydroxyl ion in the polymer matrix thus enhancing the rate.

Figure 10 shows plots of ln [Q′(t)/Q′(inf)] vs. ln [t] for latexes 1 and 3 hydrolyzed at basic pHs. Curve A of Figure 7 shows a similar plot for latex 4. Table 3 shows that the slopes obtained for these lines vary between 0.256 and 0.404. The values have been reported with their correlation coefficients and 95% confidence intervals. Slope values of 0.5 would suggest that the rate of drug transport out of the polymer matrix is controlled by Fickian diffusion.[40,41] As mentioned for self-catalyzed acid hydrolysis, slopes ranging between 0.5 and 1.0 would suggest a non-Fickian (anomalous) transport mechanism. The fact that these slopes fit neither of these cases may indicate establishment of a boundary layer caused by

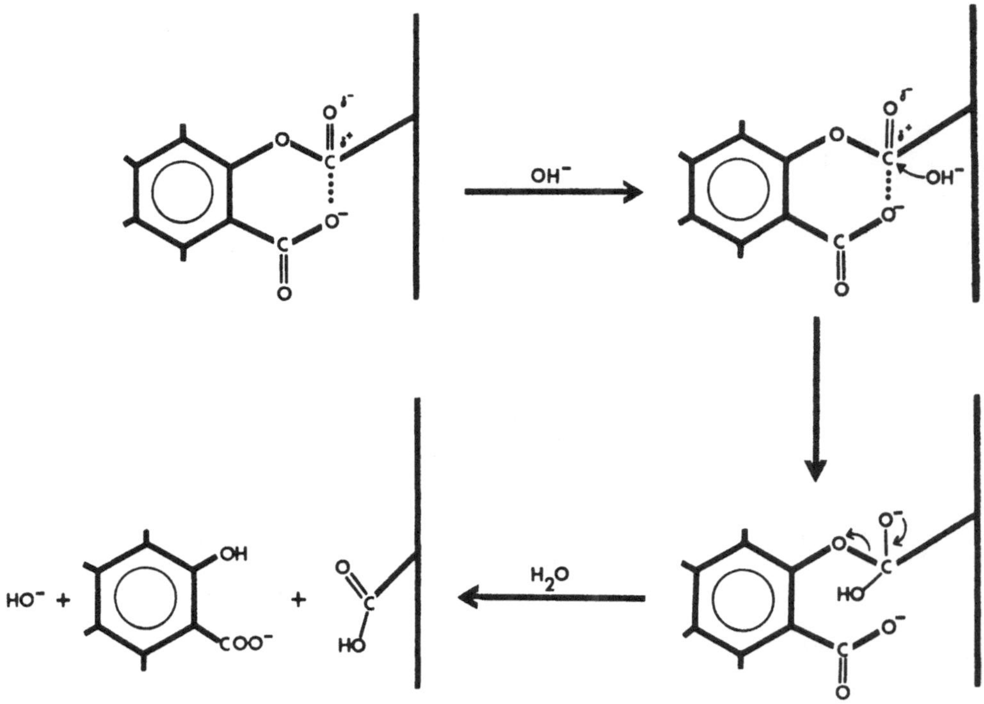

FIGURE 9. Mechanism responsible for anchimeric assistance to base-catalyzed hydrolysis of ester linkages.

the buildup of a gelatinous shell of polyacrylic acid around the particle as it hydrolyzes. This gel layer would be expected to increase in thickness with time as the reaction zone recedes toward the particle interior and could affect the rate of drug transport out into the continuous phase.

A theoretical model has been proposed to explain base-catalyzed hydrolysis of polyacrylate ester latex particles and is based on a model first described by Tarcha.[47] However, it should be noted that this model is not considered to be rigorous. Hydrolysis is believed to occur in both an interfacial reaction zone of volume V_i which decreases with time, and a gel reaction zone of volume V_g which increases with time (see Figure 11). The rate of ester hydrolysis in the interfacial reaction zone in a single particle can be expressed by:

$$-de/dt = k_i e_i [OH^-] \tag{7}$$

where e_i is the concentration of ester in the interfacial reaction zone.

The total number of moles, E, of ester hydrolyzed in the interfacial zone per hour per liter of latex is

$$-(dE/dt)_i = k_2 e_i V_i N \tag{8}$$

where N is the total number of particles per liter of latex.

Next it is assumed that e_i and V_i remain relatively constant. These assumptions seem reasonable since e_i represents the concentration of unhydrolyzed ester at the interface and V_i is a thin reaction zone whose volume is proportional to the particle surface area which changes relatively little during the early stages of hydrolysis.

Considering these assumptions and combining constants yields:

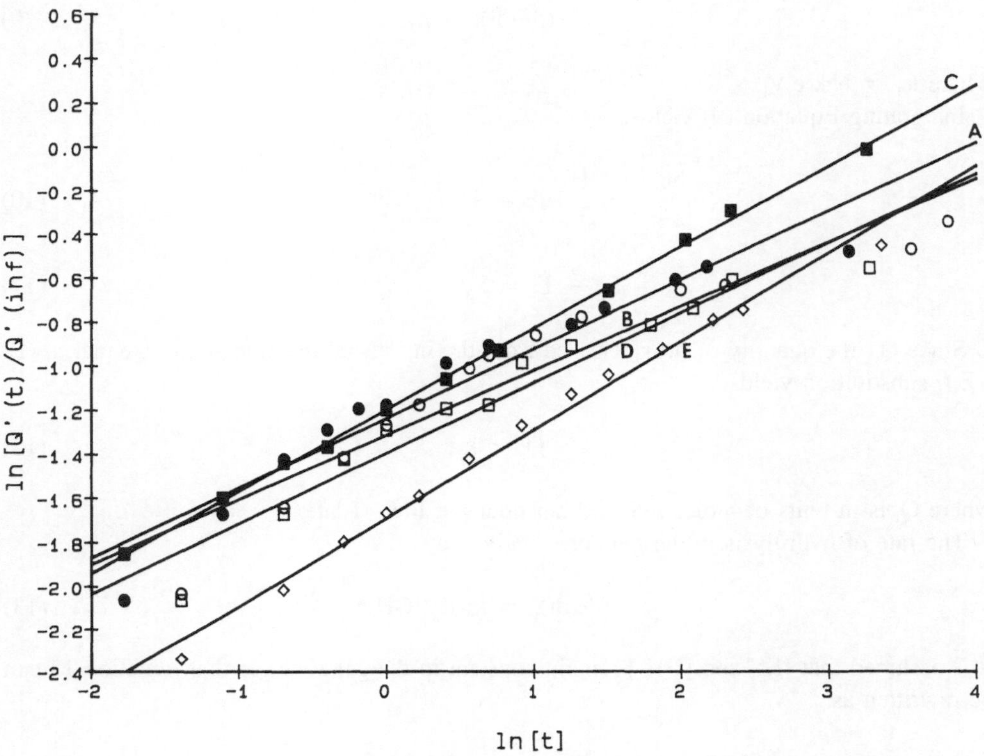

FIGURE 10. ln[Q′(t)/Q′(inf)] vs. ln[t]: Curves A and B, for base-catalyzed hydrolysis of latex 1 buffered at pH 7.4 and at 50 and 37°C, respectively; curves C and D, for base-catalyzed hydrolysis of latex 3 buffered at pH 7.4 and at 50 and 37°C, respectively; curve E, for latex 3 buffered at pH 7.8 and at 37°C.

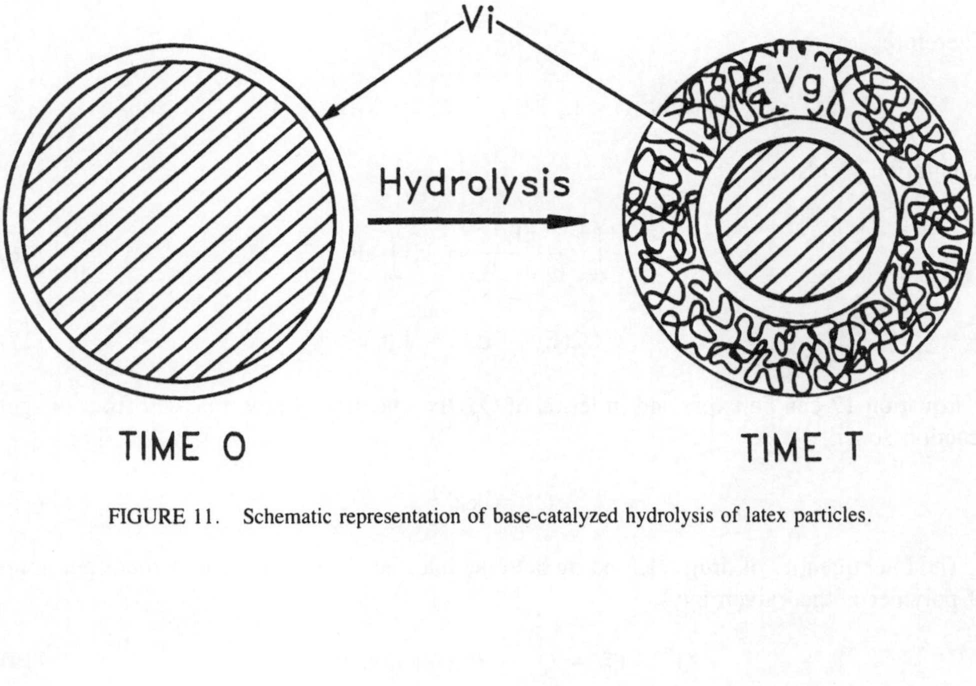

FIGURE 11. Schematic representation of base-catalyzed hydrolysis of latex particles.

$$-(dE/dt)_i = k_3 \tag{9}$$

where $k_3 = N{\cdot}k_2e_iV_i$.

Integrating Equation (9) yields:

$$-\int_{E_o}^{E_t} dE = k_3 \int_{t=0}^{t} dt \tag{10}$$

$$E_o - E_t = k_3t \tag{11}$$

Since Q_i, the quantity of drug released from the interfacial reaction zone, is equal to ($E_o - E_t$), substitution yields:

$$Q_i = k_3t \tag{12}$$

where Q_i is in units of moles released per hour per liter of latex.

The rate of hydrolysis in the gel zone is given by:

$$-(de/dt)_g = k_4(E_g)[OH]^o \tag{13}$$

It is assumed that [E_g] and [OH-] remain constant in the gel zone, so that Equation 13 can be rewritten as:

$$-(de/dt)_g = k_4 \tag{14}$$

$$-(dE/dt)_g = k_4V_g \tag{15}$$

However, $V_g \propto (E_o - E_t)$; or $V_g = K(E_o - E_t)$

Therefore:

$$-(dE/dt)_g = k_4\, K(E_o - E_t) = k_5(E_o - E_t)$$

Integration yields:

$$-\int_{E_o}^{E_t} \frac{dE}{E_o - E_t} = k_5 \int_0^t dt \tag{16}$$

$$1/2(E_o - E_t)^2 = k_5t \tag{17}$$

Equation 17 can be expressed in terms of Q_g, the quantity of drug released from the gel reaction zone:

$$Q_g = k_5t^{1/2} \tag{18}$$

The total quantity of drug released by both the interfacial and gel reaction zones per gram of polymer is then given by:

$$Q' = Q_i' + Q_g' = [k_3t + (2k_5t)^{1/2}]/n \tag{19}$$

in units of moles released per gram of polymer per hour, where n is the number of grams of polymer per liter of latex.

Superposed on the experimental data in Figures 3 (curves C, D, and E) and 5 (curve A) are theoretical curves obtained using Equation 19 and a fit-function routine in RS1 software on a Digital VAX 785 computer. The experimental and theoretical curves appear to match well as evidenced by the correlation coefficients obtained for each fit.

C. Potential In Vivo Applications

Latex 3 was buffered at pH 2.0 and maintained at 37°C to determine the drug release which might occur in the human stomach.

Salicylic acid (0.02 mg) was released per gram of latex per hour by latex 3 under these conditions. The fact that the drug molecules initially are immobilized on the polymer backbone and that the rate of hydrolysis is low at this pH, suggests that only small quantities of free salicylic acid would be released during travel through the stomach. This might eliminate aggravation of the stomach lining often caused by aspirin and could be advantageous relative to enteric coating of tablets because the dispersion could be administered as a liquid. Release of salicylic acid would then greatly accelerate after passage into the small intestine.

This is borne out by experiments at pHs of 7.4 and 7.8 in which latex 3 exhibited a R_1 of approximately 31 mg/g polymer per hour and released approximately 80% of the bound salicylic acid during the first 12 hr of hydrolysis. Rates of this magnitude suggest that therapeutic quantities of drugs might indeed be released during passage through the small intestine.

Due to the inert nature of the polymer hydrocarbon backbone, latex particles with sufficiently high molecular weights would probably pass out of the digestive system without being absorbed into the bloodstream. Similar aqueous dispersions and water-soluble polymers of poly(meth)acrylate esters are currently being used for enteric coatings of pharmaceutical preparations.[48] When fed to a variety of test animals, radioactively labeled polymers of these types with mean molecular weights of 250,000 were found to be excreted completely from the gastrointestinal tract.[49] Quantities as high as 11 g/kg body weight were found to be tolerated without producing any toxic effects.[49]

Perhaps the most exciting potential application of these colloidal drug-delivery systems is for targeted chemotherapeutic treatment of cancer. It is now well accepted that tumor-associated antigens (TAAs) are present on the surface of human cancer cells.[50] The presence of the surface receptor sites with the inception of cancer initiates the production of antibodies specific to the TAAs. Methods have been devised for the separation and purification of these antibodies, and a variety of toxins have been conjugated to them to increase their tumor-selective cytotoxicity.[50,51] In some cases these conjugates have been found to produce markedly selective in vivo cytotoxicity.[50-55]

Bonding of polymer colloid drug-delivery systems to tumor-specific antibodies (TSAs) might represent an ideal method to achieve sustained release of cytotoxic molecules at TAA sites. Among the methods available for bonding antibodies to colloidal particles[56-58] the carbodimide method[56] is probably the most convenient for the type of colloids described in this study, due to the availability of carboxyl groups at the particle-water interface.

Alternatively, targeting could be achieved by direct intratumor injection of the colloid. This could offer the advantages of less acute local toxicity, lower dose levels, and less diffusion of drug out of the tumor mass into the circulatory system. These devices could also offer treatment periods of several days or weeks which is highly desirable.[51]

By targeting treatment at the cellular level the quantity of drug required for a therapeutic dose would be greatly reduced, thus eliminating many toxic side effects associated with conventional chemotherapy. The use of more potent antitumor drugs such as 6-mercaptopurine[59,60] or adriamycin[50,61] might reduce therapeutic levels even further.

In vivo biodegradability and body clearance will require that the molecular weight of polyacrylic acid (PAA) residues left after complete drug release be less than 50,000.[62] Even faster removal of PAA would be obtained by reducing the molecular weight below 30,000.[62] Use of chain transfer agents during colloid synthesis could be used to reduce the molecular weight to the desired level. Or appropriate cyclic ketene acetal monomers could be incorporated into the polymer backbone during emulsion polymerization to provide hydrolyzable ester linkages.[63-65] Bailey has used such monomers to prepare biodegradable polyolefins.[66]

It is interesting to note that long polymer chains with a high density of carboxyl groups have been shown to produce antiviral activity, tumor growth inhibition, and interferon induction.[67-69] It is possible that PAA residues present in the bloodstream after particle degradation might even be capable of producing some beneficial side effects in the patient. The presence of sulfate and sulfonate end groups at the particle interface in these colloids may also help to improve blood compatibility since heparin and heparinoids possessing such groups are known for their anticoagulant activity.[62,70]

VII. CONCLUSION

Acrylic polymer particles 40 to 400 nm in diameter with attached pendant drug molecules have been synthesized by aqueous emulsion polymerization. This work suggests that a variety of other drug molecules could also be incorporated into the particle matrix by this method. Sustained release of drug molecules occurs during either acid- or base-catalyzed hydrolysis of the pendant ester linkages. Mathematical models, based on surface and gel reaction zone theories, have been proposed to explain the drug-release kinetics observed.

These colloidal systems offer potential as sustained release devices for oral, topical, and parenteral administration of drugs. Future work in this area will include preparation of latex particles capable of releasing antitumor drugs and synthesis of biodegradable latices. Experiments will also be performed to define, in greater detail, the various factors controlling drug release.

APPENDIX

Using Equations 8 and 9,

$$-dE/dt = kNe_i 4\pi r^2 \Delta r \tag{20}$$

Letting $K = kNe_i 4\pi\Delta r$, Equation 20 becomes:

$$-dE = Kr^2 dt \tag{21}$$

However, $r = f(E)$, and from Equation 4 it is known that:

$$cE = N \cdot 4/3\pi r^3, \quad \text{therefore:}$$

$$r = (3cE/4\pi N)^{1/3} \tag{22}$$

Substituting Equation 22 into Equation 21 yields:

$$-dE = K(3c/4\pi N)^{2/3} \cdot E^{2/3} dt \tag{23}$$

Rearrangement of Equation 23 yields:

$$dE/E^{2/3} = -K'dt \tag{24}$$

where:

$$K' = K(3c/4\pi N)^{2/3} = ke_i\Delta\, r(6\pi^{1/2}N^{1/2}c)^{2/3}$$

Mass balance requires that:

$$Q = E_o - E$$

$$dQ = -dE$$

where:

Q is the concentration of drug released (mol/ℓ), E is the concentration of ester at time t, and E_o is the concentration of ester at $t = 0$.

Substituting these into Equation 24 and integrating yields:

$$\int_0^Q \frac{-dQ}{(E_o - Q)^{2/3}} = -K' \int_0^t (dt)$$

$$3(E_o - Q)^{1/3} \Big]_0^Q = -K't$$

$$3(E_o - Q)^{1/3} - 3E_0^{1/3} = -K't$$

$$(E_o - Q)^{1/3} = \frac{3E_0^{1/3} - K't}{3} \tag{25}$$

$$E_o - Q = [(3E_0^{1/3} - K't)/3]^3$$

$$Q = (E_0^{1/3})^3 - [(3E_0^{1/3} - K't)/3]^3 \tag{26}$$

$$\text{Note:} \quad A^3 - B^3 = (A - B)(A^2 + AB + B^2) \tag{27}$$

Letting $A = E_0^{1/3}$ and $B = (E_0^{1/3} - K''t)$, where $K'' = K'/3$

Equation 26 can be factored in the form of Equation 27:

$$Q = [E_0^{1/3} - (E_0^{1/3} - K''t)][E_0^{2/3} + E_0^{1/3}(E_0^{1/3} - K''t) + (E_0^{1/3} - K''t)^2]$$

$$= K''t[E_0^{2/3} + E_0^{2/3} - E_0^{1/3}K''t + E_0^{2/3} - 2E_0^{1/3}K''t + K''^2 t^2]$$

REFERENCES

1. **Sugibayashi, K., Marimoto, Y., Nadai, T., Kato, Y., Hasegawa, A., and Arita, T.,** Drug-carrier property of albumin microspheres in chemotherapy. I. Tissue distribution of microsphere-entrapped 5-fluorouracil in mice, *Chem. Pharm. Bull.*, 27, 204, 1979.
2. **Sugibayashi, K., Morimoto, Y., Nadai, T., and Kato, Y.,** Drug-carrier property of albumin microspheres in chemotherapy. II. Preparation and tissue distribution in mice of microsphere-entrapped 5-Fluorouracil, *Chem. Pharm. Bull.*, 25, 3433, 1977.
3. **Ishizaka, T., Endo, K., and Koishi, M.,** Preparation of egg albumin microcapsules and microspheres, *J. Pharm. Sci.*, 70, 358, 1981.

4. **Widder, K., Flouret, G., and Senyei, A.,** Magnetic microspheres: synthesis of a novel parenteral drug carrier, *J. Pharm. Sci.*, 68, 79, 1979.
5. **Kramer, P. A.,** Albumin microspheres as vehicles for achieving specificity in drug delivery, *J. Pharm. Sci.*, 63, 1646, 1974.
6. **Bakan, J. A.,** Microcapsule drug delivery systems, in *Polymers in Surgery and Medicine*, Kronenthal, R., Oser, Z., and Martin E., Eds., Plenum Press, New York, 1975, 213.
7. **Deasy, P. B., Brophy, M. R., Ecanow, B., and Joy, M. M.,** Effect of ethylcellulose grade and sealant treatments on the production and in vitro release of microencapsulated sodium salicylate, *J. Pharm. Pharmacol.*, 32, 15, 1980.
8. **Luzzi, L. A., Zoglio, M. A., and Maulding, H. V.,** Preparation and evaluation of the prolonged release properties of nylon microcapsules, *J. Pharm. Sci.*, 59, 338, 1970.
9. **Luzzi, L. A.,** Microencapsulation, *J. Pharm. Sci.*, 59, 1367, 1970.
10. **Lim, F., Ed.,** *Biomedical Applications of Microencapsulation*, CRC Press, Boca Raton, Fla., 1984.
11. **Speiser, P.,** Kolloide Verteilungszustande in der Pharmazeutischen Technologie: Koazervierung, Adsorption und Mizellpolymerisation, *Progr. Colloid Polymer Sci.*, 59, 48, 1976.
12. **Couvreur, P., Kante, B., Roland, M., and Speiser, P.,** Adsorption of antineoplastic drugs to polyalkylcyanoacrylate nanoparticles and their release in calf serum, *J. Pharm. Sci.*, 68, 1521, 1979.
13. **Couvreur, P., Kante, B., Lenaerts, V., Scailteur, V., Roland, M., and Speiser, P.,** Tissue distribution of antitumor drugs associated with polyalkylcyanoacrylate nanoparticles, *J. Pharm. Sci.*, 69, 199, 1980.
14. **Couvreur, P., Kante, B., Roland, M., Guiot, P., Bauduin, P., and Speiser, P.,** Polycyanoacrylate nanocapsules as potential lysosomotropic carriers: preparation, morphological and sorptive properties, *J. Pharm. Pharmacol.*, 31, 331, 1979.
15. **Marty, J. J., Oppenheim, R. C., and Speiser, P.,** Nanoparticles — a new colloidal drug delivery system, *Pharm. Acta Helv.*, 53, 17, 1978.
16. **Oppenheim, R. C.,** Nanoparticles, in *Drug Delivery Systems*, Juliano, R., Ed., Oxford Univ. Press, 1980, 177.
17. **Kreuter, J.,** Nanoparticles and nanocapsules — new dosage forms in the nanometer size range, *Pharm. Acta Helv.*, 53, 33, 1978.
18. **Gregoriadis, G., Ed.,** *Liposome Technology*, Vol. 3, CRC Press, Boca Raton, Fla., 1984.
19. **Juliano, R.,** Drug delivery systems: a brief review, *Can. J. Physiol. Pharmacol.*, 56, 683, 1978.
20. **Fendler, J., and Romero, A.,** Liposomes as drug carriers, *Life Sci.*, 20, 1109, 1977.
21. **Rymen, B. E.,** Liposomes: possible potential in drug delivery, in *Controlled Release Delivery Systems*, Roseman, T. J., and Mansdorf, S. Z., Eds., Marcel Dekker, New York, 1983, 27.
22. **Ringsdorf, H.,** Structure and properties of pharmacologically active polymers, *J. Polymer Sci. Polymer Symp.*, 51, 135, 1975.
23. **Trouet, A.,** Carriers for bioactive materials, in *Polymeric Delivery Systems, Midl. Macromol. Monograph Series No. 5*, Kostelnik, R. J., Ed., Gordon & Breach, New York, 1978, 157.
24. **Goldberg, E. P.,** Polymeric affinity drugs, in *Polymeric Delivery Systems*, Midl. Macromol Monograph Series No. 5, Kostelnik, R. J., Ed., Gordon & Breach, New York, 1978, 227.
25. **deDuve, C., deBarsy, T., Poole, B., Trouet, A., Tulkens, P., and Van Hoof, F.,** Lysosomotropic agents, *Biochem. Pharmacol.*, 23, 2495, 1974.
26. **Kim, S. W., Petersen, R. V., and Feijen, J.,** Polymer drug delivery systems, in *Drug Design*, Ariens, J. J., Ed., Academic Press, New York, 1980, 193.
27. **Hartsough, R. R., and Gebelein, C. G.,** The release of 5-fluorouracil from 1-(N-2-ethylmethacrylcarbamoyl)-5-fluorouracil monomers, polymers and acrylate copolymers, *Polymer Mater. Sci. Eng.*, 51, 131, 1984.
28. **Schuerch, C.,** The chemical synthesis and properties of polysaccharides of biomedical interest, *Adv. Polymer Sci.*, 10, 173, 1972.
29. **Fitch, R. M., Gajria, C., and Tarcha, P. J.,** Acrylate polymer colloids: kinetics of autocatalyzed hydrolysis, *J. Colloid Interface Sci.*, 71, 107, 1979.
30. **Hopfenberg, H. B.,** Controlled release from erodible slabs, cylinders and spheres, *Pap. Meet. — Am. Chem. Soc.*, 36, 229, 1976.
31. **Banks, A. R., Firbiger, R. F., and Jones, T.,** A convenient synthesis of methacrylates, *J. Org. Chem.*, 42, 3695, 1977.
32. **Batz, H. G., Ringsdorf, H., and Ritter, H.,** Pharmacologically active polymers. VII, *Makromol. Chem.*, 175, 2229, 1974.
33. **Scholsky, K. M.,** Ph.D. Thesis, University of Connecticut, 1982.
34. **McCarvill, W. T. and Fitch, R. M.,** The surface chemistry of polystyrene latices initiated by the persulfate/bisulfite/iron system, *J. Colloid Interface Sci.*, 67, 204, 1978.
35. **McCarvill, W. T. and Fitch, R. M.,** The preparation and surface chemistry of polystyrene colloids stabilized by sulfonate surface groups, *J. Colloid Interface Sci.*, 64, 403, 1978.

36. **Vanderhoff, J. W., Van den Hul, H. J., Tausk, R. J., and Overbeck, J. Th. D.,** The preparation of monodisperse latexes with well-characterized surfaces, in *Clean Surfaces: Their Preparation and Characterization for Interfacial Studies,* Goldfinger, G., Ed., Marcel Dekker, New York, 1970, 15.
37. **Horne, R. W.,** Negative staining techniques, in *Techniques for Electron Microscopy,* 2nd ed., Kay, H., Ed., Blackwell Scientific Publishers, 1975, 328.
38. **Scholsky, K. M. and Fitch, R. M.,** Examination of small polystyrene latex particles using a negative staining technique, *J. Colloid Interface Sci.,* 104, 592, 1985.
39. **Brandrup, J. and Immergut, E. H., Eds.,** *Polymer Handbook, Part IV,* John Wiley & Sons, New York, 1975, 309.
40. **Sinclair, G. W. and Peppas, N. A.,** Analysis of non-Fickian transport in polymers using simplified exponential expressions, *J. Membr. Sci.,* 17, 329, 1984.
41. **Peppas, N. A.,** Purdue University, private communication, 1984.
42. **Tarcha, P. J., Fitch, R. M., Dumais, J. J., and Jelinski, L. W.,** Particle morphology of self-hydrolyzed acrylate polymer colloids. A ^{13}C NMR and DSC study, *J. Polymer Sci., Polym. Phys. Ed.,* 21, 2389, 1983.
43. **Weast, R. C., Ed.,** *Handbook of Chemistry and Physics,* 56th ed., CRC Press, Boca Raton, Fla., 1975, D-135.
44. **Ingold, C. K.,** *Structure and Mechanism in Organic Chemistry,* Cornell University Press, Ithaca, N.Y., 1953.
45. **March, J.,** *Advanced Organic Chemistry,* 2nd ed., McGraw-Hill, New York, 1977, 279.
46. **Turpin, E. T.,** Hydrolysis of water-dispersible resins, *J. Paint Technol.,* 47, 40, 1975.
47. **Tarcha, P. J.,** Ph.D. Thesis, University of Connecticut, 1983.
48. **Lehmann, K.,** Enteric and retard coatings of pharmaceutical preparations with aqueous acrylic resins dispersions, *Acta Pharm. Technol.,* 21, 255, 1975.
49. Eutragit™ product information bulletins, Rohm Pharma GMBH, Darmstadt, West Germany.
50. **Ghose, T., Blair, A. H., Vaughan, K., and Kulkarni, P.,** Antibody directed drug targeting in cancer therapy, 1, in *Targeted Drugs,* Goldberg, E. P., Ed., John Wiley & Sons, New York, 1983, 1.
51. **McGlaughlin, C. A. and Goldberg, E. P.,** Local chemo- and immunotherapy by intratumor drug injection: opportunities for polymer-drug compositions, in *Targeted Drugs,* Goldberg, E. P., Ed., John Wiley & Sons, New York, 1983, 231.
52. **Ghose, T. and Blair, A. H.,** Antibody-linked cytotoxic agents in the treatment of cancer: current status and future prospects, *J. Natl. Cancer Inst.,* 61, 657, 1978.
53. **Pouton, C. W.,** Drug targeting — current aspects and future prospects, *J. Clin. Hosp. Pharm.,* 10, 45, 1985.
54. **Rowland, G. F., O'Neill, G. J., and Davies, D. A.,** Supression of tumor growth in mice by a drug-antibody conjugate using a novel approach to linkage, *Nature (London),* 255, 487, 1975.
55. **Thorpe, P. E., Edwards, D. C., Davies, A. J., and Ross, W. C.,** Monoclonal antibody-toxin conjugates: aiming the magic bullet, in *Monoclonal Antibodies in Clinical Medicine,* Fabre, J. W. and McMichael, A. J., Eds., Academic Press, New York, 1982, 167.
56. **Goodfriend, T. L., Levine, L., and Fasman, G. D.,** Antibodies to bradykinin and angiotensin: a use of carbodiimides in immunology, *Science,* 144, 1344, 1969.
57. **Cuatrecasas, P.,** Protein purification by affinity chromatography. Deriviatizations of argarose and polyacrylamide beads, *J. Biol. Chem.,* 245, 3059, 1970.
58. **Molday, R. S., Dreyer, W. J., Rembaum, A., and Yen, S. P. S.,** New immuno-latex spheres: visual markers of antigens on lymphocytes for scanning electron mciroscopy, *J. Cell Biol.,* 64, 75, 1976.
59. **Brown, D.,** *Heterocylic Compounds, Fused Pyrimidines* (Part II), John Wiley & Sons, New York, 1971, 302.
60. **Hitchings, G. H. and Elion, G. B.,** Specificities of purine metabolizing enzymes and implications for chemotherapy, in *The Purines — Theory and Experiment,* Bergmann, E., Ed., Academic Press, New York, 1972, 565.
61. **Tritton, T. R. and Yee, G.,** The anticancer agent adriamycin can be actively cytotoxic without entering cells, *Science,* 217, 248, 1982.
62. **Ottenbrite, R. M.,** Structure and biological activities of some polyanionic polymers, in *Anionic Polymeric Drugs,* Donaruma, L. G., Ed., John Wiley & Sons, New York, 1980, 21.
63. **Bailey, W. J., Ni, Z., and Wu, S.,** Free radical ring-opening polymerization of 4,7 dimethyl-2-methylene-1,3-dioxepane and 5,6-benzo-2-methylene-1,3-dioxepane, *J. Polymer Sci.,* 20, 3021, 1982.
64. **Bailey, W. J., Gupad, B., Lin, Y., Ni, Z., and Wu, S.,** The use of free radical ring-opening polymerization for the synthesis of reactive oligomers, *Polymer Prepr.,* 25, 142, 1984.
65. **Bailey, W. J.,** Free radical ring-opening polymerization, *Polymer Prepr.,* 25, 210, 1984.
66. **Bailey, W. J. and Gupad, B.,** Synthesis of biodegradable polyethylene, *Polymer Prepr.,* 25, 58, 1984.

67. **Kaplan, A. M., Ottenbrite, R. M., Regelson, W., Carchman, R., Morahan, P., and Munson, A.,** Immunoadjuvant, antiviral and antitumor activity of synthetic polyanions, in *The Handbook of Cancer Immunology,* Waters, H., Ed., Garland STPM Press, New York, 1978, 135.
68. **Ottenbrite, R. M., Regelson, W., Kaplan, A. M., Carchman, R., Morahan, P., and Munson, A.,** Biological activity of polycarboxylic acid polymers, in *Polymeric Drugs,* Donaruma, L. G., Vogl, O., Eds., John Wiley & Sons, New York, 1978, 263.
69. **Ottenbrite, R. M. and Regelson, W.,** Water-soluble polymers, in *Encyclopedia of Polymer Science and Technology,* Vol. 2, John Wiley & Sons, New York, 1977, 118.
70. **Gregor, H. P., Ed.,** Anticoagulant activity of sulfonate polymers and copolymers, in *Biomedical Applications of Polymers,* Plenum Press, New York, 1975, 51.

Chapter 8

TARGETING OF DRUG ACTION TO THE CELL SURFACE BY MICROSPHERES

Kevin L. Ross and Zoltán A. Tökés

TABLE OF CONTENTS

I. INTRODUCTION TO THE CONCEPT OF MEMBRANE-TARGETED DRUG ACTION

The objective of this chapter is to illustrate how microspheres can be used to restrict drug action to the plasma membrane resulting in a novel mechanism of cytotoxicity. This approach needs to be distinguished from some other applications of microspheres for drug delivery where the fundamental mechanism of drug action is not altered and the microspheres serve only as carriers for the slow release of therapeutic agents.[1]

The concept of membrane-targeted drug action utilizes microspheres as solid supports to which the therapeutic agents are covalently linked. Thus, the surface of microspheres consists of a repetitious structure formed by the drug. If the drug is capable of interacting with the lipid bilayer or any other components of the plasma membrane, multiple interactions can occur. The additive effects of these interactions can focus the drug action to a restricted zone on the cell surface and may create a lethal effect. Drug interactions with the plasma membrane may involve intercalation with lipid molecules and thereby change membrane fluidity, block specific transport molecules, change the configuration of various receptors either by direct or indirect interactions, and alter ion pumps or channels involved in the maintenance of cellular-ionic equilibrium. Minor membrane perturbations produced by individual drug molecules may become amplified by the microsphere-mediated multiple interactions. Such an amplified action can create cellular responses which are not achieved by free drug.

The principle of cell surface-directed cytotoxic action is illustrated with the anthracycline antibiotic Adriamycin (Adr). Adr is one of the most widely used antineoplastic drugs currently available. Its importance is based on its broad spectrum of activity against both solid tumors and leukemias.[3] Early mechanistic studies of Adr suggested that DNA was the major site of action and in this regard it was demonstrated that Adr binds to DNA at a low drug-to-DNA molar ratio corresponding to the formation of an intercalation complex.[4] In addition, it was shown that Adr can cause fragmentation of the DNA strand,[5,6] as well as inhibition of DNA and RNA biosynthesis.[7,8] In further support of a DNA target, Adr was shown to accumulate in cell nuclei.[9] However, more extensive studies of the cytotoxic mechanism of the drug have suggested that DNA intercalation is not an absolute requirement for activity. For instance, it has been shown that the cytotoxic effect of some anthracyclines occurs at concentrations which do not affect DNA synthesis.[10] Examination of Adr analogs reveal that *N*-acetyldaunomycin inhibits cell mitosis while having only a weak affect on nucleic acid synthesis.[11] Still another anomaly is the analog *N*-trifluroacetyl Adr, which possesses only a weak binding affinity for DNA, yet retains considerable cytotoxic activity.[12] Therefore, several alternative sites for Adr action are being investigated and thus far one of the most promising sites is the plasma membrane. It has been demonstrated that Adr alters a variety of membrane properties, including protein glycosylation,[13] microtubule assembly,[14] membrane fluidity,[15] and cotransport.[16] Changes in lectin agglutination[17] and hormone receptors[18] have also been reported. It is not evident which of these observed effects may be the most crucial. Perhaps even the notion of a "lethal hit" as a single event necessary and sufficient to kill a cell is biologically naive.[19] It is, however, clear that Adr affects the plasma membrane, and this serves as the rationale for choosing anthracyclines as models for illustrating the concept of membrane-targeted cytotoxicity with microspheres.

The chemical and physical properties of microspheres utilized for membrane targeting can vary significantly as long as they are biocompatible and can be covalently linked with the chosen drug. Polyglutaraldehyde(Pgl) was chosen as carrier for Adr because the amino groups on the daunosamine residues readily form Schiff base condensation with aldehyde residues on the microspheres.[20] Experiments utilizing Pgl-Adr complexes with 0.45 and 0.1 μm diameters are reviewed in this chapter. However, it is important to emphasize that the

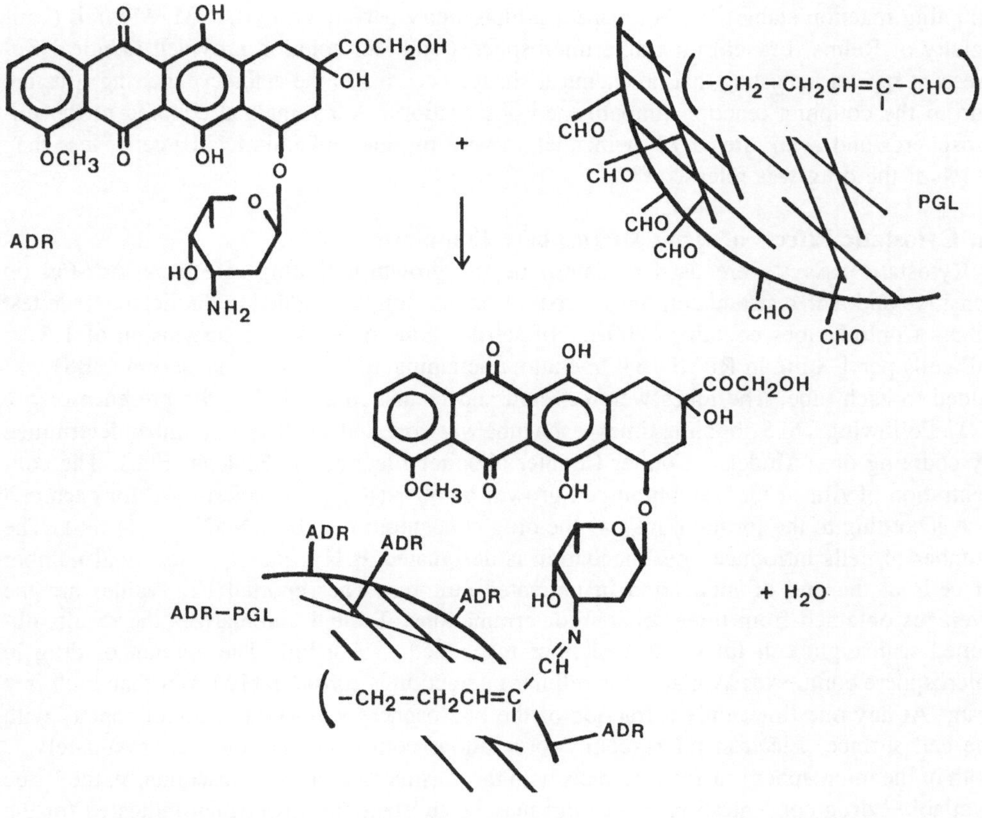

FIGURE 1. Diagram of the coupling reaction between Adr and Pgl.

"ideal" carriers have not yet been identified and extensive studies will be required to define them for each therapeutic agent.

Preparations of microsphere-Adr complexes, evaluations of their in vitro cytotoxic activity and indications for their in vivo utilization are summarized here. Experiments are reviewed which show that a variety of cell types are sensitive to membrane-targeted drug delivery. Estimates are provided as to the number of Adr molecules which may interact with the membrane and how this may result in signals leading to lysis or to differentiation at sublethal doses.

II. PREPARATION OF MICROSPHERE-ADR COMPLEXES AND THEIR IN VITRO EVALUATION

Microspheres were synthesized by polymerizing 10% glutaraldehyde at pH 11.0 in the presence of 0.1% Aerosol 604 (American Cyanamide, Wayne, N.J.), a nonionic detergent.[20] Following synthesis, the microspheres (Pgl) were washed exhaustively with water to remove unreacted glutaraldehyde. Covalent coupling of Adr was accomplished by reacting the drug with the Pgl in a 1:20 ratio (w/w) at pH 6.0 for 24 hr. Subsequent to coupling, noncovalently bound Adr was removed by treatment with 0.05% Nonidet P-40 (Particle Data Laboratories, Elmhurst, Ill.) and repeated washing with phosphate buffered saline (PBS) and 30% ethanol water. The average diameter of the microspheres under these conditions was 4500 Å.[20] Figure 1 is a diagram of Adr coupling at the polymer surface. The Schiff base condensation between amino and aldehyde groups are, in general, reversible. However, an extra double bond on the polymer β carbon position provides resonance-stabilizing energy making the coupling reaction stable.[21,32] Microspectrofluorometry performed by Dr. M. Monfait (University of Reims, France) on single microspheres and Adr solution provided identical fluorescent spectra indicating that no chemical change occurred in the anthracycline ring structure due to the coupling reaction (unpublished observation). Adr remained coupled to the microspheres and even after a 72-hr incubation with murine leukemia L1210 cells, less than 0.1% of the drug was released.[20]

A. Cytostatic Effects of Drug-Microsphere Complexes

Cytostatic assays were used to determine the growth-inhibiting effects of Adr-Pgl on sensitive and Adr-resistant cell lines. Free- or bound-drug was added to duplicate sterile test tubes. Control tubes contained 200 $\mu\ell$ of sterile saline only. A cell suspension of 1.5×10^4 cells per 1.8 mℓ in RPMI 1640 medium containing 10% fetal bovine serum (FBS) was added to each tube. The tubes were covered and incubated at 37°C in the presence of 5% CO_2. Following 3 to 5 doubling times, each tube was vortexed and the cell number determined by counting on a Model B Coulter Counter (Coulter Electronics, Hialeah, Fla.). The concentration of Adr needed to inhibit cell growth by 50% (IC_{50}) was determined for each cell line according to the formula: IC_{50} = the drug concentration at 0.5 (Nc/No − 1) + 1. The number of cells introduced per incubation is designated as N_o, and N_c is the final number of cells at the end of incubation in untreated cultures. The reported IC_{50} values are the averages obtained from three separate determinations. Table 1 summarizes the results obtained with eight cell lines. All cell lines responded to Adr-Pgl. The amount of drug in microsphere complexes available for cellular interaction is considerably lower than with free drug. At any one time, only a fraction of the microsphere surface is in direct contact with the cell surface. Electron microscopic observations confirmed that only approximately 1/20th of the microsphere surface interacts with the plasma membrane, consequently, the "true available" drug concentration in Adr-Pgl may be substantially lower than indicated for the IC_{50} values in Table 1. This suggests that the coupled drug is more effective than the listed values would indicate. Of considerable interest is the observation that cells resistant to free Adr become sensitive to the microsphere drug complexes. The response of these cells, in general, was similar to the response of the sensitive parental cell line to the free drug.

B. Determination of Cytotoxicity Using the Colony-Forming Assay

As a measure of cytotoxicity, the colony-forming ability of individual cells was assayed in soft agarose by Dr. Kathy Rogers,[22] using a modification of the procedure of Macpherson and Patterson.[23] Suspension cultures of cells in RPMI 1640 medium supplemented with 10% FBS were incubated with various concentrations of free or bound Adr for 24 hr. Following the initial incubation, a layer of 5 mℓ of 0.4% Sigma Type II agarose in RPMI 1640 medium was poured into each 60-mm petri dish and allowed to solidify for 30 min. Subsequently, the drug exposed cells (10^4) suspended in RPMI 1640 medium containing 10% FBS, 10% dialyzed horse serum, 2 m*M* pyruvate (Sigma, St. Louis, Mo.), and 0.24% agarose were spread over the previously poured agarose layer. Dishes were then incubated for 10 to 14 days in a humdified 37°C incubator under 5% CO_2. The number of colonies in the drug-treated dishes was then compared to the number of colonies in untreated dishes plated at the same cell density. Only colonies which contained a minimum of 50 cells were scored. The cloning efficiency for both the sensitive and resistant cells was greater than 80%.[22]

Exponentially growing cultures of the $L1210_S$ cell line were exposed to various concentrations of either free Adr or Adr-Pgl for a period of 24 hr. Cell survival was determined in the soft-agar colony assay and the representative dose-survival curves are shown in Figure 2A. A sigmoid survival curve was obtained for the free drug which was characterized by an initial plateau followed by a rapid decline in slope as the drug concentration increased. After 90% of the cells had been killed, the slope became more shallow. In contrast, the

Table 1
IC_{50} VALUES FOR FREE AND MICROSPHERE-BOUND ADR IN VARIOUS SENSITIVE AND RESISTANT CELL LINES

	IC_{50} nM	
Cell line	Free Adr	Adr-Pgl[a]
L1210[b]	28	15
L1210 Adr-resistant-1	2200	40
L1210 Adr-resistant-2	6100	45
S-180[c]	23	56
CCRF-CEM[d]	11	12
CCRF-CEM/VLB 500	320	20
CCRF-CEM/VLB 100	390	20
RLC[e]	18000	18

[a] The molar concentration of Adr-Pgl represents the total amount of Adr associated with the microspheres in a given volume.
[b] Murine leukemia L1210 cells were obtained from Dr. T. Moore, Roswell Park Memorial Institute, Buffalo, N.Y. Their resistant variants were generated by long-term exposure to increasing doses of Adr.
[c] Sarcoma-180 cells were provided by Dr. T. Tritton, University of Vermont, Burlington.
[d] Human leukemia cells CCRF-CEM and their resistant clones to vinblastine which are cross-resistant to Adr were obtained from Dr. W. Beck, St. Jude Children's Research Hospital, Memphis, Tenn.
[e] RLC are carcinogen-induced rat hepatocytes from Dr. Brian Carr, City of Hope National Medical Center, Duarte, Calif.

Adr-Pgl survival curve reached plateau at a lower concentration and was followed by a steady decline in slope between 10^{-10} and 10^{-7} M drug concentrations. The Adr concentrations resulting in 90, 50, and 10% survival with free and polymer-bound drug are shown in Table 2. A control consisting of KBH_4 reduced, nondrug carrying plain Pgl (P-Pgl) had no effect on viability at equivalent drug concentrations. A control dose-survival study was also done in which the free drug was removed from the medium subsequent to the 24-hr exposure time. The results produced a curve identical to that obtained when the free drug was present throughout the assay. The 50% survival concentration for $L1210_S$ was 1×10^{-7} M when free Adr was removed after 24 hr, and 0.9×10^{-7} M when it was allowed to remain throughout the assay. An additional control consisting of enough P-Pgl to couple 1×10^{-6} M Adr and 1×10^{-10} M free Adr was included. The concentration of free drug was chosen to reflect a minimum of a tenfold higher amount of drug which was known to be released from the Adr-Pgl.[20] This control was done to test whether a small amount of free Adr together with a large number of microspheres could perturb the cells enough to produce the decrease in viability. Both cell types remained 100% viable in the presence of the free Adr and P-Pgl demonstrating that covalent coupling of Adr to the microspheres was responsible for cell killing.

Dose-survival curves were also generated for the Adr-resistant $L1210_R$ cells with both the

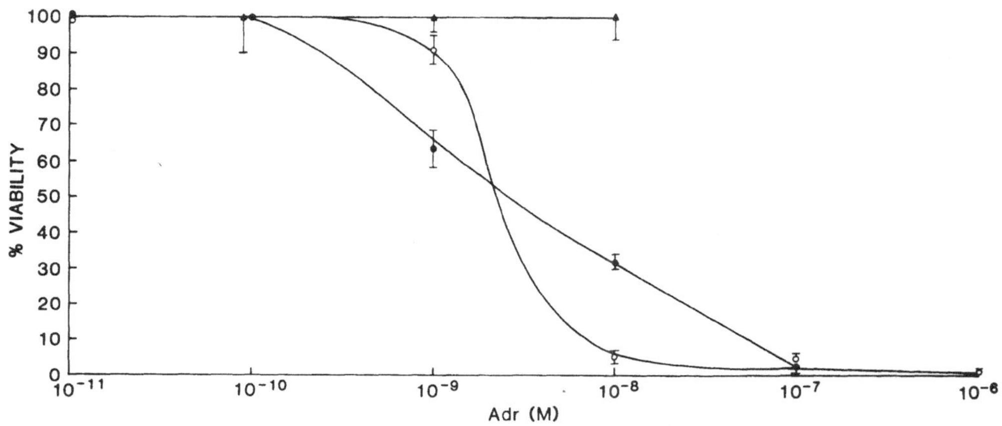

A

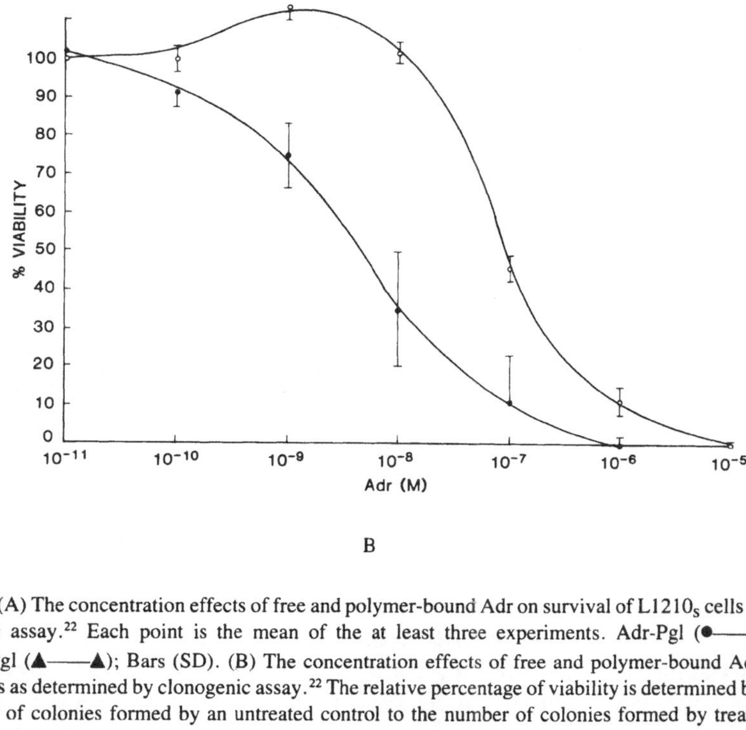

B

FIGURE 2. (A) The concentration effects of free and polymer-bound Adr on survival of $L1210_S$ cells as determined by clonogenic assay.[22] Each point is the mean of the at least three experiments. Adr-Pgl (●——●); free Adr (○——○); P-Pgl (▲——▲); Bars (SD). (B) The concentration effects of free and polymer-bound Adr on survival of $L1210_R$ cells as determined by clonogenic assay.[22] The relative percentage of viability is determined by comparison of the number of colonies formed by an untreated control to the number of colonies formed by treated cells. The apparent increase in the relative percent of viability at 10^{-9} *M* free Adr represents a growth stimulation which is routinely observed at low concentrations of free drug. Each point is the mean of at least three determinations. Adr-Pgl (●——●); free Adr (○——○); Bars (SD).

free and bound drugs. The results are shown in Figure 2B. Sigmoid curves were obtained for both the free and bound drug. However, free Adr stimulated growth of the $L1210_R$ cells in the threshold region while the Adr-Pgl did not. In addition, 140 times smaller concentration was necessary for Adr-Pgl than for free Adr to obtain 90% survival, as summarized in Table 2. The 50% survival concentration is similar for both the free and bound drug with $L1210_S$. However, a fourfold increase in Adr-Pgl is needed to decrease the survival of sensitive cells to 10% as compared to free drug. When the free and bound drugs were compared with $L1210_R$ cells, 12-fold less Adr-Pgl were needed for 50% survival, and 10-fold less for 10% survival.[22]

Table 2
SENSITIVITY OF MURINE LEUKEMIC CELL LINES TO FREE OR POLYMER-COUPLED ADR[22]

Cell line	Form of Adr	Adr concentration (*M*) resulting in 90% Survival	50% Survival	10% Survival
$L1210_S$	Free	1.0×10^{-9}	3.5×10^{-9}	9.0×10^{-9}
	Polymer bound	2.5×10^{-10}	2.5×10^{-9}	3.7×10^{-8}
$L1210_R$	Free	2.2×10^{-8}	1.0×10^{-7}	1.2×10^{6}
	Polymer bound	1.6×10^{-10}	8.0×10^{-9}	1.2×10^{-7}

In this study the clonogenic assay was used to determine cell viability after drug treatment with polymer-bound Adr. This method was selected because it has been shown to correlate well with results obtained in vivo.[23-28] There are two major objections, however, to this method of viability determination. The first is that plating a small number of cells in agarose to form colonies imposes undue stress on the cells. Since the cloning efficiency of the control cells in our studies was always greater than 80% we believe the stress due to cloning was minimal. It has also been suggested that drugs may cause a lethal injury to some cells which does not become manifested for several cell generations. This could result in an overestimate of cell kill if the clones were counted too soon and an underestimate if the colonies were counted late.[27] In our assays the estimates of cell kill did not significantly vary during an additional 3 to 4 generation times after 90% of the control colonies had reached at least 50 cells. The longer assay times should have allowed any clones, which had been sublethally damaged, enough additional time to divide so that they could be scored as colonies, thus avoiding overestimates of killing.

The shapes of the survival curves for $L1210_S$ and $L1210_R$ for both free and bound drug were characterized by a plateau region at low drug concentrations where minimal cell kill was seen. At higher concentrations, both free and bound Adr killed cells exponentially. The shoulder regions imply that the cells are able to sustain a certain degree of damage which is sublethal.[27] Also revealed by the survival curves for free Adr with both cell types is a more shallow slope at the highest drug concentrations. This is presumably due to the presence of a cell population which is resistant to free Adr but which is subject to the effects of the bound drug. It appears that more sublethal damage can be accumulated by either $L1210_S$ or $L1210_R$ cells when they are exposed to free Adr than when they are treated with the bound drug. This does not necessarily mean that the cells possess the ability to repair the damage, but does reflect the capacity of hypothetical cellular targets to absorb damage.[26,27] Hence, it follows that more targets would have to be affected by the free drug, when compared to the bound drug, before the effects became manifested. Alternatively, the same number of targets could be affected by both forms of the drug but to a significantly greater extent by the bound Adr. A third possibility is that a new set of molecules become targets to the bound drug, which are normally not affected by free Adr. The sensitivity of the cellular targets to free or bound Adr can be compared by examining the 90% survival concentrations. In the case of the $L1210_S$ cells there is an apparent fourfold increase in the sensitivity to the bound drug. In the case of the $L1210_R$ cells, the hypothetical targets are approximately 140 times more sensitive to the bound drug than they are to the free drug. The increased sensitivity of the $L1210_R$ cellular targets to the Adr-Pgl is also seen at the 50 and 10% survival concentration. The tenfold increase in effectiveness against the resistant cells is noteworthy because 90% cell kill is thought to be the minimum killing which translates into an in vivo response.[28-30]

The study using the clonogenic assay was also extended to human surgical malignant specimens.[31] A total of 48 specimens were investigated representing eight different histological types of malignancy. Two drug concentrations were used: 0.1 and 10 μg of Adr/mℓ in free or bound form. Marked increase in cytotoxicity was observed in 35% of the specimens at 0.1 μg/mℓ and in 55% of the specimens at 10 μg/mℓ due to covalent attachment of Adr to microspheres. Indistinguishable results were obtained in 45% and 25% of the specimens at 0.1 and 10 μg Adr per milliliter, respectively. Slightly decreased activity was observed only in 20% of the samples. The most striking increase in activity was obtained with lung cancer specimens where 67% showed enhanced sensitivity with Adr-Pgl at both concentration, and in ovarian cancer where 43 and 57% of the samples showed an increased sensitivity at the low and high drug concentration, respectively. These results clearly establish that the microsphere-drug complexes express cytotoxic effects on a large variety of cells and that their further therapeutic evaluation is warranted.

C. Transmission Electron Microscopic Studies

Since it was demonstrated that Adr-Pgl exerts cytotoxicity under conditions where free Adr is not released in sufficient quantities to be effective as a cytotoxic agent, the cell plasma membrane became the prime candidate for a probable site of action. Transmission electron microscopy was chosen to visualize the location of the microspheres following 12 to 24 hr incubations.[22]

L1210 cells (1×10^5 cells per milliliter) were incubated in suspension culture with either 10^{-6} *M* Adr-Pgl or the equivalent number of drug-free P-Pgl. At the conclusion of 24 hr, cells and microspheres were pelleted by centrifuging in a microfuge (Beckman Instruments, Fullerton, Calif.) for 1 min. The supernatant was removed and replaced by 1 mℓ of 5% glutaraldehyde (Pelco, Tustin, Calif.) at ambient temperature. Following a 12-hr fixation period, the samples were washed twice with 0.1 *M* cacodylate buffer (pH 7.2) and allowed to sit overnight in 2 mℓ of fresh buffer. Subsequently, samples were postfixed in 1% osmium tetroxide at 4°C for 45 min and dehydrated in a graded series of ethanol. Epon® (Pelco, Tustin, Calif.) was used for embedding. Thin sections were stained with uranyl acetate and lead citrate, and viewed with an AEI Corinth 500 transmission electron microscope (Manchester, U.K.). The locations of the microspheres were categorized as extracellular, surface-bound, cytoplasmic, or nuclear.

P-Pgl microspheres did not bind to the plasma membrane and were not endocytosed. Less than 2% of the microspheres were in contact with the cell surface. With Adr-Pgl, 20% of the microspheres were associated with the plasma membrane. Figure 3 shows Adr-Pgl in close proximity with the plasma membrane. The drug-microsphere complexes were not internalized by L1210 cells.

It is possible to estimate from the electron micrographs the area of Adr-Pgl directly in contact with the cell surface. Based on the position of 20 drug-carrying microspheres which were attached to cell surfaces, we have estimated that an average contact area was approximately 3×10^6 Å^2, representing a circle with an estimated diameter of approximately 2100Å. For the purpose of this estimation we assumed that the microspheres and the cells interacting with them represent smooth, uniform surfaces. The reservations for these assumptions are considerable since significant surface irregularities may exist, therefore the approximate surface area for the interaction represents an average minimum estimate. An assumption was also made that this equals the diameter of a circle which represents the contact area between the polymer drug and the cell surface. The approximate size of this area is 3.2×10^6 Å^2 which is approximately 1/20 of the total PGL-microsphere's surface area of 63×10^6 Å^2. Earlier an estimation was made that each microsphere carries approximately 0.8 to 1.2×10^6 drug molecules,[20] therefore each contact surface area could present as much as 40 to 60,000 molecules of Adr to the cell surface. This number probably represents the highest estimation since not all of the drug molecules might be accessible for

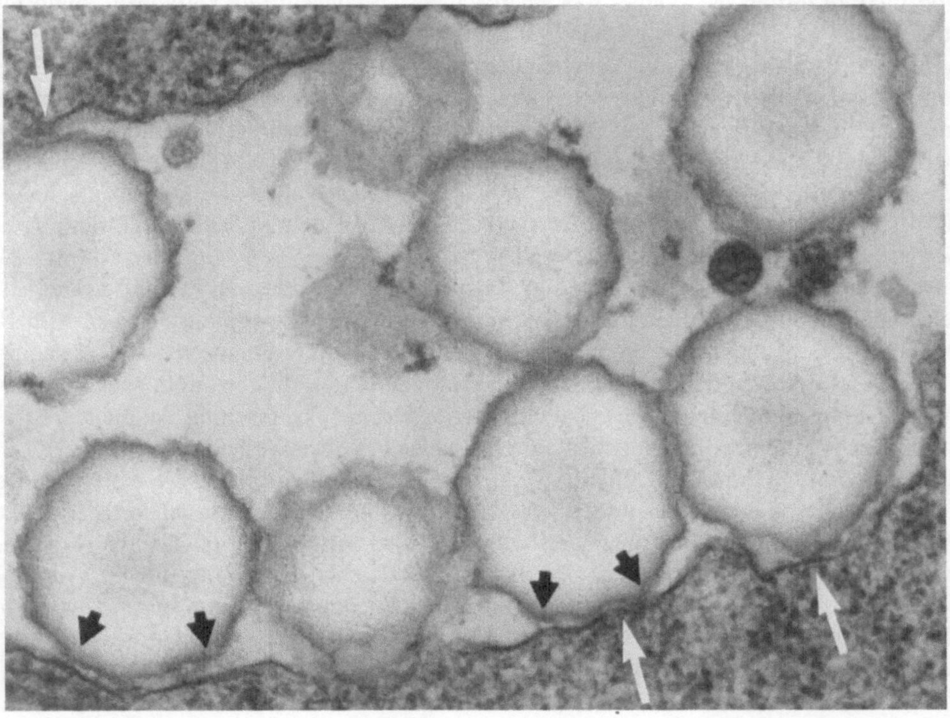

FIGURE 3. Electron micrograph of the plasma membrane of an L1210 cell treated with 10^{-6} *M* Adr-Pgl for 12 hr. The figure illustrates the position of microspheres bound to the cell surface. The contact areas between the cell membrane and the microspheres are indicated by black arrows positioned inside the microspheres, spanning an average distance of approximately 2000 Å, white arrows point to the plasma membrane. Photomicrograph by Dr. Kathy Rogers.[22] (Magnification approximately × 20,000.)

interaction with membrane components. Nevertheless, it is important to emphasize that each polymeric drug complex is capable of simultaneously presenting several thousand drug molecules to a restricted domain on the cell surface. Thus, it is proposed that the interaction results in multiple binding or "multiple-hits". The simultaneous interaction of thousands of drug molecules would infer that more than one type of membrane component may be perturbed by Adr-PGL. These components may include molecules which would not be otherwise affected by free Adr. Such a perturbation due to multiple binding would make the cells less able to absorb as much damage as they could with the free drug.

III. IN VIVO EVALUATION OF MICROSPHERE-ADR COMPLEXES

The demonstrated activity of Adr-Pgl microspheres against cultured mammalian tumor cell lines and excised human tumors prompted the question of whether these complexes could also display beneficial antitumor activity in vivo. Exploratory experiments were performed with the understanding that further studies are necessary to define the ideal carriers before a thorough evaluation of in vivo therapeutic efficacy is undertaken.

In the first set of experiments the biodistribution of 0.45 and 0.1 μm diameter drug microspheres was investigated. Special carriers were prepared for these experiments, which contained entrapped ^{125}I-labeled substances in order to enable their quantitation in tissues.

A. Preparation of ^{125}I-Labeled Adr Microspheres

Large molecular-weight components, such as proteins, when included in the polymerization mixture during the synthesis of microspheres, become firmly entrapped within the matrix of the Pgl.[32] Purified egg albumin was used as a means of labeling the microspheres. Egg albumin was labeled with ^{125}I by chloramine-T methodology.[33] Briefly, 1 mg of egg albumin was incubated for 3 min with 5 mCi of carrier-free ^{125}I and chloramine-T, 1 mg in 1 mℓ H_2O, at room temperature. The reaction was stopped with the addition of sodium metabisulfite, 0.5 mg. Free ^{125}I was separated from labeled protein with the use of a Sephadex G-25 column after which the ^{125}I-labeled protein was extensively dialyzed against saline.

Two sizes of Pgl microspheres were synthesized. Aqueous glutaraldehyde solutions of 10% and 1.5% were used to make microspheres of 0.45 and 0.10 μm diameter, respectively. The surfactant, Aerosol 604 was added at 1% of the total volume and 500 μCi of the ^{125}I-labeled egg albumin was added to each reaction vessel. The pH was adjusted to 11 with 10 *N* NaOH and the containers were purged with nitrogen. This was repeated at 1, 2, and 4 hr. The 150-mℓ plastic bottles containing 100 mℓ of reaction mixture were shaken for 24 hr at room temperature on a mechanical wrist-action shaker. The resulting Pgl microspheres were dialyzed, to remove nonpolymerized glutaraldehyde, then washed with PBS, pH 7.2, and centrifuged, five times at 6000 × *g* for 15 min. Pelleted microspheres were resuspended by vortexing or sonicating when necessary. Labeled protein at the external surfaces of the microspheres was removed by treatments with 0.25% trypsin for 30 min. The microspheres were washed and trypsinization was repeated until no further label was removed. Adr was coupled to the labeled microspheres by mixing a 20:1 microsphere to drug ratio (w/w) in PBS, at pH 6.0 for 24 hr at room temperature in a dark container. The microspheres were then washed extensively to remove noncovalently attached drug.

B. Analysis of Organ Distribution in Rats After Intravenous Injection

A volume of 0.5 mℓ containing 3×10^5 cpm of ^{125}I-labeled Adr-Pgl of either of the two sizes, 0.45 and 0.10 μm average diameter, was injected into the lateral tail veins of female Fischer 344 rats (Charles River, Portage, Del.). At time points of 15 min, 8, 24, and 40 hr, animals were anesthetized with Metofane (Pitman-Moore, Washington Crossing, N.J.), sacrificed by thoracotomy and organ samples were collected. Samples of heart, lung, liver, spleen, kidney, bone, muscle, thyroid, feces, blood, and urine were analyzed. Blood was collected from the heart and urine from the bladder, and their volumes were recorded. The heart, lung, liver, spleen, and kidney were weighed whole and then minced. Bone and muscle samples were taken from upper thigh. Feces samples were collected at upper, mid, and lower intestinal tract and mixed. Thyroid tissue was collected by removing the trachea along with adjoining tissue by making incisions above and below the thyroid. All samples were weighed and analyzed for radioactivity on a Packard 5260 gamma counter. For calculations of whole organ distribution as "percent of injected counts", the heart, lungs, liver, spleen, and kidneys were weighed as whole organs. The whole organ size for the following tissues was established from rat anatomical data tables according to the average weight of the animals (305 g in this case): bone (60 g), muscle (150 g), feces (10 g), blood (20 mℓ), and urine (3 mℓ). Counts injected and not accounted for in the whole organs analyzed were categorized as "other".

As shown in Table 3, the liver was the main receptacle for injected microspheres, collecting from 50 to 70% of injected counts. Bone, muscle, lung, and spleen followed in collecting from 4 to 15% of injected dose as whole organs. Heart and kidney were especially low in uptake of Adr-Pgl with less than 1% of injected label associated with these organs.

Few differences were found between the distribution of the 0.1 μm drug microspheres as compared to those of 0.45 μm. One notable difference was in the 15-min value for the lung which was 12% with the smaller microspheres as compared to 3% with the larger microspheres.

Changes occurring over the time course of 15 min to 40 hr, for the 0.1 μm microspheres, showed liver and lung values to decrease while muscle values increased. These trends were

Table 3
WHOLE-ORGAN DISTRIBUTION OF INTRAVENOUS INJECTIONS OF ^{125}I ADR MICROSPHERES (Adr-Pgl) OF TWO SIZES, 0.1 μm AND 0.45 μm

Whole organ	15 min	8 hr	24 hr	40 hr
0.1 μm Adr-Pgl (% of injected counts)				
Heart	0.11	0.09	0.08	0.13
Lung	12.89	4.39	2.21	2.01
Liver	63.31	57.36	53.00	48.99
Spleen	4.20	3.30	5.32	7.27
Kidney	0.37	0.52	0.41	0.63
Bone	5.46	11.23	7.05	7.45
Muscle	7.55	9.58	14.59	17.53
Feces	0.85	1.41	2.46	1.75
Thyroid	0.08	1.42	2.79	2.82
Blood	2.71	4.14	5.26	5.62
Urine	0.70	4.32	3.42	1.73
Other	1.76	2.24	3.40	4.09
0.45 μm Adr-Pgl (% of injected counts)				
Heart	0.09	0.09	0.09	0.07
Lung	3.24	4.16	1.07	1.01
Liver	55.80	68.60	68.29	63.70
Spleen	3.88	6.06	5.09	7.41
Kidney	0.19	0.43	0.31	0.29
Bone	14.35	7.10	7.56	6.55
Muscle	10.63	4.20	5.94	8.48
Feces	0.48	1.16	1.78	1.39
Thyroid	0.04	1.46	1.77	2.50
Blood	1.91	3.00	3.74	3.51
Urine	0.89	2.75	2.97	3.12
Other	8.50	0.98	1.39	1.98

Table 4
TISSUE-SPECIFIC CONCENTRATION OF ^{125}I Adr-Pgl OF TWO SIZES, 0.1 μ (A) AND 0.45 μ (B)

Tissue	15 Min	8 Hr	24 Hr	40 Hr
0.1 μ Adr-Pgl (% of injected counts per gram tissue)				
Heart	0.10	0.09	0.07	0.10
Lung	7.90	2.42	1.26	1.16
Liver	7.08	6.78	6.14	5.40
Spleen	7.00	4.45	6.56	10.09
Kidney	0.52	0.50	0.44	0.71
Bone	0.09	0.18	0.11	0.12
Muscle	0.05	0.06	0.09	0.11
0.45 μ Adr-Pgl (% of injected counts per gram tissue)				
Heart	0.07	0.06	0.07	0.05
Lung	1.66	2.44	0.64	0.60
Liver	6.22	5.57	8.78	6.56
Spleen	5.79	7.57	7.71	9.03
Kidney	0.19	0.28	0.36	0.29
Bone	0.23	0.11	0.12	0.10
Muscle	0.07	0.02	0.03	0.05

not as apparent with the larger microspheres. In both sizes, spleen, blood, and thyroid values increased with time. Only with feces and urine was an increase seen in the first 24 hr, followed by a decrease at 40 hr.

Tissue distribution can also be analyzed by calculating the specific uptake per gram tissue rather than whole organ. This diminishes the role of organ size and focuses on tissue-specific concentration. As shown in Table 4, spleen, lung, and liver had, by far, the highest amount of Adr-Pgl per gram tissue. In both sizes of microspheres, lung concentration decreases while spleen concentration increases over 40 hr.

Small spherical particles less than 7 μm in diameter are normally removed from the circulation by phagocytic cells of the reticuloendothelial system. Kupffer cells in the liver, free and fixed macrophages in spleen and bone marrow, are highly active in clearing the bloodstream of particles such as sheep red blood cells, polymeric microspheres, and lipid vesicles.[34-36] Nonspherical particles, such as fibers, and spherical particles larger than 7 μm are mechanically filtered in the lung, being that it is the first capillary bed encountered upon intravenous administration. The great majority of macrophage cells in the lung reside in the alveolar compartments, separated from the circulation by capillaries which contain an uninterrupted lining of endothelial cells. Thus, alveolar macrophages may not contribute to the observed blood clearance because they do not line the blood channels as do their counterparts in liver and spleen.

The results of this study agree well with published reports of other injected solid particles,[35,37] with 65 to 80% of injected Adr-Pgl found in liver, spleen, and bone. Lung-associated microspheres are most likely mechanically filtered aggregates. Adr forms a hydrophobic coating on the Pgl microspheres which induces some homotypic aggregation observed after drug coupling. This phenomena may explain the quantity of Adr-Pgl found in the lung even though the injection mixture was vigorously vortexed prior to the injections. Consistent with this explanation is the decrease in lung values over 40 hr which could be due to recirculation or release of entrapped Adr-Pgl.

Counts observed in the thyroid are indicative of free ^{125}I. These values increased with time and suggest that residual degradation of microsphere label occurred but remained below 3% over 40 hr. Blood values showed that the great majority of Adr-Pgl was cleared by the 15-min time point. The increase in blood values with time suggests that either microspheres are being released back into the circulation or that labeled protein is being released by the microspheres. Feces and urine values along with the total counts retained by the animals over 40 hr show that renal excretion and bile account for a loss of about 7% of the injected label over this time period.

Heart and kidney had the lowest values of all organs examined. This is significant because these two organs display the highest sensitivity to the toxic effects of free Adr. Cardiomyopathy is a dose-limiting effect of Adr treatment in the clinical setting. Any treatment modality that can limit drug exposure to heart and kidney while maintaining antitumor activity would be advantageous as long as new organ-specific toxicities are not introduced.

Distribution studies of free anthracyclines after intravenous injections show these drugs to have a rapid uptake in most tissues with especially high initial association with kidney and lung which peak at about 30 min.[42,43] All tissue levels fall sharply between 30 min and 2 hr, except spleen and lymph node which increase over 8 hr. Approximately 40% of injected dose was shown to be excreted by feces (30%) and urine (10%) within 48 hr.

C. Evaluation of the Antitumor Effect of Adr-Microspheres In Vivo

The ability of Adr microspheres to bind to and kill tumor cells in vitro without Adr detaching from the microspheres and without the microspheres being internalized demonstrated that the cytotoxic effect occurs at the level of the cell membrane upon contact with drug microspheres. Whether this cytotoxic mechanism could occur in vivo remained unknown. Organ distribution studies indicated that the highest tissue-specific concentration of Adr-Pgl after intravenous injection occurred in spleen, liver, and lung. Because contact between drug microspheres and tumor cells appears to be a critical factor, an animal tumor model involving one of these organs was desirable. One of the models chosen was the lung-colonizing MTLn3 rat-mammary adenocarcinoma line developed by Neri et al.[38] This line was established from a metastatic lung colony produced by the carcinogen-induced rat mammary adenocarcinoma line 13762 originally developed by Segaloff.[39] The MTLn3 cell line is highly active in metastasizing to the lung after either intravenous injection (experimental metastasis) or injections into the inguinal mammary fat pad (spontaneous metastasis). A high correlation was demonstrated between the actual lung tumor load and the number of visible surface lung tumors.[40] Therefore, scoring the number of surface lesions is a reliable method of assessing tumor load by this tumor in the lung. Tissue distribution studies indicate that the lung is comparatively high in exposure to both free and microsphere-bound Adr following intravenous administration. Consequently, we felt that this model could provide a reliable comparison of the in vivo antitumor activities of Adr and Adr-Pgl.

Rat MTLn3 mammary adenocarcinoma cells, passage number 22, were grown in 75 cm^2 tissue culture flasks (Costar Corp., Van Nuys, Calif.) in RPMI 1640 media with 10% FBS to a range of 70 to 90% confluency. Culture media were removed and the attached cells were washed with Dulbecco's calcium-free and magnesium-free phosphate buffered saline (DPBS). Cells were detached with the addition of 3 mℓ of 0.25% trypsin in DPBS for 15 min at 37°C. Flasks were gently tapped to remove remaining attached cells and the mixture was passed through a 5-mℓ pipette and repeated to obtain a single-cell suspension. Trypsinization was stopped by the addition of 5 mℓ of RPMI 1640 with 10% FBS. Cells were pelleted by centrifugation and resuspended in cold serum-free RPMI. Cell counts were made with the use of a hemocytomer and viabilities determined by trypan blue exclusion. Only cell populations with viabilities above 92% were used for animal injections. The cell density was adjusted to 3×10^5 viable cells per milliliter in cold serum-free RPMI for intravenous injections.

Female Fischer 344/CRBL rats, 6 to 9 weeks of age (Charles River, Portage, Del.), anesthetized with metofane (Methoxyflurane, Pitman-Moore, Inc., Washington Crossing, N.J.), were inoculated with 1.5×10^5 MTLn3 tumor cells. Tuberculin syringes (1-cc volume, 27-gauge needles) were used to deliver a volume of 0.5 mℓ intravenously into the lateral tail vein. Four treatment groups of ten animals per group were established. Treatment groups consisted of animals treated with Adr, Adr microspheres, plain drug-free microspheres, and saline. Animal groups received 0.5-mℓ i.v. doses of 2 mg/kg Adr, 2 mg/kg equivalents of Adr microspheres, 0.5 mℓ of an equivalent concentration of plain microspheres and 0.5 mℓ of sterile saline at days 2, 9, and 16 after tumor cell inoculations. At day 23 the animals were anesthetized with metofane, sacrificed by thoracotomy, and lungs were removed and fixed in Bouin's solution (30:10:2 ratio of saturated picric acid, formalin, and glacial acetic acid) for 24 hr. (Lung tumors stain yellow and are easily visualized on the pink lung parenchyma background.) Tumors were counted and measured with the use of a dissecting microscope. The number of surface lung tumors with diameters greater than 0.5 mm was recorded for each animal. Tumor counts of each group were compared by Student-Newman-Kells and t-ratio procedures of group significant difference analysis.

Tumor counts after various forms of treatments are shown in Figure 4. The average number of tumors per treatment group and range occurred as follows: saline, 61.3 (44 to 125); plain drug-free microspheres, 63.0 (21 to 85); Adr, 37.0 (21 to 47); Adr-Pgl microspheres, 25.9 (9 to 53). Differences between the means of the control treatment groups, saline and plain microspheres, as compared to treatment groups of Adr and Adr microspheres were highly significant ($p < 0.001$). Differences between means of saline as compared to plain microsphere treatment were not significant nor were differences between treatment groups of Adr and Adr microspheres. The sizes of lung-surface tumors ranged from 0.1 to 5 mm in diameter. A few animals in each group displayed a background of micrometastatic colonies of around 0.1 mm in size, but this was not consistent among animals in any particular treatment group. Only established tumors of ≥ 0.5 mm diameter were scored for comparisons. Although the range of tumor counts was larger in the controls, the average size of tumors was not different among groups. The majority of tumor sizes were between 1 and 3 mm. Small palpable tumors were found at the injection site of about one quarter of the animals, suggesting some extravasation of tumor cells upon inoculation. No tumors were detected outside of the lung in the thoracic or peritoneal cavities. There was no evidence of acute toxicities related to the various treatment modalities including the Adr microspheres.

The results establish that the microsphere drug complexes are well tolerated and retain their antitumor activity. However, further studies need to be conducted to evaluate their efficacy for the treatment of solid tumors and circulating leukemias. These studies should include Adr-resistant cells and the most feasible carriers that can be identified from in vitro studies.

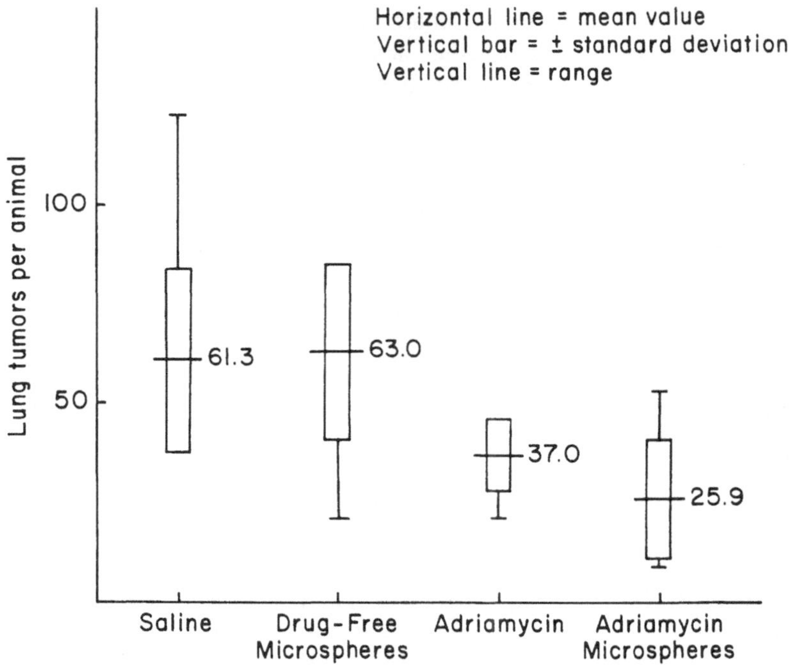

FIGURE 4. Metastatic mammary tumors in the lung of rats treated with Adr, Adr-microspheres, drug-free microspheres, or saline. (Ten animals were used for each treatment group.)

IV. DISCUSSION AND SUMMARY

The concept of using microspheres to focus drug action to the plasma membrane has now been applied in experimental systems. The result of this method of drug presentation is a novel mode of cytotoxicity which is not available to the free form of the drug. In order to further establish the novelty of the outlined approach, results of three additional sets of experiments are summarized below.

In the first set of experiments, the inactive anthracycline analog 4-demethoxy-7,9-di-epi-daunorubicin (7R-9R) was covalently coupled to Pgl microspheres. The resulting 7R-9R-Pgl complexes acquired significant cytostatic activity as evaluated with Adr-resistant and sensitive murine L1210 leukemia cells.[41] The free form of the analog had no detectable growth-inhibiting activity even at concentrations of 10^{-5} *M* with either the sensitive or resistant cells. The IC_{50} values of the microsphere-analog complexes were 2.1×10^{-7} *M* and 3.2×10^{-7} *M* with the sensitive and resistant cells, respectively. Combinations of P-Pgl and free drug failed to elicit growth inhibition, establishing that covalent attachment is essential for producing cytotoxic effects. These experiments also rule out the possibility that the observed effects are due to slow release of free-drug molecules, since they do not express activity against either sensitive or resistant cells.

L1210 cells were exposed to Adr-Pgl for 12 months in the second set of experiments in order to compare the capacity of the cells to develop resistance to free or polymer-bound Adr and to compare their biological properties.[44] Experiments were performed with parallel cell lines cultured under continuous exposure to free Adr or Adr-Pgl. The drug sensitivity of each line was regularly monitored by growth-inhibition analysis. Drug levels were adjusted and maintained at IC_{70} concentrations. Fresh drug was added at every cell passage of 3 to 4 days. After a period of 8 months, cells grown in the presence of free Adr developed a 130-fold increase in resistance to the free drug, but were unaltered in their sensitivity to polymer-bound Adr. In contrast, cells grown in the continuous presence of polymer-bound Adr did not develop resistance to either the free or polymer-bound form of the drug. These experiments further establish the stability of Adr-Pgl and the lack of substantial free-Adr release during incubation which would have resulted in some resistance to Adr. They also substantiate the different and nonoverlapping modes of cytotoxicity by these two forms of Adr. Furthermore, the experiments demonstrate that cells cannot readily modify membrane properties which could lead to increased resistance to this mode of cell surface-directed drug delivery.

In the third set of experiments drug-induced differentiation was evaluated.[45] Adr induces hemoglobin (Hb) synthesis in human leukemic K562 cells.[46] The free drug is readily taken up by these cells and distributes in several subcellular compartments, including the nuclei. Since Adr-Pgl focuses the drug action to the plasma membrane, the question was raised as to what effect the microsphere-mediated membrane targeting may have on the induction of Hb synthesis. K 562 cells were incubated for 3 to 5 days with free Adr on Adr-Pgl. No cellular internalization of the drug-microsphere complexes occurred. Hb was labeled with Fe^{55} and identified by gel electrophoresis and autoradiography. Untreated cultures contain 2 to 4% Hb-positive cells, identified by benzidine staining. These cells synthesize Hb types Gower 1 and Portland. At 50% growth inhibition with 15 n*M* free Adr, 15% of the cells were Hb positive and with 130 n*M* Adr-Pgl, 11% were positive. At 87% growth inhibition, the Hb positive cells were 37% for free Adr and 41% for Adr-Pgl. Free Adr induced the synthesis of Hb types Gower 1, Portland, and trace amounts of HbF. Microsphere-Adr complexes induced the syntehsis of Hb Gower 2 and elevated levels HbF in addition to Hb Gower 1 and Portland. These results demonstrate that the increased plasma membrane interaction with Adr, due to covalent attachment to microspheres, leads to the synthesis of a different Hb profile which is consistent with differentiation. Slow release of free drug would not be capable of such induction of Hb synthesis, again emphasizing the need for covalent attachment to microspheres. More extensive testing for the capacity to induce differentiation in other cellular models is now of considerable interest.

In summary, these three sets of experiments and the previously outlined in vivo and in vitro studies demonstrate a scope of biological activities that can be obtained by membrane targeting of anthracyclines. These activities warrant further studies to extend the observations to other drugs and microcarriers with the specific aims of defining the therapeutically useful complexes.

REFERENCES

1. **Davis, S. S., Illum, L., McVie, J. G., and Tomlinson, E., Eds.,** *Microspheres and Drug Therapy: Pharmaceutical, Immunological and Medical Aspects,* Elsevier, Amsterdam, 1984.
2. **Tökés, Z. A. and Rogers, K., and Rembaum, A.,** U.S. Patent 4, 460, 560, 1984.
3. **Blum, R. H. and Carter, S. K.,** *Ann. Intern. Med.,* 80, 249, 1974.
4. **Neidle, S.,** *Topics in Antibiotic Chemistry,* Vol. 2 (Part D), Sammers, P., Ed., Ellis Horwood, Chichester, 1978, 240.
5. **Schwartz, H. S.,** *J. Med.,* 7, 33, 1976.
6. **Lee, Y. C. and Byfield, J. E.,** *J. Natl. Cancer Inst.,* 157, 221, 1976.
7. **DiMarco, A.,** *Cancer Chemother. Rep.,* 6, 91, 1975.
8. **Misjuno, N. S., Zakis, B., and Decker, R. W.,** *Cancer Res.,* 35, 1542, 1975.
9. **Silvestrini, R., Gambarucci, C., and Dasdia, T.,** *Tumori,* 56, 137, 1970.
10. **Tobey, R. A.,** *Cancer Res.,* 32, 2720, 1972.
11. **Zunino, F., Gambetta, R., and DiMarco, A.,** *Biochem. Pharmacol.,* 24, 309, 1975.
12. **Israel, M., Modest, E. J., and Frei, E., III,** *Cancer Res.,* 35, 1365, 1975.

13. **Kessel, D.,** *Mol. Pharmacol.*, 16, 306, 1979.
14. **Na, C. and Timasheff, S.,** *Arch. Biochem. Biophys.*, 182, 147, 1977.
15. **Murphree, S. A., Tritton, T. R., Smith, P. L., and Sartorelli, A. C.,** *Biochim. Biophys. Acta*, 649, 317, 1981.
16. **Gosalvesj, M. and Blanco, M.,** *5th Int. Biophys. Congr.*, Copenhagen 5, 4, 1975.
17. **Murphree, S. A., Cunningham, L. S., Hwang, K. M., and Sartorelli, A. C.,** *Biochem. Pharmacol.*, 25, 1227, 1976.
18. **Zuckier, G. N., Tomiko, S. A., and Tritton, J. R.,** *Fed. Proc. Fed. Am. Soc. Exp. Biol.*, 40, 1877, 1981.
19. **Smuckler, E. A. and James, J. L.,** *Pharmacol. Rev.*, 36, 775, 1984.
20. **Tokes, Z. A., Rogers, K. E., and Rembaum, A.,** *Proc. Natl. Acad. Sci.*, U.S.A., 79, 2026, 1982.
21. **Monsan, P., Puzo, G., and Mazarguil, H.,** *Biochimie*, 57, 1281, 1975.
22. **Rogers, K. E.,** Ph.D. Thesis, Univ. So. Calif., Los Angeles, 1983.
23. **Macpherson, L. and Patterson, M. K., Eds.,** *Tissue Culture: Methods and Applications*, Academic Press, New York, 1973, 276.
24. **Yuhas, J. M., Toya, R. E., and Pazmino, N. H.,** *J. Natl. Cancer Inst.*, 53, 465, 1974.
25. **Von Hoff, D. D., Casper, J., Bradley, E., Sandbach, J., Jones, D., and Makuch, R.,** *Am. J. Med.*, 70, 1027, 1981.
26. **Wiesenthal, L. M., Dill, P. L., Kurnid, N. B., and Lippman, M. E.,** *Cancer Res.*, 43, 258, 1982.
27. **Drewinko, B., Roper, P. R., and Barlogie, B.,** *Eur. J. Cancer*, 15, 93, 1979.
28. **Drewinko, B., Loo, R. L., and Gottlieb, J. A.,** *Cancer Res.*, 36, 511, 1976.
29. **Drewinko, B., Green, C., and Loo, T. L.,** *Cancer Treat. Rep.*, 61, 1513, 1977.
30. **Wolberg, W. H.,** *Arch. Surg.*, 102, 344, 1971.
31. **Tokes, Z. A., Rogers, K. E., Daniels, A. M., and Daniels, J. R.,** *Proc. Am. Assoc. Cancer Res.*, 24, 255, 1983.
32. **Margel, S., Zisblatt, S., and Rembaum, A.,** *Immunol. Methods*, 28, 341, 1979.
33. **Morrison, M.,** *Methods Enzymol.*, 70, 214, 1980.
34. **Bradfield, J. W. B.,** *Microspheres and Drug Therapy*, Davies, S. S., Illum, L., McVie, J. G,. and Tomlinson, E., Eds., Elsevier, Amsterdam, 1984, 25.
35. **Tomlinson, E.,** *Int. J. Pharm. Tech. Prod. Mfr.*, 4, 49, 1983.
36. **Poste, G., Bucana, C., Raz, A., Bugelski, P., Kirsh, R., and Fidler, I. J.,** *Cancer Res.*, 42, 1412, 1982.
37. **Illum, L., Davis, S. S., Wilson, C. G., Thomas, N. W., Frier, M., and Hardy, J. G.,** *Int. J. Pharm.*, 12, 135, 1982.
38. **Neri, A., Welch, D. R., Kawaguchi, T., and Nicolson, G. L.,** *J. Natl. Cancer Inst.*, 68, 507, 1982.
39. **Segaloff, A.,** *Recent Prog. Horm. Res.*, 22, 351, 1966.
40. **Welch, D. R., Neri, A., and Nicolson, G. L.,** *Invas. Metast.*, 3, 65, 1983.
41. **Rogers, K. E. and Tokes, Z. A.,** *Biochem. Pharmacol.*, 32, 605, 1984.
42. **Yesair, D. W., Schwartzbach, E., Shuck, D., Denine, E. P., and Asbell, M. A.,** *Cancer Res.*, 32, 1177, 1972.
43. **Rusconi, A., DiFronzo, G., and DiMarco, A.,** *Cancer Chemother.*, Rep. 52, 331, 1968.
44. **Ross, K. L. and Tokes, Z. A.,** *Proc. Am. Assoc. Cancer Res.*, 27, 273, 1986.
45. **Comoe, L., Jeannesson, P., Jardillier, J. C., Ross, K. L., and Tokes, Z. A.,** *Proc. Am. Assoc. Cancer Res.*, 28, 263, 1987.
46. **Jeannesson, P., Ginot, L., Manfait, M., and Jardillier, J. C.,** *Anticancer Res.*, 4, 47, 1984.

Chapter 9

MICROSPHERES AND PHAGOCYTOSIS

Emma Fernandez-Repollet and Abraham Schwartz

TABLE OF CONTENTS

I. INTRODUCTION

Phagocytosis is the process by which single cells ingest and digest particulate material. Phylogenetically, phagocytosis is the oldest and one of the most efficient defense mechanisms of the organism against invading or foreign matter. The phagocytic process or "cellular eating" was first described by the Russian biologist Elie Metchnikoff in 1882,[1] who described his conception of the process as follows:

> "Thus it was in Messina that the great event of my scientific life took place. A zoologist till then, I suddenly became a pathologist. I entered into a new road in which my later activity was to be exerted.
>
> One day when the whole family had gone to a circus to see some extraordinary performing apes, I remained alone with my microscope, observing the life in the mobile cells of a transparent star-fish larva, when a new thought suddenly flashed across my brain. It struck me that similar cells might serve in the defense of the organism against intruders. Feeling that there was in this something of surpassing interest, I felt so excited that I began striding up and down the room and even went to the seashore to collect my thoughts.
>
> I said to myself, if my supposition was true, a splinter introduced into the body of a star-fish larva, devoid of blood-vessels or of a nervous system, should soon be surrounded by mobile cells as is to be observed in a man who runs a splinter into his finger. This was no sooner said than done.
>
> There was a small garden to our dwelling . . . I fetched from it a few rose thorns and introduced them at once under the skin of some beautiful star-fish larvae as transparent as water.
>
> I was too excited to sleep that night in the expectation of the result of my experiment, and very early the next morning I ascertained that it had fully succeeded. That experiment formed the basis of the phagocyte theory, to the development of which I devoted the next twenty-five years of my life."[2]

Metchnikoff's phagocytic theory generated a controversy regarding the role that cellular vs. humoral factors play in the host protection.[3]

Eventually, this controversy was resolved through the realization that collaboration between phagocytes and humoral factors is essential for the overall phagocytic process. Phagocytosis is a complex phenomenon, and consequently, it is not surprising that a century after Metchnikoff's phagocytic theory, a full understanding of the mechanism of phagocytosis is still lacking.

This chapter will briefly review the phagocytic process and discuss the advantages, applications, and limitations of the utilization of microspheres in the study of phagocytosis.

II. THE PHAGOCYTIC PROCESS

A. The "Professional Phagocytes"

Although several types of cells can be induced under certain circumstances to engulf particles, only neutrophils and mononuclear cells are considered to be phagocytes (Table 1). Neutrophils constitute the principal cell population in the early inflammatory response to infection, while mononuclear cells predominate in the late inflammatory exudates. Neutrophils and mononuclear cells are considered the first and second line of defense, respectively, against invading particles.

1. The Polymorphonuclear Phagocytic System

The modern term for polymorphonuclear leukocyte (PMN) with granules is granulocyte with the adjective neutrophil, eosinophil, or basophil added according to the staining properties of their respective cytoplasmic granules. Neutrophils are the predominant granulocyte cell type and represent two-thirds of the total white cell population. Their half-life in the blood circulation averages only a few hours (~6.6 hr).[4] The neutrophils measure 10 to 15 μm in diameter. The nucleus is homogeneous and the cytoplasm contains numerous (50 to 200) granules[5] (Figure 1). Histochemically a variety of enzymes have been identified in the granules including peroxidase, dehydrogenase, acid, and alkaline phosphatase.[5,6] The presence of glycogen and lipid deposits has also been demonstrated in these cells.[7]

Table 1
THE PROFESSIONAL PHAGOCYTES

System	Cell types	Location
Polymorphonuclear	Neutrophils	Blood tissues
Mononuclear	Promonocytes	Bone marrow
	Monocytes	Blood
	Macrophages	
	Histiocytes	Connective tissue
	Free and fixed macrophages	Lymphoid tissue
	Pleural and peritoneal macrophages	Serous cavities
	Alveolar macrophages	Lungs
	Kupffer cells	Liver
	Osteoclasts	Bone tissue
	Microglial cells	Nervous system

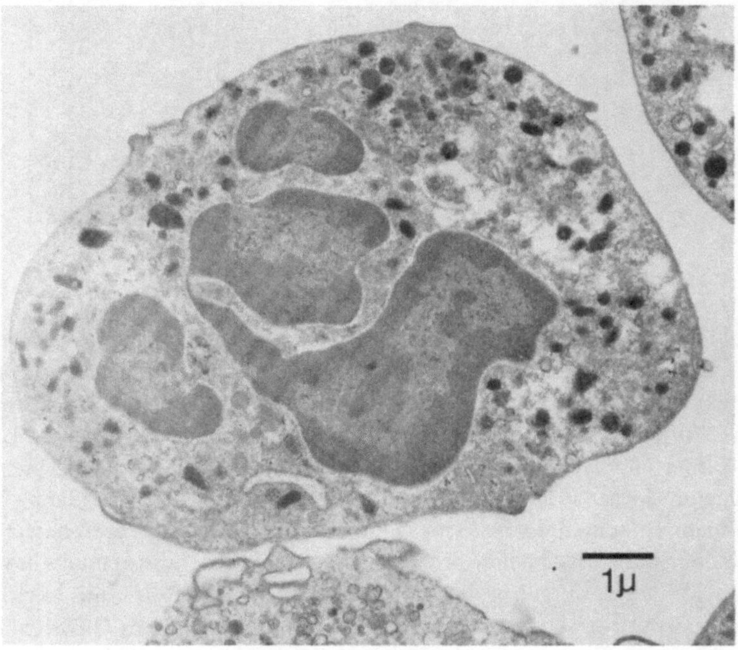

A

FIGURE 1. (A) Transmission electron micrograph of a polymorphonuclear leukocyte (PMN) stained with uranyl acetate and lead citrate (courtesy of W. Ambrose); (B) scanning electron micrograph of a PMN.

2. The Mononuclear Phagocytic System

Although the term reticuloendothelial system (RES) has been used for years in reference to the phagocytic cells of the body, today it is generally agreed that the term is inappropriate, lacking precision, and should be replaced by mononuclear phagocyte system.

The most active mononuclear phagocytes are the promocytes in the bone marrow which

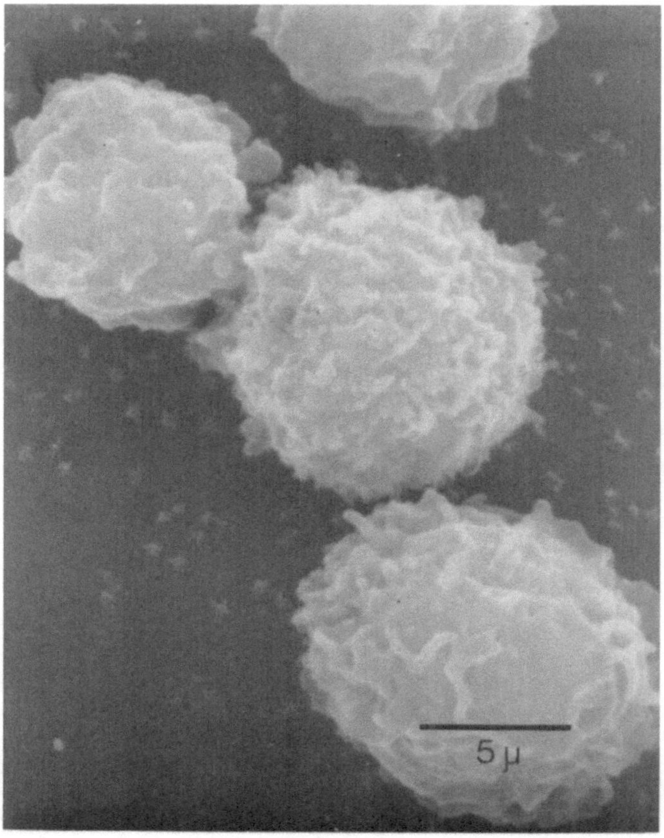

FIGURE 1B.

enter the blood as monocytes. They eventually become macrophages in tissues such as the liver (Kupffer cells), the lungs (alveolar macrophages), the lymph nodes (free and fixed macrophages), etc. (Table 1).[8] In man, there are normally 2×10^9 monocytes in the circulation. Morphological studies have revealed that a monocyte measures 12 to 15 μm in diameter.[8,9] It has a grayish-blue cytoplasm and a kidney-shaped nucleus. In contrast, tissue macrophages are larger (15 to 18 μm) and also differ from the circulating monocytes in having remnants of engulfed material (Figure 2). Macrophages are characterized by the presence of large numbers of heterogeneous dense granules. These granules have been shown to contain acid hydrolase and probably represent secondary lysosomes emerging from the fusion of digestive or phagocytic vacuoles with primary lysosomes.[10] During phagocytosis, numerous digestive vacuoles and phagosomes develop in the cytoplasm. The ingested material acts as substrate for the lysosomal enzymes contained in the granules. Macrophages from different tissues have different proportions of hydrolytic enzymes.

B. The Phases of Phagocytosis

The phagocytic process can be divided into three general phases: the attachment of foreign particles to the surface of the phagocytic cell, the ingestion of such particles by the phagocytes, and the digestion of the ingested material within the cell (Figure 3).[11]

1. Attachment Phase

The first phase of phagocytosis is defined as the contact and attachment of the particle with the surface of the phagocyte. Prior to this event, however, the circulating phagocytes

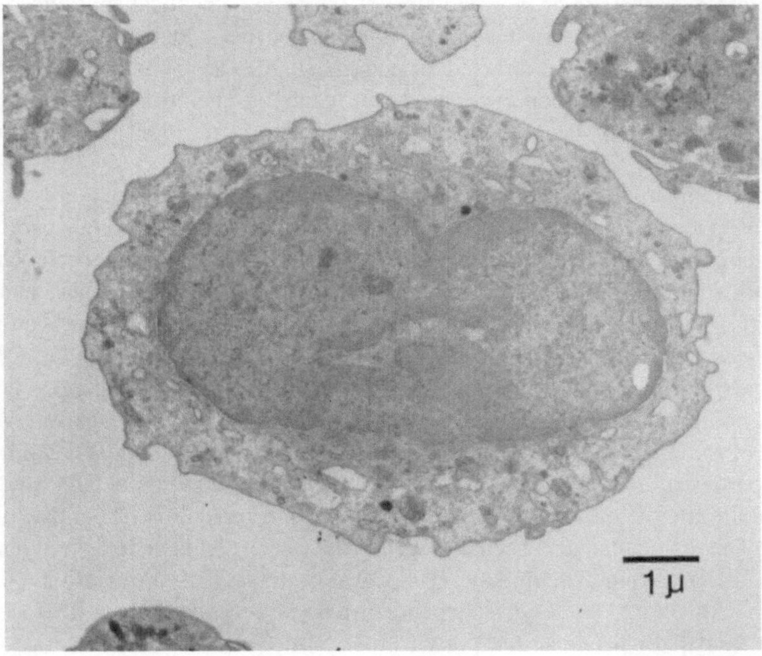

FIGURE 2. Transmission electron micrograph of a macrophage (Mϕ) stained with uranyl acetate and lead citrate (courtesy of W. Ambrose).

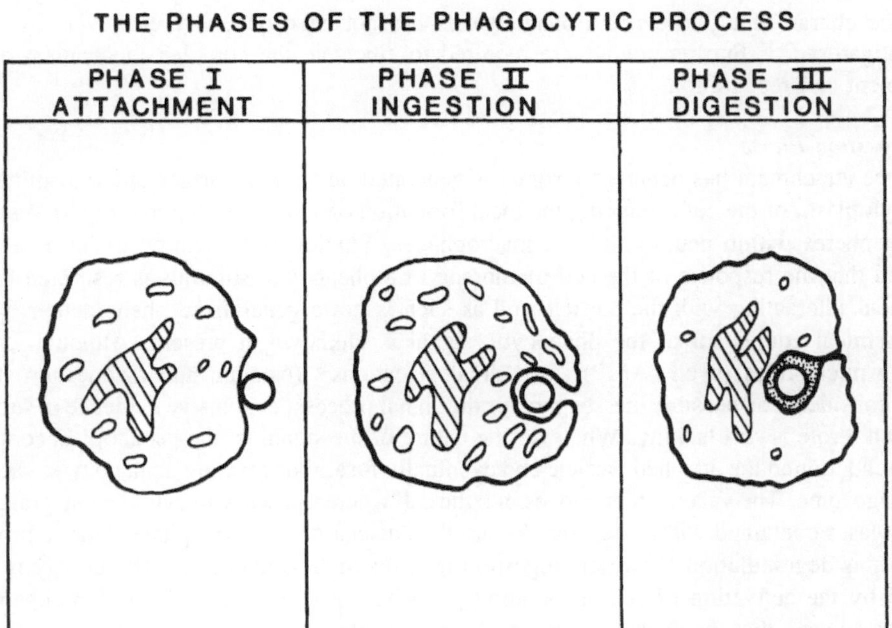

FIGURE 3. Illustration of the phagocytic process involving attachment (phase I), ingestion (phase II), and digestion (phase III).

move, or migrate, to the site of injury in response to certain chemical substances (primarily complement factors: C_3a, C_5a) that attract the phagocytes and cause them to accumulate.[12-14] This unidirectional and directed migration is known as chemotaxis. Neutrophils are much more responsive than monocytes to the influence of chemotaxic factors.[11] Although chemotactic migration has been extensively studied, the exact mechanism of the process remains unknown.

The factors regulating recognition and binding during this first phase are also unclear. Experimental evidence indicates that the attachment or binding of foreign particles to the phagocytes is greatly enhanced by the coating or opsonization of the particle surface with serum factors such as immunoglobulins IgG and IgG_3, specific antibodies, and the third component (C_3) of the complement system.[15,16] Based on these data, it has been postulated that phagocytes have surface receptors for the specific antibodies and/or the C_3 fragment which play a significant role in immunologically mediated phagocytosis. Opsonization by specific antibodies has been suggested to mediate the specificity of the self vs. nonself recognition of the phagocytes.[17] On the other hand, it has also been postulated that certain surface properties of the invading particles, such as charge, hydrophobicity, and chemical composition are the factors determining whether or not a particle is recognized and phagocytized.[17-22] These so-called nonspecific receptors have been distinguished both functionally[23] and metabolically[24] from the immune (Fc, C_3) receptors. This observation is especially important in explaining the serum-independent recognition and ingestion of a variety of synthetic or inert particles since the presence of specific receptors for this type of particle on the phagocyte cell surface is extremely unlikely. Further support in favor of surface physiochemical properties as determinants of phagocytosis emerges from studies in which complement (C_3b) increased the hydrophobicity of antibody-sensitized bacteria, and enhanced its phagocytosis.[17,20] A similar causal association between hydrophobicity or net surface charge of a particle and phagocytosis has not been observed, however, by other investigators.[25-26] Further studies are required to elucidate the complex mechanism of attachment and recognition.

2. Ingestion Phase

Once attachment has occurred a signal is generated on the cell surface and transmitted to the cytoplasm of the cell, inducing the local formation of small pseudopods. Figure 4 shows microspheres within neutrophils and macrophages. Studies by Griffin et al. have demonstrated that the response of the cell membrane to a phagocytic stimulus is restricted to the segment interacting with the particle and as such is not a generalized phenomenon.[27] The biochemical equivalent of the phagocytic signal is unknown at present. Although cyclic purine nucleotides (cyclic-AMP)[28] and divalent cations[29] (calcium and magnesium) have been considered as possible messengers for the signal process, conclusive evidence in support of such a role is still lacking. Whatever the nature of the signal is, the pseudopods continue to extend around the attached particle and eventually fuse, thus forming a phagocytic vacuole or phagosome. The vacuole then moves into the cell where it fuses with cytoplasmic granules. Hydrolases contained within the granules are then discharged into the phagosome, a process known as degranulation.[30] Particle ingestion requires metabolic energy. The energy is provided by the activation of ATP-generating processes, primarily glycolysis.[31] Biochemical studies suggest that the energy is utilized by contractile proteins (actin, myosin, and actin-binding protein) to provide the motile activity necessary for this phagocytic stage.[32]

3. Digestion Phase

The digestive stage of phagocytosis includes the killing and degradation of bacteria as the microorganisms are engulfed within the phagocytic vacuole. After the phagosome forms there is an abrupt fall in pH inside the vacuole.[33] This acidification can destroy or inactivate

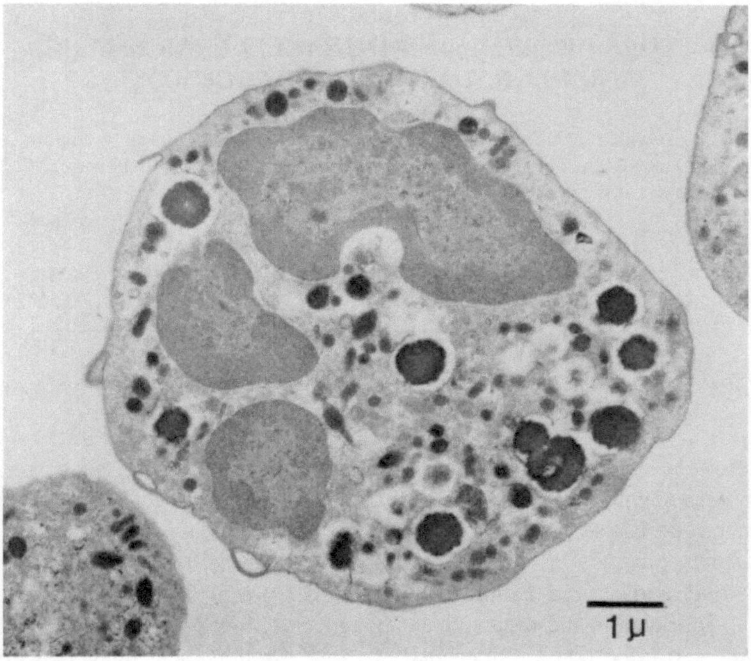

A

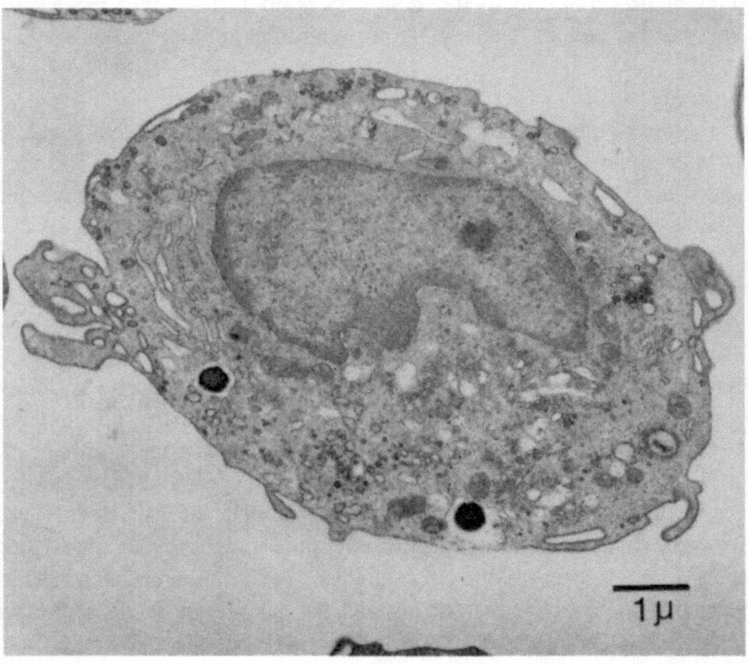

B

FIGURE 4. Transmission electron micrographs of rat peritoneal, (A) PMN and (B) macrophage which have ingested polyacrylic microbeads (courtesy of W. Ambrose).

Table 2
FACTORS DETERMINING INTRACELLULAR KILLING AND DEGRADATION BY PHAGOCYTES

Acid pH within the phagocytic vacuole or phagosome	bactericidal by itself
Lysozyme	glucosaminidase which can attack cell walls of certain bacteria, especially if they are opsonized
Phagocytin	cationic protein found in neutrophil granules with bactericidal activity against many Gram-positive and Gram-negative bacteria
Lactoferrin	an iron-binding protein present in neutrophils that inhibits growth of microorganisms
Hydrogen peroxidase	potent bactericidal agent whose activity is potentiated by myeloperoxidase with halide ions
Lysosomal hydrolytic enzymes	acid phosphatases, acid nucleases, proteases, lipases, polysaccharidase, etc.

certain bacteria and microorganisms.[11] In addition, the acidity of the phagosome activates many lysosomal enzymes capable of hydrolyzing proteins, carbohydrates, and lipids. The granules of neutrophils also contain large amounts of bactericidal or antimicrobial substances such as lysozyme, phagocytin, and lactoferrin.[5] Hydrogen peroxide, especially in the presence of myeloperoxidase and halide ions, is another potent bactericidal agent present in neutrophils.[34] Similarly in the mononuclear phagocytes, low pH, lysozyme, and hydrogen peroxidase are also important antimicrobial factors.[35] Mononuclear phagocytes do not contain, however, phagocytin nor lactoferrin. Despite effective killing, degradation of bacteria or microorganisms is only partial in macrophages as evidenced by the long-term persistence of ingested material in secondary lysosomes of these cells.[36] This morphological feature is not observed in neutrophils due to the relatively short life span of these cells. Table 2 summarizes the various factors determining antimicrobial activity of phagocytes.

C. Types of Phagocytosis

Three general types of phagocytosis has been described: spontaneous, normal, and immune phagocytosis.[35] Serum proteins are not required for the spontaneous phagocytic process while nonspecific and specific serum proteins are necessary for normal and immune phagocytosis, respectively. These serum proteins markedly increase the rate of phagocytosis by interacting with specific receptors on the phagocyte surface. Quantitative rather than qualitative differences seem to distinguish specific from nonspecific opsonization.

1. Spontaneous Phagocytosis

Serum-independent ingestion of a variety of inert particles has been demonstrated in experimental studies.[18,37,38] Relatively few phagocytes participate in spontaneous phagocytosis and few particles are ingested. The rate of phagocytosis can be enhanced, however, by increasing (1) the number of collisions between phagocytes and particles,[18] (2) the size and density of phagocytes,[18] and (3) the absolute number of particles or phagocytes.[40] Spontaneous phagocytosis is also facilitated by trapping the particle against a suitable rough surface (alveolar walls, surface of other phagocytes, fibrin, collagen fibers, etc.) a phenomenon known as surface phagocytosis.[41] This process has proven to be of great importance in controlling acute bacterial infections before specific antibodies have had time to form.[36]

2. Normal Phagocytosis

Extensive experimental evidence indicates that addition of normal serum enhances phagocytosis.[15-16] This effect is mediated by the so-called normal serum opsonins or natural antibody.[42] Such nonspecific opsonizing activity appears to reside in the alpha-2 globulin fraction, but most natural antibody is IgM.[43] Binding between natural antibody and foreign particles is of low affinity, suggesting a relatively poor fit. Nevertheless, natural antibodies (IgM), acting in conjunction with complement (C_3, C_5) as co-opsonin, are of great relevance in the handling of pathogens during the early, prespecific-antibody stages of microbial or bacterial invasion. Moreover, normal serum opsonins may initially permit phagocytes to recognize foreign particles, which can be opsonized.

3. Immune Phagocytosis

The phagocytic activity of phagocytes is dramatically enhanced by specific-immune antibody and further accelerated by complement.[11,15,16,43] Specific opsonization implies binding of the particle surface to a specific antibody, induced by previous exposure to the antigen and directed against antigenic determinants on the particle surface. Receptors for both antibodies and complement has been identified on the surface of phagocytic cells.[15,16,27] Antibody attachment is mediated by the Fc portion of the IgG molecule.[38] As a result of antigen binding, IgG antibody develops increased binding affinity for specific sites (receptors) on the surface of phagocytes. These antigen-antibody complexes at the particle surface are also capable of binding complement which, once bound, enhances phagocytosis.[11,43] Specific antibodies can facilitate phagocytosis not only by opsonizing but also by agglutinating particles, and in this way immobilizing them. Agglutination is very important when particles are too small to be phagocytized.[18,20,44] Another manner in which opsonization can immobilize a particle is through immune adherence. This entails the attachment of particles, carrying the C_3b fragment, to red blood cells, platelets, and macrophages, making it easier for phagocytic cells to engulf them.[45] Immune adherence is probably a major pathway leading to phagocytosis of foreign particles after coating with opsonizing antibody and complement.

It is evident from the previous discussion that phagocytosis is a complex process, depending upon three factors: the surface properties of the particle, the surface properties of the phagocyte, and the composition of the medium in which phagocytosis occurs. Although each of these variables exerts a profound effect on the overall phagocytic process, in the remainder of this chapter we discuss in detail only the influence exerted by modifications in the physiochemical surface properties of the particle when synthetic microspheres are used as the phagocytic stimulus.

III. PHYSICOCHEMICAL ASPECTS OF PHAGOCYTOSIS

A. Surface Tension

The major factor affecting the engulfment of any particle by phagocytic cells is the overall free energy of the process. Van Oss[46] describes this factor in terms of engulfment. In general, the overall free-energy change, ΔF_{net}, for the process of engulfment per unit area, independent of the particle shape, may be represented by

$$\Delta F = g_{PS} - g_{SW} \tag{1}$$

where g_{sw} is the interfacial tension of the particle and the suspending medium, and g_{ps} is the interfacial tension between the phagocyte and the particle. Total engulfment of the particle can occur only when $\Delta F < 0$.

The most convenient method for determining the interfacial tension is through the use of Young's equation where a liquid in contact with a solid surface under equilibrium conditions forms a unique angle, θ_e, with the solid such that;

$$g_{SV} - g_{SL} = g_{LV} \cos \theta_e \tag{2}$$

where g_{SV} is the solid/vapor interfacial tension, g_{sl} is the solid/liquid interfacial tension, g_{lv} is the liquid/vapor interfacial tension, and θ_e is the equilibrium contact angle (Figure 5).

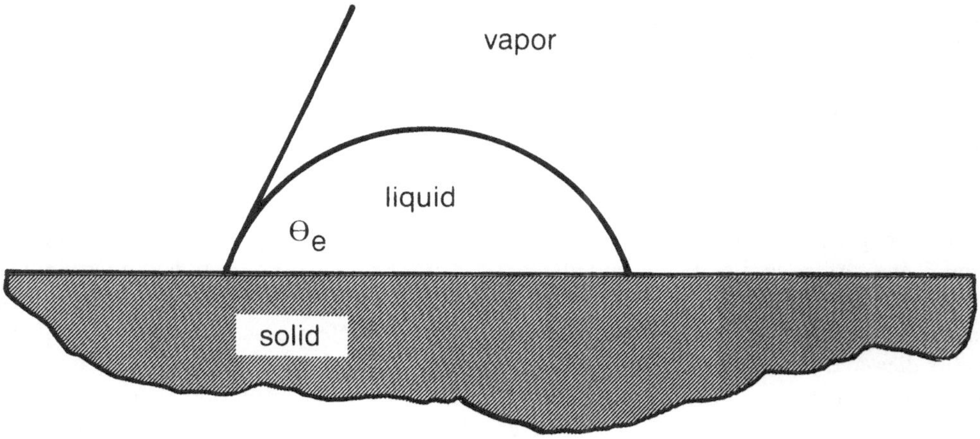

FIGURE 5. Illustration of the relationship between the contact angle and the interfacial energies of the particular components of liquid, solid, and air.

The contact angle measurement can be made under static or dynamic conditions. The static measurement is made by placing a drop of the liquid medium on the solid surface and sighting a telescope mounted on a goniometer along the surface plane of the solid and measuring the contact angle of the drop.[47] Alternatively, the tangent contact angle may be measured in a dynamic fashion by immersing and withdrawing the sample into and out of a liquid medium at a constant rate while measuring the forces exerted on the sample by the medium. The measurements can be conveniently done with an Instron. This method of contact angle measurement yields additional information in that advancing and receding contact angles are determined. These two contact angles usually are not equal. The variance between them is generally stable and commonly referred to as a contact-angle hysteresis. The differences in these contact angles are caused by the different forces exerted under dynamic conditions of advancing or receding the meniscus of the liquid over the solid surface due to heterogeneties in the surface. The low-energy components of the heterogeneous surface determining the advancing angle, and the higher energy components determine the receding angle.[48]

Much of the work on various phagocytic cells and bacteria with respect to contact-angle measurements have been done in the static mode.[49] Such determinations for the phagocytic cells may be done by depositing or growing confluent flat monolayers of the cells on glass coverslips. The contact-angle measurements on such monolayers have been found most reproducible when the monolayers have been allowed to air dry to the point where the surface glossiness disappears and a matted texture replaces it prior to the measurement. Van Oss has measured the contact angles of various mammalian species, as listed in Table 3.[50]

The contact angles of the microspheres can be measured in a similar manner as those of the PMNs by coating glass coverslips with layers of the microspheres. Styrene latex microspheres, 0.8 μm in diameter, have been found to have a contact angle of 66°.[51] The contact angles of other hydrophobic polymeric materials have similar high contact angles as styrene, e.g., polymethyl methacrylate (θ = 62°). Such contact angles can be related to critical surface tensions, e.g., polystyrene = 35 dyn/cm.[52] Polymers which are thought to be hydrophilic have lower contact angles, e.g., polyhydroxyethyl methacrylate (θ = 18°). Coleman et al.[53] have found that copolymerization of polar and apolar monomers such as methyl methacrylate and hydroxyethyl methacrylate can result in intermediate contact angles, e.g., for 50/50 MMA/HEMA, θ = 30°.

As a simplification, the relationship between the overall free energy of engulfment, ΔF_{net}, and the interfacial energies may be reduced to considering contact angles of the phagocytic

Table 3
CONTACT ANGLES OF PHAGOCYTIC CELLS

Cell type	Species	Contact angle
Polymorpholeukocytes	Human	18°
Macrophages	Guinea pig	21°
Neutrophils	Monkey	18°
	Rabbit	17.5°
	Chicken	17°
	Mouse	15°
	Guinea pig	14°
	Rat	13°

Table 4
ELECTROKINETIC POTENTIALS OF HUMAN BLOOD CELLS

Cell type	Potential (mV)
Platelets	11
Polymorpholeukocytes	12
Lymphocytes	16
Erythrocytes	18

cell and the particle. If θ particle >θ phagocyte, then engulfment is favored. If the two contact angles are equal, then only simple attachment may be expected with a rare case of engulfment taking place. If θ particle <θ phagocyte, then engulfment is no longer favored.[54]

B. Particle Size and Energetics

One of the major energetic factors to consider about phagocytosis with respect to small particles is their translational kinetic energy. This Brownian, or thermal energy is the same for all single particles, no matter what their size or shape. This translational energy may be represented by 3/2 kT where k is the Boltzmann's constant = 1.38×10^{-16} erg/degree, and T is the absolute temperature expressed in degrees Kelvin. Therefore, at the physiological temperature of 37°C, all particles have a translational energy of 6.42×10^{-14} erg. The important aspect of translational energy of a particle with respect to phagocytosis is its energy per unit area, i.e., the larger the particle, the lower its energy per unit area. This suggests that there is a critical size below which the energy per unit area is too high to be overcome by interfacial energy, and thus the particle will resist attachment and engulfment by phagocytic cells. This resistance to engulfment holds true of particles smaller than the critical size even when they have high interfacial energies.

C. Collision Factors

The ability of phagocytic cells to attach and engulf particles is directly related to the probability that they will collide with the particles. This probability is determined by factors such as concentration, size, and density of both the phagocytic cells and the particles. Increasing these factors tends to increase the phagocytic activity. For instance, Deierkauf has shown that the phagocytosis of 0.48 μm polystyrene latex particles by rabbit PMNs follows a linear relationship as a function of the latex concentration.[55] It has also been shown that if the number of phagocytic cells is held constant, a great percentage of them engage in the phagocytic activity when the number of particles is increased. Clumping of either the cells or particles also tends to increase the rate and degree of the phagocytic process by increasing the opportunity for collisions.

D. Surface Charge and Shape

The surface charge of phagocytic cells with respect to particles can strongly influence their potential for collision and interaction. If both the phagocyte and particle are of the same charge, they will tend to repel each other, thus limiting interaction. This is even true of the peripheral blood cells interacting with each other because they have similar net charges, as seen in Table 4.

For example, the electrokinetic potential of all human blood cells is >10 mV, i.e., platelets, PMNs, lymphocytes, and erythrocytes are 11, 12, 16, and 18 mV, respectively.[56]

These potentials keep the spherical or disk-shaped cells from approaching each other any closer than 50 to 80 Å which is too far for cellular interactions.[57] However, this repulsion can be overcome by protrusions from the cells, e.g., microvilli or pseudopods, provided that they have a curvature of radius <500 Å. Platelets are a good example of cells that can be activated to extend long pseudopods which result in formation of aggregates of cells.

It follows that if the particle is also carrying the same charge as the phagocytic cells, then the chance of attachment will be greatly reduced, whereas, particles of the opposite charge will have a high tendency toward attachment and engulfment. For example, polyanions with a negative charge such as albumin and polyglutamic acid were shown to have a strong inhibitory effect on phagocytosis of latex by rabbit PMNs, whereas polylysine was found to stimulate phagocytosis.[55,58] Such charges in either case would lower the interfacial tension and contact angle relative to electrically neutral particles.

IV. MICROSPHERES AS PHAGOCYTIC STIMULUS

A. General Aspects

Research on phagocytosis has attracted many investigators since first described almost a century ago. A variety of substances and particles have been used as the phagocytic stimulus in the evaluation of this phenomenon throughout these years. In early studies, the uptake of vital dyes was utilized to identify phagocytic cells.[59] Experimental evidence demonstrated later that these substances were also pinocytosed by other type of cells (i.e., fibroblasts) and therefore did not constitute a reliable marker.[60] Colloidal suspensions (i.e., carbon, iron oxide) were then utilized to assess phagocytic activity.[61,62] The use of these particles represented an improvement in terms of cell specificity, visualization, and quantitation of the phagocytic process.[63] Despite these advantages, colloids are not too stable in suspension,[64] lack size uniformity,[65] and cause the aggregation of phagocytes.[18,36,40] All of these factors are known to influence phagocytosis. More recently, synthetic particles of different physicochemical properties (i.e., latex, silica, acrylic) have been used extensively in studying both qualitatively and quantitatively the phagocytic process.[18,66,67] Commercially available microspheres can be prepared highly uniform in diameter and surface area. Moreover, chemical modifications of the surfaces of the microspheres (addition of hydroxy, carboxyl, amide, or other function groups) enable binding of protein, dyes, fluorescent molecules, and pharmacological agents.[64,65,68] These modifications have been of great importance in evaluating the influence of physicochemical surface properties on phagocytosis, identifying, and isolating subpopulations of cells based on their phagocytic activity, and developing more sophisticated and accurate methods to quantitate phagocytic activity in individual cells or in a particular cell population. Thus, microspheres have emerged as an important tool in the study of phagocytosis and the factors influencing the process.

B. Types and Properties of Microspheres

The synthesis of microspheres can be accomplished with a variety of monomers and initiation schemes. Many of these are covered in other chapters of this book. In addition, the surface of the microspheres may be modified after their synthesis to modify their chemical and physical surface properties and interactions with respect to cellular interaction.

1. Polyaromatic Microspheres

Of the aromatic microspheres, polystyrene latex is the most widely known and used (Figure 6). Polystyrene microspheres have been commercially available for many years and have served, in many cases, as sizing standards for electron microscopes, particle counters, flow cytometers, and similar instruments. Polystyrene microspheres have been useful as standards with these instruments because they may be made extremely uniform. The term

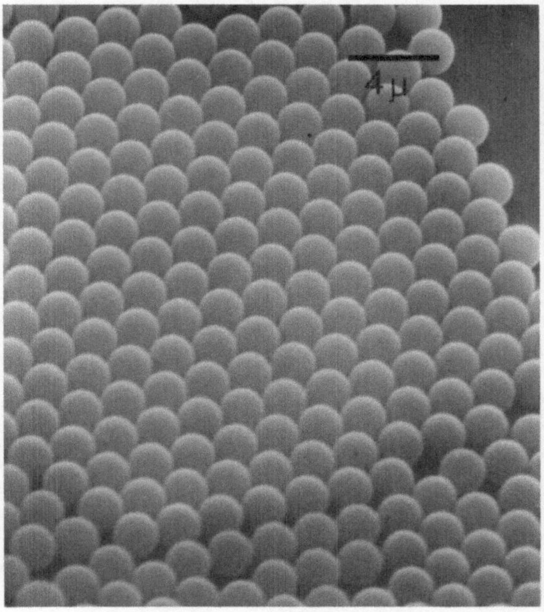

FIGURE 6. Scanning electron micrograph of commercial polyvinyl toluene microspheres (Dow Diagnostics).

"monodispersed" is used in regard to these particles and refers to variation in diameter of less than 1%.[69] It is the uniformity of size and chemical composition of these microspheres which is attractive in the design of phagocytic studies.

The choice of polystyrene microspheres for phagocytosis research was most likely dictated by considerations of convenience and availability, rather than physiological design. The inert behavior of these particles with respect to their suspension in physiological media made them ideal for the early phagocytic studies. In addition, these particles have the advantage over naturally occurring particles of being easily visualized by light or electron microscopy due to their uniformity, or by immunofluorescent or flow cytometry techniques, especially when the polystyrene microspheres contain fluorescent dyes. Examination of the physical and chemical properties of polystyrene suggests, that as a bulk material, it would invoke a cellular response from its very hydrophobic nature and the lack of an ionic charge. Similar results would be expected from microspheres produced from similar aromatic monomers, i.e., polyvinyl toluene which is also hydrophobic and of neutral charge. However, analysis of residual charge on polystyrene microspheres has been reported, and surfactant, as well as the initiator associated with the end groups was found present even after extensive clean-up procedures involving exhaustive dialysis and treatments with ion-exchange resins.[68]

Another advantage of using polyaromatic microspheres in phagocytic studies is that the surface properties can be modified by direct chemical reactions. For instance, one method of chemically modifying the benzene ring of polystyrene or polyvinyl toluene particles which has been used commercially involves introduction of nitro groups on the ring with fuming nitric acid. These can be further modified to amines and diazonium ion groups to covalently link various proteins directly to the microspheres.[70] In addition, styrene can be copolymerized with other monomers which contain specific functional groups, such as methacrylic acid. These carboxyl groups can, in turn, be activated by reagents such as water-soluble carbodiimides to covalently link proteins to the surface of the particle for opsonization.[71]

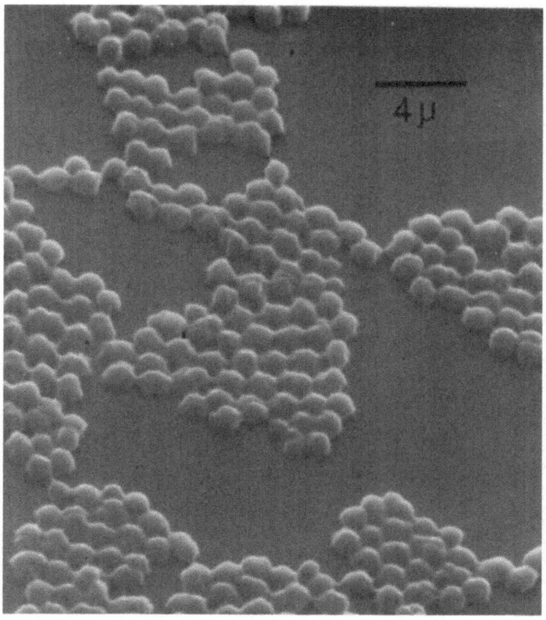

FIGURE 7. Scanning electron micrograph of polyhydroxyethyl methacrylate/methacrylic acid microspheres.

2. Polyacrylate Microspheres

Microspheres have also been synthesized from acrylic monomers (Figure 7). The most extensively studied acrylic microspheres are those based on 2-hydroxy ethyl methacrylate.[10,36,40,64] This methacrylate monomer can be copolymerized into microspheres with a variety of neutral and ionically charged monomers resulting in microbeads containing almost any desired chemical functionality. This functionality can be conjugated to various proteins as with the case of the aromatic copolymer microspheres.[72] Although these microspheres can be formed by chemical initiation, initiation by gamma radiation is preferred since it does not leave any residue from the initiator. Polyacrylate microspheres do have some of the same limitations as the aromatic microspheres in that their maximum size is about 1 μm while still being singlet microspheres in suspension. Attempts to grow larger particles results in doublets and multiplets fused together. The uniformity of these acrylate microspheres is also less than the aromatic microspheres with coefficients of variations (CV) in diameter being 10 to 20%, as opposed to the CV of 1 to 3% for the hydrophobic aromatic microspheres. In addition, they do not have the shape stability of the aromatic microspheres, even when excessively cross-linked. This is because they are hydrophilic and allow free diffusion of water throughout their structure. Such hydrophilic polyacrylic microspheres are not stabilized by the interfacial forces which are present in the case of the aromatic microspheres. This may be seen by scanning electron microscopy where air-dried acrylate microspheres appear more flattened. These physical properties of hydrophilicity and shape are important considerations in regard to phagocytosis and activation of phagocytes, as previously mentioned.

3. Polyvinyl Pyridine Microspheres

Microspheres with properties which lie between the aromatic and acrylate materials with respect to hydrophilicity and polarity are those formed from vinyl pyridine and comonomers.[73] Again, these microspheres can be formed by chemical or radiation initiation, but are not as limited in terms of the size (Figure 8). With the proper cosolvents, e.g., acetone or methanol, polyvinyl pyridine microspheres can be grown as large as 10 to 12 μm in one

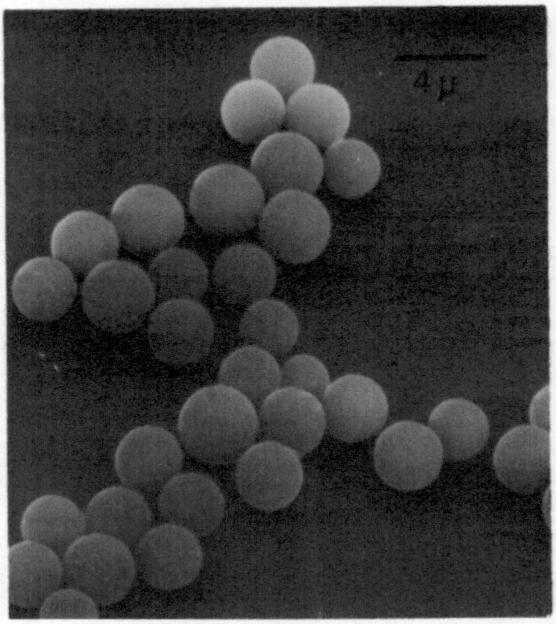

FIGURE 8. Scanning electron micrograph of polyvinyl pyridine microspheres.

step. In addition, their coefficient of variation in diameter can be reduced to a range of 5 to 10%. Their shape stability is similar to the aromatic microspheres in that they do not have to be critically point dried to retain a spherical appearance. In phagocytic studies monitored with transmission electron-microscopic investigations, these polyvinyl pyridine microspheres are advantageous in that they readily take up the common electron-opaque staining materials, e.g., lead and uranium salts. This is because the pyridine functionality readily coordinates with such metal ions. This property can be used to an advantage by prestaining the microspheres with gold or platinum salts and reducing with a borohydride prior to the introduction to the phagocytic cells.[74] Such a treatment will also increase the hydrophobicity of the microspheres and enhance the activation of the cells toward them.

V. QUANTITATION OF THE PHAGOCYTIC PROCESS

Microspheres have been used extensively to measure phagocytosis. Their uniform size, versatility in physicochemical properties, and fluorescent intensity greatly facilitate quantitation. Moreover, although polystyrene microspheres are ingested by serum-independent phagocytosis, their surface can be modified to study serum-dependent phagocytosis. Numerous methods have been developed for measurement of phagocytic uptake of microspheres including clearance studies, microscopic and radio-labeled techniques, spectrofluorometry, and more recently, flow cytometry.

A. In Vivo and In Vitro Studies

Studies evaluating the phagocytic process have been performed under both in vivo and in vitro conditions.[63,75,76] Each one of these approaches is characterized by certain limitations and advantages. In vivo methods, for example, allow the examination of the response of the whole organism to foreign particles without altering the composition of the medium in which phagocytosis occurs or the participation of the different phagocyte populations. This technique, however, is restricted to the measurement of the clearance of intravenously infused particles (carbon, latex microspheres, lipid emulsions, etc.) from the circulation of experimental animals or humans.[63,76] Reticuloendothelial phagocytic activity is expressed as the rate of disappearance of particles from the bloodstream. This method does not mimic a natural process insofar as foreign particles rarely enter the circulation in large numbers. Moreover, changes in blood flow to the splanchnic bed, Kupffer cell number and activity, particle size and composition, and serum properties such as pH and calcium concentration can alter the clearance of particles.[63] Injection of high concentrations of microspheres has also been associated with a temporary blockade of the fixed- or tissue-phagocyte system while small doses of particles measure hepatic blood flow rather than phagocytic function.[63,76,77] To these limitations one must add the impossibility of quantitating or assessing the contribution of the different phagocytic cell populations to the overall phagocytic response. Thus, in vivo clearance studies require a careful interpretation of results and an understanding of all factors that may influence this process.

Most of the studies concerning the quantitation of phagocytosis are performed under in vitro conditions. This offers the advantages of studying homogenous phagocyte populations, controlling the effects of different serum constituents on phagocyte function, and examining the different phases of the phagocytic process.[75,76] Neutrophils, monocytes, and macrophages are the phagocytic cells most widely used for in vitro studies. Neutrophils are the predominant leukocytes in whole blood and can be easily obtained by sedimenting erythrocytes with a dextran solution and centrifuging the leukocyte-rich supernatant at a moderate speed. After several washes with heparinized saline the cells are resuspended in a balanced salt solution. Although PBS (pH 7.2) has been extensively used, a balanced salt solution is more effective in preserving the structure and function of cells. A review on the composition of the media and sera employed in phagocytosis experiments has been published.[75] Neutrophils can also be obtained from acute inflammatory peritoneal exudates. High yields of neutrophils have been reported in guinea pigs and rats 1 day after the intraperitoneal injection of sodium caseinate, glycogen, or polysaccharides.[75] Resident macrophages, on the other hand, are usually harvested from the peritoneal or pleural cavity by washing the respective serous cavity with PBS. Intraperitoneal injections of the experimental animal with starch, glycogen, or mineral oil enhances the yield of macrophages in exudates harvested 4 to 6 days after injection.[78,79] Following harvesting the percentage of phagocytic cells in suspension is determined and adjusted to approximately 1 to 2 $\times$ 10^7 cells per milliliter which is the optimal cell concentration for in vitro experiments of phagocytosis. The viability of the phagocytic cell is assessed at the beginning and end of the experimental procedure by dye exclusion of trypan blue in at least 100 cells. After isolation, cells can also be cultured for in vitro studies. Thus, phagocytosis can be measured in freely suspended as well as attached (monolayer system) phagocytes. It should be mentioned that in the monolayer system the extent of cell motility plays a significant role in cell-particle contact while in shaking suspension systems this is mainly determined by collision probabilities.[63,80] In addition, in suspended cells the entire cell surface is exposed to the mcirospheres while in attached cells some membrane surface remains inaccessible to the particles and uptake may therefore be reduced.[67]

For most phagocytic tests, phagocytes are incubated in the presence of microspheres in a medium containing or lacking serum for opsonization. The mixture is continuously rotated at 37°C. At the end of the experimental period the monolayer or cell suspension is rinsed free of microspheres and the percentage of phagocytes ingesting the particles and/or the number of particles per phagocytic cell assessed.

B. Measurement of Phagocytosis

Methods of quantifying phagocytosis determine the intracellular accumulation of microspheres and/or their extracellular disappearance. Since attachment of particles to phagocytic cells does not necessarily lead to ingestion, attempts have been made to differentiate between these two aspects of microsphere uptake. Light and transmission electron microscopy,[22] utilization of large microspheres,[81] and treatment of fixed phagocyte cells with xylene, chloroform, or dioxane to dissolve extracellular polystyrene particles[60] are among the techniques that have been utilized for this purpose. It should be mentioned that even sophisticated methods such as flow cytometry do not distinguish between attachment and ingestion, and thus measure the sum of both processes.

1. In Vivo Clearances

Several types of microspheres have been used to assess in vivo phagocytosis.[63,76] Among the microspheres most commonly used in experimental animals are fluorescent and radiolabeled beads. In all instances, the microsphere suspension is injected intravenously. Particles are removed primarily by the liver and the spleen.[62] The rate of disappearance of fluorescent particles from the circulation is measured spectrofluorimetrically while that of radiolabeled microspheres is measured in a gamma counter. The logarithms of values obtained for blood radioactivity or fluorescence are plotted against time. The slope of the line describing this correlation is equivalent to the rate of phagocytosis (K) or the uncorrected phagocytic index. The corrected phagocytic index (∂) is calculated using the following formula:

$$\partial = \text{cube root of K} \times (\text{BW/LW} + \text{SW})$$

where BW, LW, and SW represent body, liver, and spleen weights, respectively.[63,76]

2. Radioactive Counting

The use of radiolabeled microspheres is one of the various methods available to quantify phagocytosis in cell suspensions or monolayer systems.[82,83] The average amount of substrate phagocytized per cell or per milligram cell protein is calculated from the specific activity of each microsphere and the cell number or protein mass.[63] The specific activity of microspheres is expressed as millicuries per gram (mCi/g) or as counts per minute per bead (cpm/bead). Radioactivity is measured in a standard gamma counter and cell protein content is determined by the method of Lowry.[84] When only relative indices of phagocytosis are required, the uptake is expressed as mean radioactive counts per minute (cpm) per microgram of phagocyte protein (cpm/μg of protein) rather than as mass ingested per cell.

Experimental evidence indicates that the binding of an isotope to microspheres is stable and that the amount of free isotope remaining after washing is negligible.[82] In addition, release of radioactivity into the medium has not been detected after microsphere ingestion.[83] Thus, in most instances measured radioactivity is directly proportional to the number of microspheres ingested. Radioactive counting does not distinguish between particle attachment and ingestion, and therefore yields phagocytic indices consistently higher than those calculated by microscopic techniques (*vide infra*).[63] Bulk radioactive measurements give average results for the population but do not provide information on the existence of subpopulations.

3. Microscopy

Phagocytosis of microspheres has been extensively quantitated by microscopic or visual methods.[63,75] After the in vivo or in vitro assay, an aliquot of the microspheres-cell suspension is centrifuged and washed several times in order to remove any free particles. Phagocytic cells are then fixed in a solution of 2% glutaraldehyde in incubation medium adjusted to pH 7.4 at room temperature for 30 min, stained with Giemsa and dried by airstream flow. The percentage of cells with ingested microspheres (percent phagocytosis) is determined from counts of at least 200 phagocytic cells. The mean number of microspheres per cell (phagocytic capacity) is measured by counting the number of microspheres per cell in approximately 100 cells that have phagocytized. Phagocytosis of fluorescent microspheres is examined by phase and fluorescence microscopy.[85,86] It is evident that this counting procedure is extremely time consuming, thus, only a small number of cells from the total population can be analyzed. In addition, with this technique it is difficult to quantitate the number of ingested microspheres when they exceed 4 to 6 beads per cell as well as to differentiate between attached and ingested particles.[63,87]

In several studies, transmission electron microscopy[22,67,88] has been used to obtain a more accurate estimate of internalized microspheres. After fixation, dehydration, and staining of the cell pellet with uranyl acetate and lead citrate, serial sections (20 μm apart) of the material are studied until at least 50 cells per pellet are examined. These data are then compared to those obtained by light microscopy in whole cells from the same original preparation.

4. Spectrophotometry

Polystyrene microspheres have been of great importance in developing methods to quantitate phagocytosis. These particles can be extracted in a solubilized form from the phagocytic cells and spectrophotometrically analyzed. In studies by Roberts and Quarted, sedimented phagocytic cells were extracted (after removal of adherent latex) with dioxane overnight at room temperature.[66] Following mixing and centrifugation, the solubilized polystyrene was quantitated in the supernatant by measuring the absorbance at 255 nm. This assay has been applied to monolayer systems[89,90] and has been used in kinetic studies.[63,91] Transmission electron microscopy has also been performed to confirm the internalization of polystyrene. This technique, however, is limited by the time-consuming extraction step, the possibility of extracting substances that will also absorb at the polystyrene wavelength, and the exclusion of other latex microspheres such as polyvinylpyrrolidone which are not dissolved by dioxane.[92]

5. Spectrofluorometry

The binding of fluorescent dyes to polystyrene microspheres has provided an alternative method to quantitate the phagocytic capacity or the number of microspheres endocytosed per phagocytic cell. As reported by Schroeder and Kinden, phagocytosis can be assessed using fluorescent carboxylated microspheres (Fluoresbrite™) and a computer-centered spectrofluorimeter.[67,92] After incubation, centrifugation, and washing, sedimented cells are resuspended in PBS (pH 7.2) and placed in a quartz cuvette for determination of absorbance-corrected fluorescence using a computer-centered spectrofluorimeter developed by Holland et al.[93] The absorbance-corrected fluorescence is then compared with a standard curve of Fluoresbrite™ microspheres in PBS alone, to quantitate the number of microspheres per cell. This method has proved to be sensitive and linear over a wide range of time, microspheres per cell ratios, and microsphere sizes.[67] Despite its sensitivity and quickness, this technique does not allow the assessment of percent phagocytosis or analysis of the heterogeneity of the phagocytic response within a cell population.

6. Flow Cytometry

In recent years phagocytosis has been measured by flow cytometric techniques using fluorescent microspheres as the phagocytic stimulus. Flow cytometry has grown from a method of counting microscopic particles to a sophisticated technology that can rapidly analyze various physical and biochemical properties of individual cells at rates of 1 to 5 $\times$ 10^3 per sec.[94-97] In commercial systems, such as the Fluorescent Activated Cell Sorter (FACS™), cells are analyzed and separated on the basis of fluorescence, size, and viability.

Figure 9 illustrates the basic principles of operation of this particular system. Briefly, suspended cells (previously stained with fluorescent or absorption dyes) are injected into a flow chamber and induced to pass, one at a time (hydrodynamic focusing), through a high-power laser beam. As individual cells pass through the beam, several variables are measured.

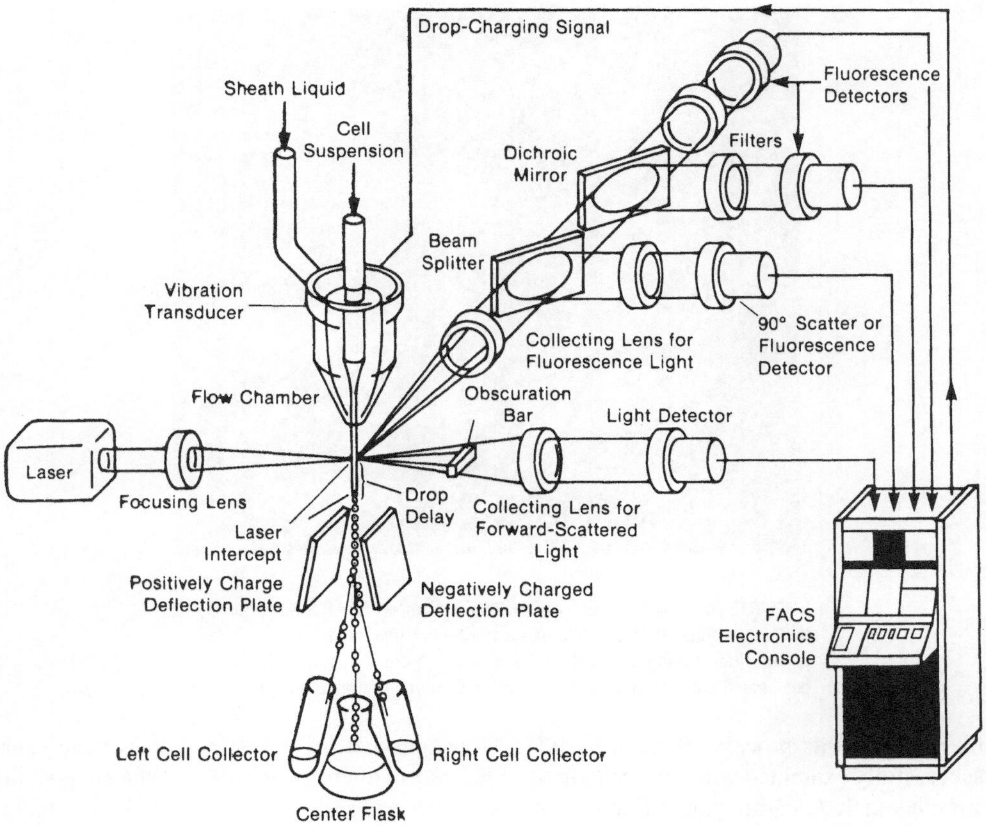

FIGURE 9. Illustration of the basic principles of operation of a FACS cell sorter.

Light scattered at low forward angles (<5° with respect to the incident laser beam) provides an index of cell size, while light scattered at larger angles gives information about internal structure or cytoplasmic granularity.

At the same time, the intensity, color, or polarization of fluorescence emitted from attached and ingested fluorescent microspheres, or from particular cellular constituents to which fluorescent dye molecules have been attached, is also measured. The scattered light and/or fluorescent optical signals are converted by special photodetectors into electrical signals which are then processed by the instrument.

In certain systems it is also possible to measure fluorescence at more than one wavelength, absorption of light by cells, cell volume, and surface area. Electronic analysis and recording of measurements may be performed with or without cell sorting.

When sorting is done, the stream carrying the cells is broken up into well-defined drops by the vibrations of an ultrasonic transducer immediately after passing the laser beam. The rate of droplet formation is approximately 40,000/sec. If a cell meets certain operator-preset criteria, a pulse of electricity is applied to the stream, charging it positive or negative. This causes the droplet containing the cell of interest to be charged and subsequently deflected as it passes through an electric field into the appropriate collector. Uncharged droplets pass undeflected into a different reservoir. The probability that two cells will appear in a single droplet is minimized by the difference between the rate at which cells emerge (5000/sec) and that of droplet formation (40,000/sec). Collection purities are reported to be greater than 95% and cell viability is usually not affected by passage through the instrument.

Application of flow cytometry to quantitate phagocytosis has added a new dimension to

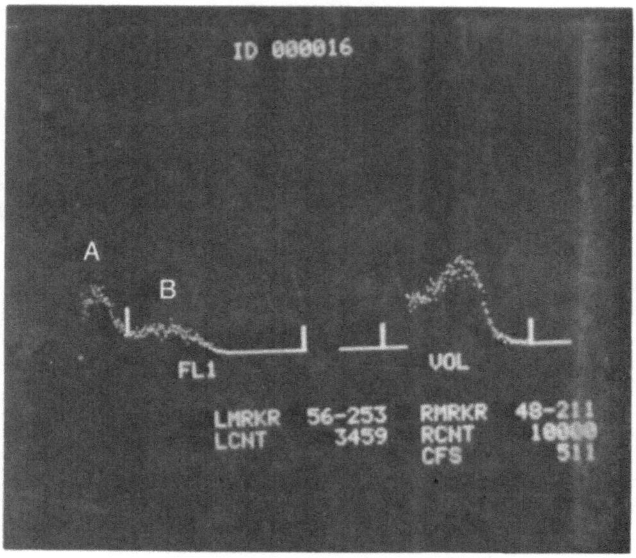

FIGURE 10. Histograms from a flow cytometer (FACS Analyzer®) of the volume and fluorescence of rat peritoneal phagocytes which have ingested fluorescent polyacrylic microspheres. Note (A) the autofluorescence peak gated out to the left of (B) the cells containing fluorescent microspheres.

the study of this process. Both in vitro[86,98-101] and in vivo[85,102-103] phagocytosis have been successfully examined with this technology. Fluorescent microspheres used as the phagocytic stimulus include carboxylated fluorescent latex microspheres, rhodamine polyvinyl toluene latex microspheres, and fluorescent methacrylate microspheres. In these studies phagocytosis of the fluorescent particles has been examined in a variety of phagocytic cells such as rat peritoneal macrophages,[102-103] rat[85] and hamster[98-99] pulmonary macrophages, rabbit pulmonary cells,[100] and human neutrophils.[86]

Analysis of the phagocytic cell suspensions by flow cytometry has provided data regarding the homogeneity of the cell population, the percentage of cells phagocytizing the fluorescent microspheres, and the average number of beads per phagocytic cell. These data have been calculated from cell size (light scatter) and fluorescence histograms. A typical example of light scatter and fluorescence histograms of rat peritoneal macrophages after phagocytosis of fluorescent methacrylate microspheres is shown in Figure 10. Flow cytometry data can also be presented as a two-dimensional dot plot in which two parameters such as light scatter and fluorescence are correlated (Figure 11). This facilitates the identification of subpopulations of cells.

Obviously, flow cytometry offers numerous advantages in quantifying the phagocytic process. First, it allows rapid determination of the percentage of cells having ingested microspheres as well as the actual number of cell-associated spheres. This provides important information concerning the variability in particle uptake among individual cells. Furthermore, with this system the number of cell-associated particles that can be accurately estimated averages 47 spheres per cell, while by visual assays only 4 to 6 particles per cell can be clearly distinguished.[98] Statistical precision is also enhanced due to the large number of cells evaluated. In addition, flow cytometry permits the evaluation of small quantities of cells and eliminates the need to remove free particles prior to analysis. Since multiple variables can be measured in a single determination, phagocytosis can be correlated with other cell properties. Moreover, continued cell viability after analysis and sorting allows further morphological and biochemical studies on a particular phagocytic cell subpopulation. Recent

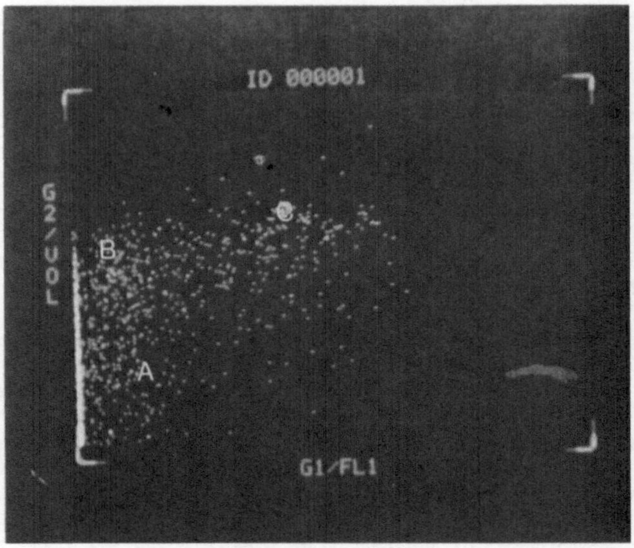

FIGURE 11. The dot plot from a flow cytometer (FACS Analyzer ™) of the volume and fluorescence of rat peritoneal phagocytes which have ingested fluorescent polyacrylic microspheres. Note (A) the population of the smaller cells (erythrocytes) which are not fluorescent, (B) the larger phagocytic cells which have not taken up fluorescent microspheres, and (C) those which have.

studies have also demonstrated applications of flow cytometric techniques in the determination of receptor molecule density on the phagocyte surface by using antireceptor antibodies labeled with a different dye (fluorescein isothiocyanate) than the latex microsphere (rhodamine).[104] In addition, flow cytometry has been used to assess monocyte-macrophage differentiation and activation[105] and to measure, simultaneously, phagocytosis and phagosomal pH by correlating the red-to-green fluorescence ratio of acridine orange-stained neutrophils and the fluorescein-isothiocyanate (FITC) fluorescence of the phagocytes.[106]

This survey of the use of flow cytometry in the quantitation of the phagocytic process clearly demonstrates the potential value of this technology in biomedical research and clinical diagnosis. Nevertheless, some limitations associated with this innovative technique should also be considered.[96] First, when flow cytometric determinations are performed in heterogeneous samples, coincident counts and peak overlap of neighboring populations may occur, affecting the interpretation of results. In addition, the assessment of cell-to-cell variability in particle uptake can be altered by the presence of other cell types as well as by cell and particle aggregates. This can be minimized by using highly purified cell suspensions and monodispersed microspheres, and by removing cell aggregates prior to analysis. Clumps of cells and smaller cell types can also be excluded from analysis by their different axial light loss signals. Although sorting speed appears to be fast it may also be a limiting factor when the cell population of interest is very small in number or when the experimental design requires further biochemical or functional studies in the sorted cell. Furthermore, as previously mentioned, flow cytometric analysis does not differentiate between attached and internalized particles and therefore measures the sum of both processes. Thus, several experimental manipulations must be performed prior to analysis to evaluate each process independently. Finally, the high cost of this instrumentation has limited its availability for research and clinical diagnosis. However, with the advent of air-cooled lasers, the substitution of lasers with mercury arc lamps and the decreased cost of computer equipment have recently made this technology simpler to utilize as well as less expensive. This will further stimulate and facilitate the applications of flow cytometric techniques to the study of phagocytosis.

REFERENCES

1. **Metchnikoff, E.,** *Untersuchungen Uber Die Intracellulare Verdauung Bei Wirbellosen Thieren,* Vienna, 1883.
2. **Metchnikoff, O.,** *Life of Elie Metchnikoff 1845—1916,* Houghton Mifflin, Boston, 1921.
3. **Wright, A. E. and Douglaus, S. R.,** Further observation on the role of the blood fluids in connection with phagocytosis, *Proc. R. Soc. London,* 73, 128, 1904.
4. **Cartwright, G. E., Athens, J. W., and Wintrobe, M. M.,** The kinetics of granulopoiesis in normal man, *Blood,* 24, 780, 1964.
5. **Bainton, D. F.,** The origin, content and fate of polymorphonuclear leukocyte granules, in *Phagocytic Mechanisms in Health and Disease,* Williams, R. C. and Fudenberg, H. H., Eds., Intercontinental Medical Book Corp., New York, 1972, 123.
6. **Baggiolini, M., Hirsch, J. G., and de Duve, C.,** Further biochemical and morphological studies of granule fractions from rabbit heterophil leukocytes, *J. Cell Biol.,* 45, 586, 1970.
7. **Scott, R. E. and Horn, R. G.,** Ultrastructural aspects of neutrophil granulocyte development in humans, *Lab. Invest.,* 23, 202, 1970.
8. **Van Furth, R.,** Origin and kinetics of monocytes and macrophages, *Semin. Hematol.,* 7, 125, 1970.
9. **Cohn, Z. A.,** The structure and function of monocytes and macrophages, *Adv. Immunol.,* 9, 163, 1968.
10. **Nichols, B. A.,** Ultrastructure and cytochemistry of mononuclear phagocytes, in *Mononuclear Phagocytes,* Vol. 2, Van Furth, R., Ed., Blackwell Scientific, Oxford, 1970.
11. **Stossel, T. P.,** Phagocytosis I-III, *New Engl. J. Med.,* 290, 717, 774, 833, 1974.
12. **Sorkin, E., Stecher, V. J., and Borel, J. F.,** Chemotaxis of leukocytes and inflammation, *Ser. Hematol.,* 3, 131, 1970.
13. **Sandberg, A. L., Snyderman, R., and Frank, M. M.,** Production of chemotactic activity by guinea pig immunoglobulins following activation of the C_3 complement shunt pathway, *J. Immunol.,* 108, 1227, 1972.
14. **Ward, P. A., Chapitis, J., and Conroy, M. C.,** Generation of bacterial proteinases of leukotactic factors from human serum, and human C_3 and C_5, *J. Immunol.,* 110, 1003, 1973.
15. **Huber, H., Polley, M. J., and Linscott, W. D.,** Human monocytes: distinct receptor sites for the third component of complement and for immunoglobulin G, *Science,* 162, 1281, 1968.
16. **Johnston, R. B., Klemperer, M. R., and Alper, C. A.,** The enhancement of bacterial phagocytosis by serum: the role of complement components and two cofactors, *J. Exp. Med.,* 129, 1275, 1969.
17. **Van Oss, C. J.,** Phagocytosis, in *Principles of Immunology,* Rose, N. R., Milgram, F., and Van Oss, C. J., Eds., Macmillan, New York, 1973, chap. 10.
18. **Fenn, W. O.,** The phagocytosis of solid particles, *J. Gen. Physiol.,* 3, 439, 1921.
19. **Mudd, S., McCutcheon, M., and Lucké, B.,** Phagocytosis, *Physiol. Rev.,* 14, 210, 1934.
20. **Van Oss, C. J.,** Phagocytosis as a surface phenomenon, *Annu. Rev. Microbiol.,* 32, 19, 1978.
21. **Beukers, H., Deierkauf, F. A., Blom, C. P., Deierkauf, M., and Riemersma, J. C.,** Effects of albumin on the phagocytosis of polystyrene spherules by rabbit polymophonuclear leukocytes, *J. Cell Physiol.,* 97, 29, 1978.
22. **Beukers, H., Deierkauf, F. A., Blom, C. P., Deierkauf, M., Scheffers, C. C., and Riemersma, J. C.,** Latex phagocytosis by polymophonuclear leukocytes: role of sialic acid groups, *Chem. Biol. Interact.,* 33, 91, 1980.
23. **Griffin, F. M., Griffin, J. A., Leider, J. E., and Silverstein, S. C.,** Studies on the mechanisms of phagocytosis. I. Requirements for circumferential attachment of particle bound ligands to specific receptors of the macrophage plasma membrane, *J. Exp. Med.,* 142, 1263, 1975.
24. **Michl, J., Ohlbaum, D. J., and Silverstein, S. C.,** 2-Deoxyglucose selectively inhibits Fc and complement receptor mediated phagocytosis in mouse peritoneal macrophages. II. Dissociation of the inhibitory effects of 2-deoxyglucose on phagocytosis and ATP generation, *J. Exp. Med.,* 144, 1484, 1976.
25. **Kozel, T. R., Reiss, E., and Cherniak, R.,** Concomitant but not causal association between surface charge and inhibition of phagocytosis by cryptococcal polysaccharide, *Infect. Immun.,* 29, 295, 1980.
26. **Rabinovitch, M.,** Phagocytosis; the engulfment stage, *Semin. Hematol.,* 5, 134, 1968.
27. **Griffin, F. M. and Silverstein, S. C.,** Segmental response of the macrophage plasma membrane to a phagocytic stimulus, *J. Exp. Med.,* 139, 323, 1974.
28. **Muschel, R. J., Rosen, N., Rosen, O. M., and Bloom, B. R.,** Modulation of Fc-mediated phagocytosis by cyclic AMP and insulin in a macrophage-live cell line, *J. Immunol.,* 119, 1813, 1977.
29. **Stossel, T. P.,** Quantitative studies of phagocytosis: kinetic effects of cations and heat-labile opsonin, *J. Cell Biol.,* 58, 346, 1973.
30. **Hirsch, J. G. and Cohn, Z. A.,** Degranulation of polymorphonuclear leukocytes following phagocytosis of microorganisms, *J. Exp. Med.,* 112, 1005, 1960.
31. **Karnovsky, M. L.,** Metabolic basis of phagocytic activity, *Physiol. Rev.,* 42, 143, 1962.
32. **Stossel, T. P.,** Contractile proteins in phagocytosis: an example of cell surface-to-cytoplasm communication, *Fed. Proc.,* 36, 2181, 1977.

33. **Mandell, G. L.,** Intraphagosomal pH of human polymorphonuclear neutrophils, *Proc. Soc. Exp. Biol. Med.*, 134, 447, 1970.
34. **Klebanoff, S. J. and Hamon, C. B.,** Role of myeloperoxidase-mediated antimicrobial systems in intact leukocytes, *RES J. Reticuloendothel. Soc.*, 12, 170, 1972.
35. **Klebanoff, S. J. and Hamon, C. B.,** Antimicrobial systems of mononuclear phagocytes, in *Mononuclear Phagocytes*, Vol. 2, Van Furth, R., Ed., Blackwell Scientific, Oxford, 1970.
36. **Carpenter, P. L.,** *Immunology and Serology*, W. B. Saunders, Philadelphia, 1965, chap. 11.
37. **Weisman, R. A. and Korn, E. D.,** Phagocytosis of latex beads by aeanthamoeba, *Biochem.*, 6, 485, 1967.
38. **Silverstein, S. C., Steinman, R. M., and Cohn, Z. A.,** Endocytosis, *Annu. Rev. Biochem.*, 46, 669, 1977.
39. **Hanks, J. H.,** Quantitative aspects of phagocytosis as influenced by the number of bacteria and leukocytes, *J. Immunol.*, 38, 159, 1940.
40. **Fenn, W. D.,** Temperature coefficient of phagocytosis, *J. Gen. Physiol.*, 4, 331, 1922.
41. **Wood, W. B. and Smith, M. R.,** Intercellular surface phagocytosis, *Science*, 106, 86, 1947.
42. **Wright, A. E. and Douglas, S. R.,** An experimental investigation of the role of the body fluids in connection with phagocytosis, *Proc. R. Soc. London*, 72, 357, 1903.
43. **Stossel, T. P.,** Phagocytosis, in *The Granulocyte: Function and Clinical Utilization*, Alan R. Liss, New York, 1977, 87.
44. **van Oss, C. J.,** Precipitation and agglutination, in *Principals of Immunology*, Rose, N. R., Milgram, F., and van Oss, C. J., Eds., MacMillan, New York, 1973, chap. 6.
45. **Mayer, M. M.,** Mechanism of cytolysis by complement, *Proc. Nat. Acad. Sci. U.S.A.*, 69, 954, 1972.
46. **Van Oss, C. J.,** Phagocytosis as a Surface Phenomenon, *Ann. Rev. Microbiol.*, 32, 19, 1978.
47. **Van Oss, C. J. and Gillman, C. F.,** *RES J. Reticuloendothel. Soc.*, 12, 283, 1972.
48. **Neumann, A. W. and Good, R. J.,** personal communication, 1972.
49. **Van Oss, C. J., Gillman, C. F., and Neumann, A. W.,** *Phagocytic Engulfment and Cell Adhesiveness*, Marcel Dekker, New York, 1975, 9.
50. **Van Oss, C. J., Gillman, C. F., and Neumann, A. W.,** Phagocytic Engulfment and Cell Adhesiveness, Marcel Dekker, New York, 9, 1975.
51. **Neumann, A. W. van Oss, C. J., and Szekely,** personal communication, 1973.
52. **Shafrin, E. G.,** Critical Surface Tension of Polymers, in *Polymer Handbook*, Brandrup, J. and Immergut, E. H., Eds., Interscience, New York, 1966, 111.
53. **Coleman, D. L., Gregonis, D. E., and Andrade, J. D.,** Blood-materials interactions: the minimum interfacial free energy and optimum polar/apolar ratio hypothesis, *J. Biomed. Mater. Res.*, 16, 381, 1982.
54. **Neumann, A. W., Good, R. J., and Hope, C. J.,** *J. Colloid Interface Sci.*, 49, 291, 1974.
55. **Deierkauf, F. A., Beukers, H., Deierkauf, M., and Riemersma, J. C.,** *J. Cell Physiol.*, 92, 169, 1973.
56. **Van Oss, C. J., Good, C. F., Neumann, A. W.,** *J. Electroanal. Chem.*, 37, 387, 1972.
57. **Good, R. J.,** *J. Theoret. Biol.*, 37, 413, 1972.
58. **Beukers, H., Deierkauf, F. A., Blom, C. P., Deierkauf, M., and Riemersma, J. C.,** *J. Cell Physicol.*, 97, 29, 1978.
59. **Florey, H. W.,** *General Pathology*, Lloyd-Luke Medical Books, London, 1970.
60. **van Furth, R. and Diesselhoff-Den Dulk, M. M. C.,** Method to prove ingestion of particles by macrophages with light microscopy, *Scand. J. Immunol.*, 12, 265, 1980.
61. **Biozzi, G., Benacerraf, B., Stiffel, C., and Halpern, B. N.,** Etude quantitative du l'activité granulopexique du systéme réticulo-endothélial chez la souris, *C.R. Soc. Biol. Paris*, 148, 431, 1954.
62. **Stiffel, C., Mouton, D., and Biozzi, G.,** Kinetics of the phagocytic function of reticuloendothelial macrophages in vivo, in *Mononuclear Phagocytes*, van Furth, R., Ed., Blackwell Scientific, Oxford, 1970.
63. **Kanet, R. I. and Brain, J. D.,** Methods to quantify endocytosis; a review, *RES J. Reticuloendothel. Soc.*, 27, 201, 1980.
64. **Molday, R. S., Dreyer, W. J., Rembaum, A., Yen, S. P. S.,** New immunolatex spheres: visual markers of antigens on lymphocytes for scanning electron microscopy, *J. Cell Biol.*, 64, 75, 1975.
65. **Palzer, R. J. and Walton, J.,** Fluorescent microsphere techniques for studying early events in cell hybridization, *J. Cell Biol.*, 75, 381, 1977.
66. **Roberts, J. and Quarted, J. H.,** Particle uptake by PMN leukocytes and Ehrlich ascites-carcinoma cells, *Biochem. J.*, 89, 150, 1963.
67. **Schroeder, F. and Kinden, D. A.,** Measurement of phagocytosis using fluorescent latex beads, *J. Biochem. Biophys. Methods*, 8, 15, 1983.
68. **Rembaum, A.,** Microspheres as immuno-reagents for cell identification in *Flow Cytometry and Sorting*, Melamed, M. R., Mullaney, P. F., and Mendelsohn, M. J., Eds., John Wiley & Sons, New York, 1979, chap. 18.
69. **Vanderhoff, J. W., Micale, F. J., El-Assaer, M. S.,** U.S. Patent 4,247,434, 1981.

70. **Tenoso, H. J. and Smith, D. B.,** Covalent bonding of antibodiers to polystyrene latex beads: a concept, NASA Tech. Brief B72-10006; NASA TSP 72-10006, Washington, D.C., 1972.
71. **Bangs, L. B.,** *Uniform latex particles,* Seragen Diagnostics Inc., Indianapolis, Ind., 1984.
72. **Molday, R. S.,** Immunolatex spheres as cell surface markers for scanning electron microscopy, in *Principles and Techniques of Scanning Electron Microscopy,* Hayat, A., Ed., Van Nos Reinhold, New York, 1978, chap. 18.
73. **Rembaum, A., Gupta, A., and Volksen, W.,** U.S. Patent 4,170,685, 1979.
74. **Rembaum, A. and Volksen, W.,** U.S. Patent 4,123,396, 1978.
75. **van Furth, R., van Zwet, T. L., and Leijh, P. C. J.,** In vitro determination of phagocytosis and intracellular killing by polymorphonuclear and mononuclear phagocytes, in *Cellular Immunology,* Vol. 2, Weir, D. M., Ed., Blackwell Scientific, Oxford, 1978, chap. 32.
76. **Stuart, A. E., Habeshaw, J. A., and Davidson, E. A.,** Phagocytes in vitro, in *Cellular Immunology,* Vol. 2, Weir, D. M., Ed., Blackwell Scientific, Oxford, 1978, chap. 31.
77. **Dobson, E. I. and Jones, H. B.,** The behavior of intravenously injected particulate material, *Acta Med. Scand.,* 144, 273, 1952.
78. **Meltzer, M. S.,** Peritoneal mononuclear phagocytes from small animals, in *Methods for Studying Mononuclear Phagocytes,* Academic Press, New York, 1981, chap. 87.
79. **Cianciolo, G. J. and Snyderman, R.,** *Quantitation of the Inflammatory Accumulation of Mononuclear Phagocytes In Vivo,* Academic Press, New York, 1981, chap. 88.
80. **Fenn, W. O.,** The theoretical response of living cells to contact with solid bodies, *J. Gen. Physiol.,* 4, 373, 1922.
81. **Rimland, D. and Hand, W. L.,** The effect of ethanol on adherence and phagocytosis by rabbit alveolar macrophages, *J. Lab. Clin. Med.,* 95, 918, 1980.
82. **Al-Ibrahim, M. S., Chandra, R., Kishore, R., Valentine, F. T., and Lawrence, H. S.,** A micromethod for evaluating the phagocytic activity of human macrophages by ingestion of radio-labeled polystyrene particles, *J. Immunol. Methods,* 10, 207, 1976.
83. **Al-Ibrahim, Valentine, F. T., and Lawrence, S.,** Activated lymphocytes depress phagocytosis of latex particles by human monocyte-macrophages, *Cell. Immunol.,* 41, 217, 1978.
84. **Lowry, O. H., Rosenbrough, N. J., Farr, A. L., and Randall, R. J.,** Protein measurement with the folin phenal reagent, *J. Biol. Chem.,* 193, 265, 1951.
85. **Steinkamp, J. A., Wilson, J. S., Saunders, G. C., and Stewart, C. C.,** Phagocytosis; flow cytometric quantitation with fluroescent microspheres, *Science,* 215, 64, 1982.
86. **Dunn, P. A. and Tyrer, H. W.,** Quantitation of neutrophil phagocytosis, using fluorescent latex microbeads: correlation of microscopy and flow cytometry, *J. Lab. Clin. Med.,* 98, 374, 1981.
87. **Kanet, R. I. and Brain, J. D.,** Phagocytosis: quantification of rates and intercellular heterogeneity, *J. Appl. Physiol.,* 42, 432, 1977.
88. **Deierkauf, F. A., Beukers, H., Deierkauf, M., and Riemersma, J. C.,** Phagocytosis by rabbit polymorphonuclear leukocytes: the effect of albumin and polyamino acids on latex uptake, *Cell Physiol.,* 92, 169, 1977.
89. **Friend, K., Eksdedt, R. D., and Duncan, J. L.,** Effect of concanavalin A on phagocytosis by mouse peritoneal macrophages, *RES J. Reticuloendothel. Soc.,* 17, 10, 1975.
90. **Tsan, M. F. and Berlin, R. D.,** Effect of phagocytosis on membrane transport of nonelectrolytes, *J. Exp. Med.,* 134, 1016, 1971.
91. **Vray, B., Saint-Guillam, M., Leloup, R., and Hoebeke, J.,** Kinetic and morphologic studies of rat macrophage phagocytosis, *RES J. Reticuloendothel. Soc.,* 29, 307, 1981.
92. **Schroeder, F. and Kier, A. B.,** Lipid composition alters phagocytosis of fluorescent latex beads, *J. Immunol. Methods,* 57, 363, 1983.
93. **Holland, J. F., Teets, R. E., and Timnick, A.,** Correction of right-angle fluorescein measurements for the absorption of excitation radiation, *Anal. Chem.,* 49, 706, 1977.
94. **Steinkamp, J. A.,** Flow Cytometry, *Rev. Sci. Instrum.,* 55, 1375, 1984.
95. **Herzenberg, L. A. and Herzenberg, L. A.,** Analysis and separation using the fluroescence activated cell sorter (FACS), in *Cellular Immunology,* Vol. 2, Weir, D. M., Ed., Blackwell Scientific, Oxford, chap. 22.
96. **Ault, K. A.,** Clinical applications of fluorescence-activated cell sorting techniques, *Diagn. Immunol.,* 1, 2, 1983.
97. **Laerum, O. D., and Farsund, T.,** Clinical application of flow cytometry: a review, *Cytometry,* 2, 1, 1981.
98. **Paiod, R. J. and Brain, J. D.,** Uptake of latex particles by macrophages: characterization using flow cytometry, *Am. J. Physiol.,* 245, C220, 1983.
99. **Paiod, R. J. and Brain, J. D.,** Uptake of latex particles by pulmonary macrophages: role of calcium, *Am. J. Physiol.,* 245, C227, 1983.

100. **Shellito, J., Murphey, S., and Warner, N.,** Flow cytometry analysis of lung cells from normal and acid-treated rabbits, *Am. Rev. Respir. Dis.*, 124, 333, 1981.
101. **Fujikawa-Yamamato, K. and Wada, M.,** Flow cytometry of the phagocytosis of fluorescent microspheres in V79 cells, *Cell Struct. Funct.*, 8, 373, 1983.
102. **Fernandez-Repollet, E., Mittler, R. S., Tiffany, S. and Schwartz, A.,** In vivo effects of prostaglandin E_2 and arachidonic acid on phagocytosis of fluorescent methacrylate microbeads by rat peritoneal macrophages, *J. Histochem. Cytochem.*, 30, 466, 1982.
103. **Fernandez-Repollet, E., Opava-Stetzer, S., Tiffany, S., and Schwartz, A.,** Effects of endogenous antidiuretic hormone (ADH) on macrophage phagocytosis, *J. Histochem. Cytochem.*, 31, 956, 1983.
104. **Kirchanski, S. J., Price, B. J., Davidovits, G. and Hoffman, R.,** Flow cytometry measurements of membrane C3b receptor modulation during phagocytosis, *Biotechniques*, 2, 1983.
105. **Haskill, S. and Becker, S.,** Flow cytometric analysis of macrophage heterogeneity and differentiation: utilization of electronic cell volume and fluroescent substrates corresponding to common macrophage markers, *RES J. Reticuloendothel. Soc.*, 32, 273, 1982.
106. **Basse, C. F., Laerum, O. D., Glette, J., Hopen, G., Haneberg, B., and Solberg, C. O.,** Simultaneous measurement of phagocytosis and phagosomal pH by flow cytometry: role of polymophonuclear neutrophilic leukocyte granules in phagosome acidification, *Cytometry*, 4, 254, 1983.

Chapter 10

THE USE OF MICROSPHERES IN THE STUDY OF CELL MOTILITY

Robert A. Bloodgood

TABLE OF CONTENTS

"Two days ago I looked down through my sparkling microscope. Opened up the diaphragm for more light and there it was. I said my my, what's this, it moves. Bumping pinpoint bouncing merrily, wow, did a standing broadjump of ten microns just then. This is most extraordinary."

From *The Mad Molecule* by J. P. Donleavy

I. INTRODUCTION

Microspheres and a variety of other inert particulate materials have been utilized to visualize motile events associated with cells; in the present context cell motility is defined to include whole cell movements, movements of cellular organelles, and movements associated with the cell surface. Microspheres have also been utilized to study cell adhesion, endocytosis, and to determine the distribution of cell surface molecules; these areas will not be dealt with in this chapter. One of the more innovative uses of microspheres in cell biology has been to enable researchers to visualize in the light microscope sites of force transduction and the movement of macromolecules or macromolecular complexes that might not otherwise be observable by routine observation. This review is subdivided into the applications of microspheres to studies of prokaryotic and eukaryotic motility systems. For the sake of completeness, information is included that has been obtained using marker objects that may not be considered strictly as microspheres.

II. PROKARYOTIC CELL MOTILITY

Prokaryotic microorganisms exhibit a wide array of motile behaviors, generally thought to be mechanistically distinct from those forms of cell motility observed among the eukaryotes. Prokaryotic motility has been differentiated into three classes: flagellar motility, spirochete motility, and gliding motility; in all these cases, the energy appears to be provided by protonmotive force. Microspheres and other inert particles have been utilized to study all three forms of prokaryotic motility.

A. Bacterial Flagellar Motility

Many bacteria possess motile flagella, although these organelles differ dramatically in their size, biochemical composition, and structure from eukaryotic cilia and flagella. The mechanism by which bacterial flagella generate force has long been controversial; many observers assumed that bacterial flagella propagated waves even though the simplicity of the structures and the apparent lack of any enzymatic activity made this a questionable hypothesis.[1,2,3] Only in 1973 was the fog of confusion surrounding bacterial flagellar motility lifted by a piece of brilliant insight[4] that built on previously published observations on the effect of antiflagellar antibodies on flagellar motility. The elegant but surprising answer is that bacterial flagella are relatively fixed helical structures that rotate at their bases. Because direct observation cannot distinguish rigid rotation from helical wave propagation, it remained for Silverman and Simon[5] to utilize two clever tricks in order to demonstrate directly bacterial flagellar rotation. If two polyhook mutant cells of *Escherichia coli* are connected via their flagellar hooks using an antibody, the two members of each pair rotate in opposite directions. If a polyhook or straight filament mutant cell is tethered to a glass surface, the cell body rotates. When antibodies are utilized to attach polystyrene microspheres (0.7 μm in diameter) to the surface of bacterial flagellar filaments on straight flagella mutants, the attached microspheres are observed in the light microscope to rotate around an invisible axis representing the position of the flagellar filament while the cell body rotates in the opposite direction (Figure 1). Berg and coworkers[6] reported additional information on flagellar rotation obtained from analyzing the movement of polystyrene microspheres attached to flagellar

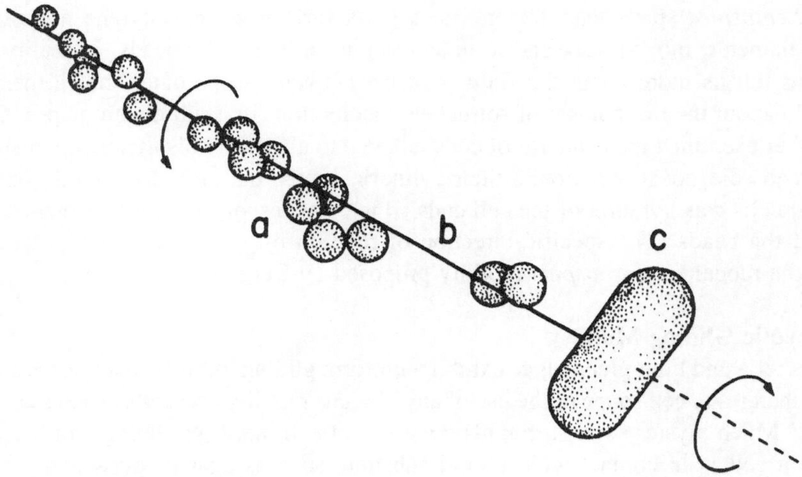

FIGURE 1. Diagram illustrating the use of polystyrene microspheres to visualize the rotation of the flagellar filament in straight flagellar mutants of *E. coli*. The flagellar filament is not visible in the light microscope; the microspheres and the cell body rotate around the same axis in opposite directions. (From Berg, H. C., *Sci. Am.*, 233, 36, 1975. With permission.)

hooks on bacterial mutants; these observations suggested that the rotation rate of the flagellar motor is nearly independent of the dynamic load when the load is light.

B. Spirochete Motility

Spirochetes can exhibit a remarkable variety of twisting, bending and flexing movements in addition to directional translocation through a liquid medium or along a solid surface.[7] Although there has been ample speculation over the years about the mechanism behind these movements, there is little in the way of conclusive experimental evidence. Spirochetes are composed of a protoplasmic cylinder with attached axial filaments that are remarkably similar to bacterial flagella; this complex is then enclosed within an external sheath.[8] In analogy to the situation with flagellated bacteria, Berg[9] proposed that spirochete motility occurs because of the rotary movements of the axial filaments within the compartment between the protoplasmic cylinder and the outer sheath. It is hypothesized that, in most spirochetes, the protoplasmic cylinder rotates relative to the flexible external sheath ''screwing'' the organism through the medium. Berg et al.[10] proposed a rather different model for the motility of *Leptospira*. In this case, the protoplasmic cylinder and the external sheath rotate in the same direction (opposite to the direction of rotation of the axial filament) and the torque is generated by the gyrations of the ends of the cell rather than by the circumferential roll of the outer sheath as is thought to be the case for other spirochetes. The *Leptospira* model predicts that polystyrene microspheres ''attached anywhere on the surface of the cell should rotate about the axis of the protoplasmic cylinder in a direction opposite to that of the anterior spiral''. When Charon et al.[11] attempted to test this prediction, they found a surprising result. Antibody-coated polystyrene microspheres attached to motile cells were always moved toward the back end of the cell, at a velocity that was dependent upon the size of the microsphere. When the cell reversed its direction of movement, the microspheres were moved to the other end of the cell. If several microspheres were initially located at several positions along a single motile cell, they all ended up at the rear end of the cell. Nonmotile mutants of *Leptospira* exhibited no microsphere movement. The authors hypothesize that the antibody-coated bead is attaching to a particular outer-membrane sheath protein, which is then being moved through the membrane because of the viscous drag of the medium on the bead. *Leptospira* spirochetes tethered to a glass surface via polystyrene microspheres (0.26 μm diameter) moved back and forth across the surface of the beads. In reality, these observations tell us more about the fluidity of the antigens in the outer sheath membrane than they do about the mechanism of spirochete locomotion. In a subsequent paper, Charon et al.[12] further examined the behavior of cells tethered to glass via polystyrene microspheres. When the cells did not rotate around their cylindrical axis, the only discernible motion of the tethered cells was gyration of the cell ends which was responsible for the rotation of the cell around the beads. The specific direction of rotation of spiral- and hook-shaped ends supported the model for *Leptospira* motility proposed by Berg et al.[10]

C. Prokaryotic Gliding Motility

Many bacteria and blue-green algae exhibit a uniform gliding motility that does not appear to involve changes in cell shape or the use of any obvious motility organelles such as bacterial flagella.[13-16] Much mystery still surrounds this form of cell motility; gliding motility occurs only when a cell is in contact with a solid substrate such as agar or glass and, in many cases, is associated with release of mucus or slime from the cells involved. Similar types of gliding motility have also been observed among eukaryotic organisms. Many early workers, especially those studying the gliding of *Oscillatoria* and similar filamentous cyanobacteria, have utilized India ink and carmine particles in order to visualize the movements of what is presumed to be mucus associated with the cell surface.[13,17-20] When the filament glides forward, the particles remain stationary relative to the substrate; if the filament is prevented from moving, particles move backward at a velocity similar to the gliding velocity. In some cases, different particles on the same filament were observed to move in opposite directions (i.e., toward each other). In other prokaryotic organisms, especially some that do not appear to secrete much mucus material, particle movement can occur on a gliding cell.

Pate and Chang[21] observed the movements of polystyrene microspheres (0.76 μm diameter) along the surface of gliding cells of *Cytophaga johnsonae* and *Flexibacter columnaris*. *Cytophaga* possesses a typical Gram-negative cell envelope consisting of a cytoplasmic membrane, a peptidoglycan layer, and an outer membrane composed of lipopolysaccharide and protein. *Flexibacter* has a similar structure but, in addition, certain species (*F. polymorphus*) possess unusual surface features that resemble goblets;[22] the presence of these surface specializations does not directly correlate with the presence of gliding motility. Pate and Chang[21] reported a variety of microsphere movements: (1) they could move up one side of the elongated cell, around the end and down the other side; (2) they could stop and reverse direction; and (3) they could remain at one end of the cell while rotating in position. Movements of different microspheres on the same cell were independent of one another. All environmental conditions or inhibitors (such as cyanide and a variety of proton conductors) that stopped gliding motility also stopped microsphere motility. Nongliding mutants failed to exhibit microsphere movement. These same authors observed ring-shape structures in the wall of *Cytophaga* and hypothesized that these structures are analogous to the ring structure of the basal bodies associated with bacterial flagella; they argued that rotation of these structures may be the mechanism for gliding motility and microsphere movements.

Lapidus and Berg[23] further studied the gliding motility of *Cytophaga* sp. Strain U67 using polystyrene microspheres of various diameters (Figure 2). They observed that microspheres moved at 2 μm/sec along cells that were actively gliding, at rest, or in suspension; this is the same velocity observed for a whole cell gliding along a glass surface and for movement of a cell along an immobilized microsphere. Microspheres moved in either direction along the cell surface and two microspheres on the same cell were observed to pass by each other in close proximity while moving in opposite directions. The rate of movement was independent of microsphere size over the range from 0.13 to 1.3 μm in diameter. Depletion of

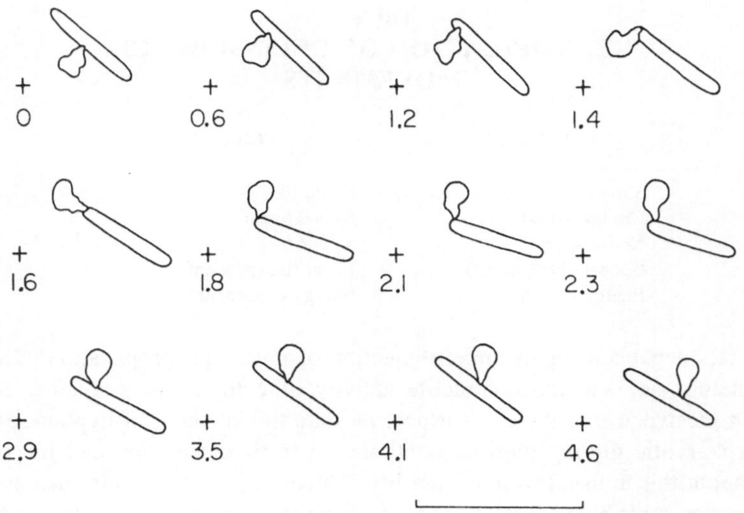

FIGURE 2. Drawings from a movie showing the movement of an aggregate of 0.13 μm diameter polystyrene microspheres along the surface of a single *Cytophaga* cell in suspension. The elapsed time is indicated in seconds. (From Lapidus, I. R. and Berg, H. C., *J. Bacteriol.*, 151, 384, 1982. With permission.)

oxygen resulted in the reversible inhibition of both cell gliding and microsphere movement. The observations of Lapidus and Berg[23] were generally in agreement with those of Pate and Chang[21] except that Lapidus and Berg were not able to find the ring-shaped structures reported by Pate and Chang using a rather different isolation procedure. Lapidus and Berg[23] argued that their observations support a model in which substrate adhesion sites within the outer membrane of *Cytophaga* move along tracks that are part of the rigid framework of the cell wall.

There is a striking similarity between the detailed characteristics of microsphere movement on *Cytophaga* observed by Lapidus and Berg[23] and the observations of microsphere movement associated with the *Chlamydomonas* flagellar membrane reported by Bloodgood and co-workers[71-73,76] (see Section III.A.2 below). In both systems:

1. Microsphere movement, whole cell gliding on a glass substrate, and cell movement along an immobilized microsphere occur at approximately 2 μm/sec.
2. Microsphere movements are bidirectional.
3. There is a load independence of the velocity of microsphere movement over a wide range.
4. Movement of different microspheres on the same cell is independent of one another.

However, it must be kept in mind that these phenomenological similarities do not necessarily imply a similar underlying mechanism; one form of motility is associated with the plasma membrane of a eukaryotic cell while the other is associated with the outer membrane of a Gram-negative prokaryote. For each of these systems, analysis of mutant cell lines has demonstrated that gliding motility and microsphere movement are two manifestations of the same energy-transducing system.[21,23,78]

In contrast to the results cited above, Dworkin et al.[24] failed to observe any polystyrene microsphere movement on gliding cells of the prokaryote *Myxococcus xanthus* and argued that gliding motility in this species results from a gradient of surface tension created by asymmetric secretion of a surfactant-like molecule.[113] A number of other models for prokaryotic gliding motility have been discussed in an excellent review by Castenholz.[153] These

Table 1
CLASSIFICATION OF CELL-SURFACE MOVEMENTS

Class I	Class II
Slow (0.01 — 0.07 μm/sec)	Rapid (0.5 — 8.0 μm/sec)
Unidirectional	Bidirectional
Nonsaltatory	Saltatory
Cooperative (global)	Local (independent)
Energy-dependent	Energy-dependent

include models dependent upon forceful ejection of mucilage, propagation of waves along the outer membrane, and the contractile activity of cytoplasmic filaments coupled with differential adhesiveness of the cell surface. Despite this plethora of hypotheses, the mechanism of prokaryotic gliding motility continues to resist elucidation and much remains to be learned about this unusual form of motility. The use of inert markers such as polystyrene microspheres has contributed much to characterizing this form of cell motility and has allowed a comparison of its characteristics in different cell types. There may well be more than one underlying mechanism involved.

III. EUKARYOTIC CELL MOTILITY

A. Cell Surface Movements

One of the principal applications of microspheres and other inert particulate markers to the field of cell motility has been in relation to the study of eukaryotic cell surface dynamics. This approach has contributed greatly to our increasing awareness of the highly dynamic nature of plasma membranes. In particular, use of marker particles attached to the surface of eukaryotic cells has allowed direct visualization in the light microscope of movements occurring in the plane of the plasma membrane. Implicit in these studies has been the assumption that the object being observed is mechanically coupled to the cell surface and that the movements of the attached marker object reflect movements of peripheral, or integral membrane components (proteins, glycoproteins, glycolipids), or else bulk flow of the entire membrane, or at least of the phospholipid bilayer. Caution must be exercised in the interpretation of these types of observations; the presence of the marker object may in some way perturb the normal dynamics of the plasma membrane. Put another way: do the movements of the attached markers truly reflect movements of membrane components occurring, albeit unseen, in the absence of the marker objects? Investigators utilizing microspheres tended to take this for granted; yet, in another system of plasma membrane dynamics, capping of cross-linked ligands, it is generally felt that the redistribution is being induced by a perturbation of the cell surface (cross-linking of membrane proteins by multivalent ligands such as lectins or antibodies).[25,26] However, some workers have interpreted the capping phenomenon as resulting from the action of a continuous membrane or lipid flow occurring in the plasma membrane but not apparent until visualized by artificial means.[27-29] For the sake of this review, the published observations on cell surface movements visualized using microspheres or other artificial markers will be divided into two classes; the characteristics of these two classes of surface movements are listed in Table 1. Class I movements tend to be slow, unidirectional, and cooperative while Class II movements are much faster (by as much as two orders of magnitude), bidirectional, saltatory and independent of one another. Inevitably, some observations do not fit neatly into either class. Although the characteristics of the capping phenomenon best place it within Class I, it must be recognized that there may be multiple mechanisms underlying the various surface movements that are here being arbitrarily classified into two categories.

1. Slow (Class I) Surface Movements

A number of early cytologists used inert materials to examine the cell surface during cell movement. This approach was applied, in particular, to the study of the motility of free-living amoebae because of the ease with which these preparations could be obtained before the advent of mammalian cell culture techniques. Mast[30] used carmine, india ink, and carbon particles in order to determine that the free surface of motile amoebae moved forward continuously as in a rolling movement. His observations agreed with the earlier studies of Jennings[31] who stated that "the surface of this species moves like the surface of a sack rolling down an inclined plane". However, controversy over the mechanism of amoeboid movement has persisted to this day and observations on the movements of the surface membrane have figured prominently in the controversy. One school of thought holds that the amoeba moves by recycling its plasma membrane into the cell; new membrane is added at the leading end of the pseudopod and old membrane is removed at the trailing end of the cell, presumably by endocytosis. The other major view is that the plasma membrane is rather stable, turns over slowly, and moves forward with the amoeba on the dorsal surface and rearward on the ventral surface, as described by Jennings[31] and Mast.[30] The first model predicts that membrane markers should remain stationary relative to the substrate but exhibit rearward movement relative to one end of the moving cell. The latter hypothesis (membrane-rolling or tank-tread model) predicts forward movement of membrane markers on the free surface of the cell at a velocity relative to the substrate that would be twice the velocity of movement of the cell; relative to the cell, the upper surface would move forward and the lower surface would shift backwards. Griffin and Allen[32] observed the forward movement of carmine, ink, or carbon particles along the surface of a variety of amoebae. In particular they noted that, in certain species of anterior-flattened amoebae, particles on the upper surface moved toward the front of the cell at approximately twice the rate that the amoebae moved. Similar observations were made by Abe.[33] On the other hand, Shaffer,[34,35] studying amoebae of cellular slime molds, observed that a variety of particles (carmine, carbon, cation-exchange resin beads, anion-exchange resin beads) bound to the amoeba surface did not move relative to the substrate, and hence accumulated at the rear of moving cells. His interpretation of these results was that the cell turned over its membrane as it moved, internalizing membrane at the rear, and inserting membrane at the front of the cell. Goldacre[36] was a strong adherent to this view but found himself saddled with the wrong observations; he consistently observed small carmine and carbon particles moving forward along motile amoebae at about twice the speed of locomotion of the amoebae. He sought to discount these movements as due to electrophoresis along the cell surface and preferred to put weight on his observations of oil droplets and glass fibers accumulating at the posterior end of motile cells as evidence that the membrane did not exhibit any forward movement. Chapman-Andresen[151] observed that polystyrene microspheres accumulate at the rear end (uroid) of moving *Amoeba proteus*. However, in experiments studying the movement of crystals of calcium oxalate applied to the surface of *A. proteus,* Czarska and Grebecki[150] observed that the small particles moved forward symmetrically on all sides of the advancing pseudopod. Most particles did not accumulate at the advancing tip, but rather the distance between particles at different positions on the pseudopod appeared to oscillate. However, these authors also observed the accumulation of large aggregates of the crystals at the rear end of motile cells. The authors proposed a theory of amoeboid movement based on folding and unfolding of the plasma membrane in the absence of significant membrane turnover and claimed that neither the turnover theory espoused by Shaffer[34,35] and Goldacre[36] or the rolling bag theory of Jennings[31] could explain both the forward movement of small particles and the accumulation of aggregates at the tail. As will be seen below, the observations on amoebae showing forward movement of surface-attached particles are at odds with studies on the movement of the upper cell surface of tissue culture cells.

Michael Abercrombie has inspired much of the modern work on use of particulate markers to study the dynamics of the cell surface during motility of cultured mammalian cells, primarily fibroblasts.[37-40] This work has been reviewed by Harris[40] and will be described only briefly. Migrating fibroblasts pick up particles from the substrate at the leading edge (ruffling membrane) and move these particles rearward on both the dorsal (upper) and ventral (lower) cell surfaces at velocities of 1 to 4 μm/min (0.02 to 0.07 μm/sec). The particles (anion-exchange resin beads, glass particles, colloidal gold, and carbon particles) eventually accumulate either in the vicinity of the nucleus or at the trailing margin of the cell. The most important observation is that all particles move rearward on motile cells; there is no evidence of bidirectional or saltatory movements. Cells need not undergo any net locomotion in order to exhibit particle transport away from areas of ruffling membrane. These authors interpret these observations to indicate that there is a continuous flow of the plasma membrane away from the ruffling membrane along both the upper and lower surfaces of the cell. The ruffling membrane is presumed to be the side of insertion of new plasma-membrane components, presumably by vesicle exocytosis. A correlate of this hypothesis is that membrane is continuously being removed from the surface at the rear of these cells and being recirculated through the cytoplasm to the leading edge. Proponents of this hypothesis also proposed that the membrane flow visualized using inert markers was propelled by a centripetal pull of cortical cytoskeletal elements and even suggested that these same cytoskeletal forces might also serve to drive the processes of reassembly of membrane along the leading edge.[27,35,40,57] Dembo and Harris[43] did a careful mathematical analysis of actual particle movements on migrating chick-heart fibroblasts and concluded that the particles were being jointly carried along either by the flow of the membrane as a whole or by the flow of some submembrane material. The observations on fibroblasts are not in agreement with the "tank tread" hypothesis for cell movement that resulted from the observations of Mast[30] and others on amoebae. In addition, recent studies on whole-cell locomotion have not tended to favor the recirculating membrane model for cell motility that resulted from the observations on fibroblasts using particulate markers[44,45] and have tended to invoke an active role of the cytoskeleton[41,42] for virtually all eukaryotic whole-cell movements.

One interpretation of the apparently conflicting observations made on the movements of cell-surface markers may be that there are multiple mechanisms for whole-cell locomotion. The forward movement of dorsally located particles on certain freshwater amoebae may truly represent a tank-tread form of locomotion while the rearward movement of particles on both upper and lower surfaces of slime-mold amoebae and cultured fibroblasts may represent a very different mode of locomotion, perhaps characterized by rapid membrane turnover.

A special case of particle movements on the plasma membrane of mammalian cells involves observations on the movement of colloidal gold particles along the microspikes (filapodia) of 3T3 fibroblasts[46] and polystyrene microspheres along the microspikes of neuroblastoma cells.[47] In both cases, the movement is inward toward the cell body. These microspikes are filled with oriented actin filaments and movements of the gold particles on the 3T3 microspikes were completely inhibited by cytochalasin B, although careful study of detergent-extracted neuroblastoma microspikes revealed no direct association between the membrane-particle attachment sites and the underlying actin filaments. The polarity of the filaments in the actin bundle should result in a myosin-based force being directed outward along the microspikes. Since this is opposite to what is observed, it is more likely that bulk membrane flow of the plasma membrane over the microspikes back toward the cell body (perhaps due to incorporation of new membrane components at the tip of the microspikes) is responsible for the movements of the gold particles.

DiPasquale and Bell[48] observed movements of polystyrene microspheres and concanavalin A-derivatized erythrocytes along the surface of epithelial cells in culture. These particles were picked up by the leading edge and moved back along the free surface of the cell to the region of the nucleus; particle movement was inhibited by cytocholasin B (2 μg/mℓ).[49] One of the main points of this paper is that the marker objects did not adhere directly to the bulk of the cell surface; the only adhesive sites were on the leading edge of the cell. As the particle moved rearward, presumably the adhesion site moved back along with it, allowing the mechanical association to be maintained. Kubota[50] reported the continuous movement of carbon particles from anterior to posterior along the surface of migrating endodermal cells; the particles moved at about 40 μm/min (0.67 μm/sec) relative to the cell but did not move relative to the substrate. Kubota[50] concluded that the cell surface is continuously flowing from anterior to posterior and that this surface movement is responsible for the whole-cell locomotion. It is interesting to note that the continuous movement of carbon particles backward along the cell surface occurred even when the cell was not migrating.

Godman et al.[51] studied the effect of cytochalasin D on the induction and subsequent migration of cell surface blebs on cultured mammalian cells. They observed that 0.8 μm diameter polystyrene microspheres attached to the HeLa cell surface exhibited irregular patterns of movement. Although the overall vector of movement tended to be toward the cell apex (thickest portion of the cell occupied by the nucleus), no clustering of microspheres was observed. However, treatment with either cytochalasin D (0.25 μg/mℓ) or colcemid (0.6 μg/mℓ) resulted in directed centripetal movement of the microspheres and their clustering at the cell apex. The directed movement of microspheres induced by colcemid treatment occurred at 3 to 4 μm/min.

Albrecht-Buhler and coworkers[52-54] developed a method for quantitating the movements exhibited by gold particles associated with the surface of fibroblasts. They utilized this method to compare the motion of gold particles on the surface of normal and transformed 3T3 fibroblasts and observed that their motion was considerably slower on virus-transformed lines than on normal cell lines. Using data from the particle movements,they calculated that the viscosity of the plasma membrane of transformed 3T3 cells was twice that of the plasma membrane of the normal cells.

Marcus[55] has studied cell-surface dynamics of HeLa cells infected with myxovirus. The virus induces synthesis and insertion into the host-cell plasma membrane of a viral hemagglutinin, which binds to the surface of red blood cells. Red blood cells bound to the cell periphery soon after viral infection are actively transported centripetally on the cell surface, presumably due to the continuous local insertion of new plasma-membrane components.

Several interesting studies utilizing particulate markers have emphasized the dynamic nature of neuronal plasma membranes.[56] Bray[57] utilized glass and carmine particles to visualize membrane movements during axonal outgrowth. Particles associated with the axonal plasma membrane exhibited rapid, local saltatory movements in both directions. However, during axonal outgrowth, particles on the axon did not change their position relative to each other or to the cell body, suggesting to Bray that new membrane is only inserted at the growing tip (growth cone), presumably utilizing membranous vesicles transported out to the growing tip by axoplasmic flow. In contrast to particles on the axon membrane proper, particles on the growth-cone plasma membrane were observed to move backwards toward the cell body until they were left behind the progressively advancing growth cone. Koda and Partlow[58] labeled the axonal plasma membrane of isolated sympathetic ganglion neurons in culture using concanavalin A-derivatized erythrocytes and polystyrene microspheres. They observed that these markers exhibited retrograde movement (movement toward the cell body) on both growing and nongrowing axons at velocities of 11 to 84 μm/hr (0.18 to 1.40 μm/sec). These workers postulated that the movement of the axonal plasma membrane either reflected passive flow of the membrane driven by the insertion of new membrane at the growth cone or the action of cytoskeletal elements within the axon acting upon membrane components. They were unable to explain the discrepancy of their data with that of Bray.[57]

Roberts and Ward[59,60] have described an unusual system of plasma-membrane movement associated with the pseudopodia of the amoeboid spermatozoa of the roundworm, *Caenorhabditis*. Polystyrene microspheres move randomly on the surface of the unstimulated spermatid; once the transformation to a spermatozoan occurs, all microsphere movement on the cell body stops while microspheres are observed to move unidirectionally away from the tip of the pseudopod to the site of the junction of the pseudopod with the rest of the cell surface. These observation have been confirmed using labeled lectins and labeled phospholipids. There appears to be a constant insertion of new membrane at the distal tip of the pseudopod and a continuous flow of plasma membrane back toward the cell body; this unidirectional membrane flow is responsible for the movement of the spermatozoan. Despite intensive effort, no evidence was found for traditional cytoskeletal elements (actin, myosin, microtubules) within the pseudopod. Mutant spermatozoa from fertilization-defective mutants of *Caenorhabditis* were observed to exhibit random movements of polystyrene microspheres over the pseudopod and the cell surface.

Capping is a phenomenon originally described on lymphocytes, in which the cross-linking of a population of cell surface proteins by a multivalent ligand (usually a lectin or antibody) results in an energy-dependent redistribution of the ligand, and presumably the membrane proteins to which the ligand is bound, to one pole of the cell.[25,26] The observations of Abercrombie and Harris and associates on the movements of particulate markers associated with the surface of fibroblasts (discussed above) have been utilized by Harris[27] to formulate a theory to explain capping of cell surface ligands due to continuous plasma membrane flow. He postulated that the bulk membrane flow resulted from contractile forces exerted on the plasma membrane by the cytoskeleton. A somewhat similar theory based on lipid flow has been postulated by Bretscher.[28,29] This theory argues that diffusional movement of uncross-linked receptors resists the lipid flow while larger, cross-linked aggregates are swept up in the flow and are carried to one end of the cell. Mathematical analysis of the motion of polymorphonuclear leukocytes (PMNS) by Dembo et al.[44] has led them to argue against the Harris and Bretscher models for capping. In place of this, they postulate that cytoskeletal linkages directly mediate the movement of cell-surface receptors; indeed, there is a large body of literature implicating cytoskeletal proteins (actin, myosin, alpha-actinin) in at least some cases of capping,[26] Several studies have utilized microspheres and other marker objects in order to study the induced redistribution of cell-surface molecules. For instance, Maher and Molday[61] studied the redistribution of lectin-derivatized methacrylate microspheres attached to the surface of mouse neuroblastoma cells.

Polystyrene microspheres have been utilized by a number of workers to study the redistribution of adhesion sites on various leukocytes. Smith and Hollers[62] observed that polystyrene microspheres coated with albumin bound uniformly to the surface of human neutrophils but redistributed to the uropod (trailing end of the cell) within 20 min after stimulation of the cells with a chemotactic peptide; this redistribution was blocked by cytocholasin B and by the protease inhibitor, *p*-tosyl-1-lysine chloromethyl ketone (TLCK). The authors interpret the redistribution of the microspheres to indicate the redistribution of the chemotactic-peptide receptors. Abnormal neutrophils exhibiting defects in motility and chemotaxis failed to exhibit this redistribution of polystyrene microsphere-binding sites in response to chemotactic-peptide stimulation.[63] Normal neonatal neutrophils were also observed to be unable to redistribute the polystyrene-microsphere adhesion sites in response to the chemotactic peptide, f-met-leu-phe.[64] Berlin and Oliver[65] demonstrated that colchicine induced a capping of concanavalin A receptors on neutrophils; similarly, cell-surface adhesion sites for polystyrene microspheres went from a uniform distribution on untreated cells to a highly asymmetric distribution on the colchicine-treated cells that was coincident with the concanavalin A receptors. Yahara and Edelman[66] found that concanavalin A-derivatized polystyrene microspheres bound uniformly to mouse lymphocytes; upon addition of colchicine, the microspheres redistributed into a single localized site (cap) on the cell surface.

Peng and co-workers[67-69] demonstrated that positively charged (polylysine-coated and polyornithine-coated) polystyrene microspheres induced redistribution of acetylcholine receptors to the site of bead-membrane contact using cultured *Xenopus* muscle cells; uncharged and negatively charged beads failed to induce any redistribution. These workers also reported the induction of post-synaptic structures and the accumulation of microfilaments at the sites of contact of the microspheres with the cell surface. This work suggests that positively charged microspheres can act as multivalent ligands in order to induce redistribution of surface macromolecules and raises the possibility that inert marker objects used to visualize cell-surface movements in other systems, may, in fact, be perturbing the organization of the cell surface and its association with the cytoskeleton and may even act as multivalent ligands inducing capping-like phenomena.

As suggested by Abercrombie,[38] most of the slow cell-surface movements described in this section probably reflect the bulk movement of large regions of the plasma membrane driven either by membrane recycling[27-29] or by plasma membrane-cytoskeletal interactions.[26,41,42] See Section III.A.3 for a more complete discussion of the possible mechanisms responsible for cell-surface motility.

2. *Rapid (Class II) Surface Movements*

The observations in the previous section have utilized microspheres and other particulate markers to help provide a view of the cell surface as a very dynamic structure. However, in general, these movements were slow, unidirectional, and could be easily interpreted as manifestations of bulk-membrane movement or bulk-lipid flow. More recently, a second category of cell-surface movements has come to light — one involving much more rapid and complex movements that are less compatible with a simple explanation. These observations on rapid, saltatory cell-surface motility associated with highly asymmetric cell extensions (cilia, flagella, axopodia, reticulipodia) illustrate the fact that cell-surface dynamic events can be very local.

The most extensively studied of this class of movements is associated with the flagellar surface of *Chlamydomonas,* a unicellular, eukaryotic green algal cell.[70,71]. Polystyrene microspheres adhere to the surface of both vegetative and gametic flagella and are rapidly (2 μm/sec) transported along the flagellar surface in a saltatory and bidirectional manner[72-74] (Figure 3). The most striking feature of these movements is that they are local; as many as 6 to 8 microspheres can be observed moving along the same flagellum and the movements of these microspheres are totally independent. Moving microspheres can stop and start, and reverse direction anywhere along the length of the flagellum; however, during the course of a saltation, the velocity is relatively constant. The number of inward- and outward-directed saltations are approximately the same on the flagella of vegetative cells;[72] However, Hoffman and Goodenough[74] have reported a difference between vegetative and gametic cells in the periods of time microspheres spend on the proximal vs. distal halves of flagella. This surface motility is a property only of the flagellar plasma membrane; mutants without cell walls exhibit binding of microspheres to the general cell-body plasma membrane, but no movement is observed. Moreover, these particle movements are independent of the nature of the object being translocated. Neutral microspheres or microspheres derivatized with bovine serum albumin (BSA), concanavalin A, hydroxyl groups, carboxyl groups, or amino groups are all bound to and translocated along the flagellar surface at the same velocity, although there are some differences in the degree of binding; even nonmotile mutants of *E. coli* can be used as marker particles.[72] The movements exhibited by these markers are energy dependent and require that the flagellum be attached to a live and metabolically active cell. The microsphere movements have been quantitated [73] by measuring the percentage of bound

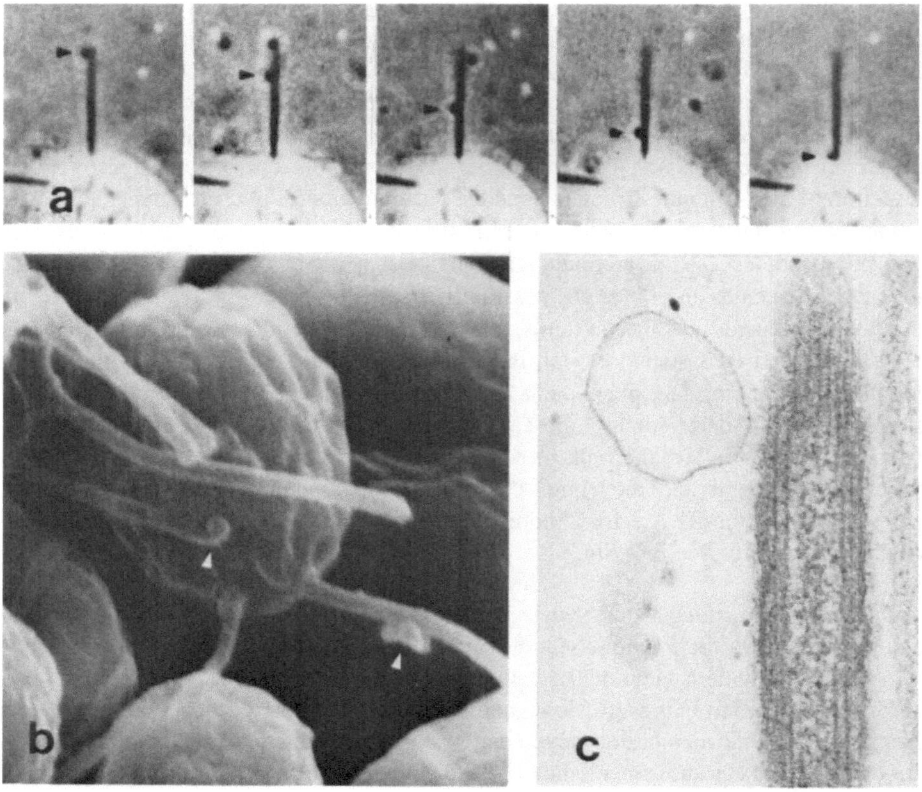

FIGURE 3. Demonstration of flagellar surface motility on *Chlamydomonas reinhardtii* strain pf-18 flagella using polystyrene microspheres. (A) Transport of a 0.357 μm diameter polystyrene microsphere along the length of the flagellum at a velocity of 2 μm/sec. The time interval between adjacent phase contrast micrographs is 0.9 sec. (Magnification × 1950.) (From Bloodgood, R. A., *J. Cell Biol.*, 75, 983, 1977. With permission.) (B) Scanning electron micrograph of *Chlamydomonas* fixed while actively moving 0.35 μm diameter polystyrene microspheres (arrows) along the flagellar surface. (Magnification × 8500.) (From Bloodgood, R. A., Leffler, E. M., and Bojczuk, A. T., *J. Cell Biol.*, 82, 664, 1979. With permission.) (C) Transmission electron micrograph of a 0.278 μm diameter polystyrene microsphere associated with the surface of the flagellar plasma membrane. The microsphere has been coated with bovine serum albumin so that it is visible in the electron microscope; this treatment does not alter the ability of microspheres to be transported along the flagellum. (Magnification ×64,000.) (From Bloodgood, R. A., *J. Cell Biol.*, 75, 983, 1977. With permission.)

microspheres in movement at the time of observation; this figure is consistently observed by a number of independent observers to be approximately 60%. This figure could represent the fact that each bead spends 60% of its time in motion on average or it could indicate that 60% of the bound microspheres are constantly in motion while 40% never move. The former appears to be the correct interpretation; long-term observations of individual microspheres show that every micropshere that becomes mechanically associated with the flagellar surface exhibits periods of movement. The microsphere motility is calcium dependent.[73] The microsphere movements can be reversibly inhibited by a number of treatments or conditions (Table 2); these treatments result in a decline in the percentage of attached microspheres exhibiting movement. No change in average velocity has been observed; a bound microsphere either moves at the maximum velocity (2 μm/sec) or not at all.

Clearly, the motor whose activity is being visualized using polystyrene microspheres has not evolved for the purpose of transporting polystyrene microspheres. Two physiological manifestations of flagellar surface motility have been elucidated: (1) whole-cell locomotion

Table 2
REVERSIBLE INHIBITORS OF FLAGELLAR SURFACE MOTILITY

Reversible inhibitor	Ref.
Fast	
Lidocaine	79
Trifluoperazine	80
Low calcium	73
High salt (NaCl or KCl)	73
Methyl xanthines	73
Cold (4°C)	73
Slow	
Cycloheximide	73
Chloral hydrate	73
Tunicamycin	71

by gliding motility,[75,76] and (2) flagellar reorientation during mating.[71,74,77] Lewin[78] has isolated nongliding mutants of *Chlamydomonas* that also fail to exhibit any movement of microspheres; this result indicates that gliding and microsphere movement are manifestations of the same energy-transduction system (motor). The first step in mating of *Chlamydomonas* gametes involves adhesion and alignment of the flagella on the two cells; a number of treatments (high salt, lidocaine, trifluoperazine) inhibit both microsphere movement and mating,[79,80] suggesting that the motor responsible for microsphere movement is also necessary for successful mating. In fact, it was over 30 years ago that Lewin[75] first suggested that *Chlamydomonas* gliding motility (or creeping, as he called it then) was also operative in flagellar alignment during mating. Collectively, these manifestations of energy transduction at the flagellar surface are referred to as ''flagellar surface motility''.

The three most important problems associated with understanding the mechanism of flagellar surface motility are

1. To identify the energy-transducer involved (presumably located within the flagellum)
2. To identify the surface-exposed components responsible for adhesion of the flagellar surface to a microsphere, to a solid substrate during gliding, or to another flagellum during mating-associated reorientation
3. To elucidate the manner in which the receptor molecules are coupled transmembrane to the motor (energy transducing system)

Preliminary studies using proteolytic modification of the flagellar surface and use of immobilized iodination systems have identified a high molecular-weight flagellar glycoprotein (or group of closely migrating glycoproteins) with an apparent molecular weight of 350,000 as the best candidate for the surface-exposed adhesive molecule that may couple the externally bound substrate (be it microsphere, glass surface, or another flagellar surface) to the motor.[81,82] The identity of the energy transducer is less clear. There is no myosin within cilia and flagella; the only actin-like component is present in small amounts and appears to be associated with the outer dynein arms.[83] Dynein-containing structures within the axoneme do not appear to be involved since mutant cell strains that lack either inner or outer dyein arms still exhibit normal flagellar surface motility. Currently, the best candidate for the energy transducer is a low molecular-weight, calcium-specific ATPase that is present within the flagellum and appears to be associated with the membrane and the matrix com-

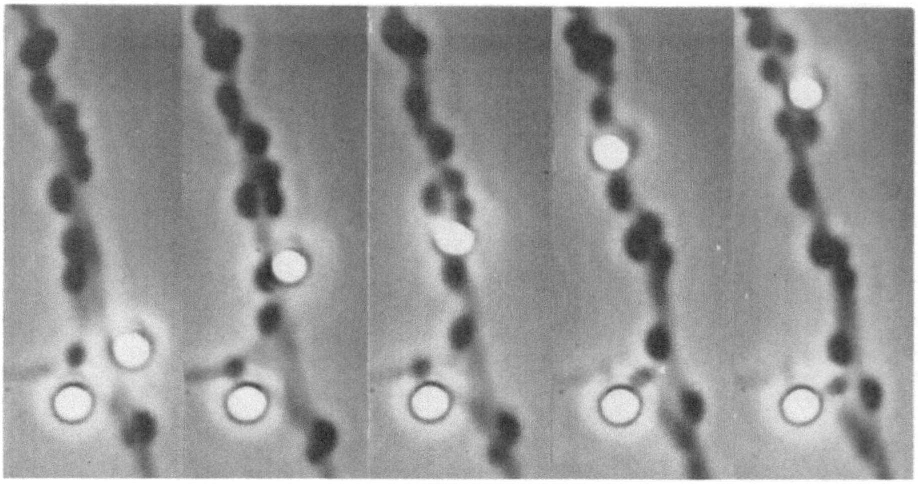

FIGURE 4. Phase contrast videomicrographs of a 1.23 μm diameter polystyrene microsphere moving along the surface of an *Allogromia laticollaris reticulopodium*. Time interval between frames = 0.1 sec. (Magnification ×3800.) (Courtesy of Dr. Samuel Bowser, New York State Department of Health.)

partments.[84,85] A complete understanding of the molecules involved in flagellar surface motility will only come from a detailed analysis of mutant cell lines with defects in flagellar surface adhesiveness and flagellar surface motility; for these types of genetic studies, *Chlamydomonas* is an excellent choice as an experimental system.[78]

The observations discussed above, although most thoroughly documented in *Chlamydomonas*, are not unique to the flagella of this organism. Similar observations of polystyrene microsphere movement have been made using sea urchin embryo cilia.[86] In this system, microspheres move outward along the ciliary membrane with velocities (8.2 ± 1.7 μm/sec) much greater than those associated with inward-directed movements (2.4 ± 0.2 μm/sec). This is in contrast to the situation in the *Chlamydomonas* flagellum, where outward-directed movements occur at the same velocity as inward-directed movements (1.7 ± 0.2 μm/sec vs. 1.8 ± 0.5 μm/sec). The role played by this ciliary surface motility in sea urchin embryos could not be either of those already mentioned for *Chlamydomonas* — gliding motility or mating. An intriguing possibility raised by Rosenbaum[166] is that ciliary and flagellar surface motility may represent a method for facilitating the transport of proteins or protein aggregates to the growing tips of cilia and flagella where the bulk of axonemal growth occurs; indeed, surface motility is exhibited during flagellar assembly (regeneration) and disassembly (chemical-induced resorption).[72] However, this explanation does not account for the continued expression of surface motility in fully grown cilia and flagella, where there is relatively low turnover of axonemal components.

Rapid, bidirectional movements of membrane-bound organelles occur within the axopodia of Heliozoa and the reticulipodia of Foraminifera.[87-90] Polystyrene microspheres and other inert marker objects placed on the external surface of the plasma membrane of Heliozoan axopodia and Foraminiferan reticulopodia also exhibit rapid movements of similar velocity.[89-96]

In *Allogromia*, microspheres move in both directions along the reticulopodia at similar velocities (4.4 ± 2.0 μm/sec outward and 5.8 ± 2.1 μm/sec inward)[95] (Figure 4). The movement properties of surface-associated polystyrene microspheres appear to be very similar to those of the organelles moving within the cytoplasm of the reticulopodia;[95,96] cytoplasmic organelle movement occurs along tracks that are composed of microtubules[96,97] and it appears that the paths of polystyrene microspheres associated with the overlying plasma membrane

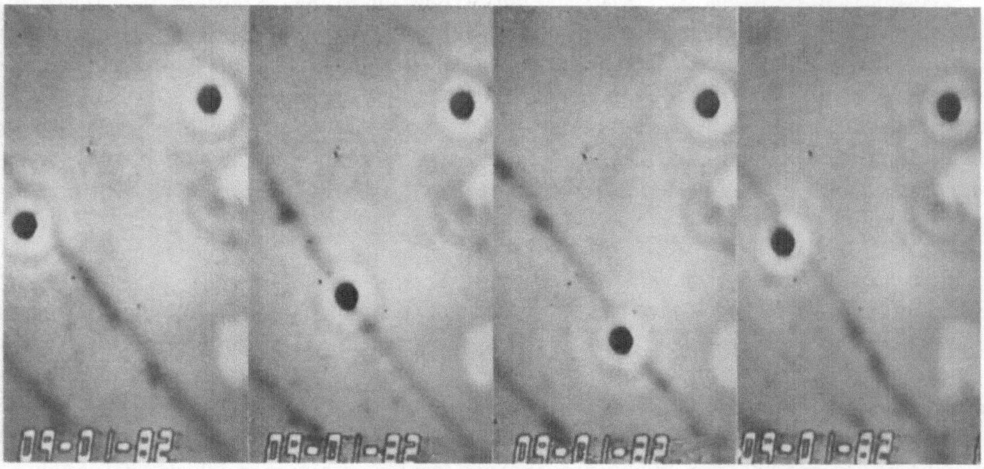

FIGURE 5. Phase contrast videomicrographs of a 0.45 μm polystyrene microsphere moving along the surface of the membrane on a "two-dimensionalized" reticulopodium of *Allogromia* sp. Note that the microsphere is moving in a path that corresponds to that of an intracellular fibril (shown by electron microscopy to be composed of microtubules); the microsphere reverses its direction of movement between the 3rd and 4th frames. Time interval between frames is 0.5 s. (Magnification × 10,000.) (Courtesy of Dr. Samuel Bowser, New York State Department of Health.)

follow closely the location of microtubules underlying the membrane[96] (Figure 5). When membrane blebs are experimentally induced on the reticulopodia, polystyrene-microsphere movement is interrupted at sites where the plasma membrane is pulled away from the underlying microtubules.[98] Both surface motility and cytoplasmic organelle motility are inhibited by colchicine, 4°C and erythro-9-3-(2-hydroxynonyl)adenine (EHNA; an inhibitor of dynein ATPase activity and of certain other enzymes).[96,99]

In the Heliozoan *Echinosphaerium,* microspheres are observed to move only outward along the axopodia at 1.8 ± 0.1 μm/sec;[91,92] however, Troyer[89] reported bidirectional movement of surface-attached objects along the axopodia of a centrohelidian, *Heterophrys,* although he observed that a number of the characteristics of the inward-directed movements differed from those of the outward-directed movements.[100] In the case of *Echinosphaerium nucleofilum,* Suzaki and Shigenaka[93] observed that the surface and cytoplasmic movements were rather different in terms of their sensitivity to various inhibitors; a wide variety of metabolic and other inhibitors affected cytoplasmic organelle movement while only colchicine had any effect on the surface movement of polystyrene microspheres.

There are observations in the literature that suggest that surface motility in Protists, such as occurs on the axopodia of Heliozoa and the reticulopodia of Foraminifera, may be involved in prey ingestion. Grell[101] observed that: "In Heliozoa, Radiolaria, and Foraminifera the food organisms are not taken up directly into the cell body but stick to the axopodia or rhizopodia. They are then carried to the cell by surface cytoplasmic streaming." Leidy[102] observed that food organisms that are brought into contact with the axopodia of the Helizoan *Actinophrys* "glide slowly along them to their base"; Kitching[103] confirmed this observation for *Actinophrys* and Suzaki et al.[104] have reported it in *Echinosphaerium*. These proximally directed movements of food organisms are in contrast to the unidirectional distal-oriented movements of microspheres observed on the surface of *Echinosphaerium* axopodia.[91,92] Cachon and Cachon[105] have suggested that Heliozoan axopodial surface motility may be utilized for whole-cell gliding motility. Choanoflagellates are marine protozoa that possess a feeding apparatus which is a collar composed of processes resembling axopodia. Ellis[106] observed that food organisms (bacteria) were transported up and down the surface of these processes.

Concanavalin A-Sepharose® beads and polystyrene microspheres have been used to study the surface dynamics of gregarine protozoa during gliding motility,[107,108] These markers move backward along the surface of gliding cells and accumulate at the rear of the cell; the velocity of marker movement is independent of size. Trifluoperazine, chlorpromazine, and cytochalasin B inhibit both gliding motility and the movement of microspheres. This suggests that gliding motility and movement of the microspheres are two manifestations of the same motor, as has been demonstrated for *Chlamydomonas*[76,78] and *Cytophaga*[21,23] using non-gliding mutants.

3. Possible Mechanisms for Surface Motility

A number of hypotheses have been proposed to explain the nondiffusional movements of components on the cell surface. These fall into four general categories.

a. Membrane-Associated Cytoskeleton Hypothesis

This model suggests that elements of the cytoskeleton mechanically interact with plasma-membrane components (presumably proteins) and that redistribution of cytoskeletal structures results in movement of membrane domains or of individual membrane proteins. Actin filaments are the best candidates for the structures that associate mechanically with the inside of the plasma membrane in mammalian cells, either directly or via a linking component such as spectrin; microtubules may be a more likely candidate in the case of cilia, flagella, axopodia, and reticulopodia. The energy transduction is presumed to result from a cytoskeletal-associated energy transducer such as myosin or dynein. Certain slow cell surface movements, especially those falling into the category of capping (see Section III.A.1), are likely to be driven by cytoskeletal interactions.[26] This mechanism is also a distinct possibility for explaining certain forms of rapid cell-surface motility (Class I in Table 1).

b. Membrane/Lipid Flow Hypothesis

Bretscher[28,29] and Harris[27] have proposed that slow cell-surface movements on mammalian cells may represent a unidirectional bulk flow of the entire plasma membrane, or at least the lipid bilayer, driven by insertion of membrane at one portion of the cell surface and removal of plasma membrane (presumably by endocytosis) at the opposite end of the cell. This model need not necessarily invoke an association of the cytoskeleton with the plasma membrane and is especially useful in situations such as the *Caenorhabditis* spermatozoan pseudopod,[59,60] where traditional cytoskeletal elements seem to be absent. Although this mechanism might explain many of the slow cell-surface movements (Class I in Table 1), it is an unsatisfactory explanation for rapid, bidirectional cell-surface movements (Class II in Table 1).

c. Surf-Riding Hypothesis

Hewitt[109] and Berlin and Oliver[110] proposed that traveling deformations of the plasma membrane (waves) could be responsible for entraining plasma-membrane proteins and moving them in the plane of the membrane; they suggested this mechanism primarily in relation to capping-like movements. Bowser and Bloodgood[94] reported observations obtained using Heliozoan axopodia and Foraminiferan reticulopodia, where moving plasma-membrane waves occur due to the movement of cytoplasmic organelles; there was no evidence that movement of membrane-surface markers (polystyrene microspheres) correlated with the movement of these membrane deformations. There still exists the possibility that this hypothesis may apply to the slower movements associated with mammalian cell surfaces (Class I in Table 1), but there is little evidence for the existence of propagated membrane waves in these systems.

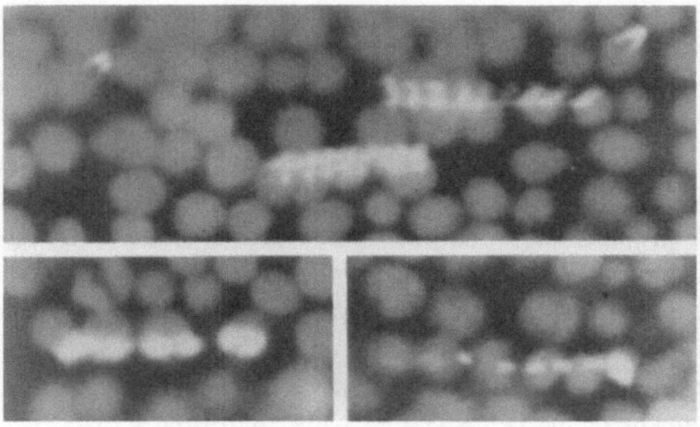

FIGURE 6. Fluorescence micrographs of fluorescent Covaspheres (Covalent Technologies. Inc., Ann Arbor, Mich.) derivatized with heavy meromyosin, moving along the chloroplast rows in an internode cell of *Nitella* that has been cut open. Multiple 1-sec exposures were taken of the same preparation. (Magnification × 1000.) (From Sheetz, M. P. and Spudich, J. A., *Nature (London)*, 303, 31, 1983. With permission.)

d. Electro-Osmosis and Surface-Tension Hypotheses

A number of authors have proposed purely surface chemical phenomena based on surface tension[111] or electro-osmotic flow of fluid[112] to explain cell-surface motility. It is doubtful that these explanations can be reconciled with the characteristics reported for the rapid, local bidirectional cell surface movements (Class II in Table 1) such as occur on the *Chlamydomonas* flagellum.[71,72]

B. Cytoplasmic Motility

Movement of a wide variety of cytoplasmic organelles has been observed within many different cell types;[114] with the exception of centrioles and of certain pigment granules, these transported organelles are generally membrane bound. In many cases, these movements are saltatory, as defined by Rebhun.[115] In general, cytoskeletal systems (either actin and myosin or microtubules and dynein) have been implicated in these movements.[114] Only in recent years have microspheres been utilized to study cytoplasmic organelle motility but they have proven to be quite useful because of their easy detectability by light microscopy (either directly using phase contrast or differential interference contrast optics or after being fluorescently labeled) and the fact that the surface of the microspheres can be controlled (in terms of charge or by adsorption or covalent linkage of specific proteins).

For many years, cell biologists have sought to establish an in vitro system for reproducing cytoplasmic organelle movements. This was only recently accomplished by Sheetz and Spudich[116,117] in an elegant manner utilizing myosin-coated fluorescent microspheres and bundles of actin filaments from the giant internode cells of the alga, *Nitella*. The cells were carefully cut open lengthwise with microscissors to expose the surfaces of the large actin bundles that are thought to be responsible for the dramatic, unidirectional streaming (60 μm/sec) that is a hallmark of this cell. The two sides of the cell exhibit streaming in opposite directions and have been shown to have actin filaments oriented in opposite directions.[118] The authors layered onto this exposed surface 0.7 μm diameter fluorescent *Covaspheres* (Covalent Technologies Inc., Ann Arbor, Mich.) that had been coated with rabbit muscle myosin. The beads were observed to move unidirectionally at velocities of 0.5 to 10 μm/sec (Figure 6); the direction of movement for the two sides of the cell matched that which was expected based on the known polarity of the actin-filament bundles. The velocity of the bead movement was independent of bead size over a range from 0.6 μm Covaspheres to 120 μm agarose beads. If the myosin bound to the beads was inactivated with *N*-ethylmaleimide (NEM), no movement was observed. The movement was dependent upon the presence ATP and magnesium, but calcium was not required. These researchers[117] have shown that different myosin species (skeletal muscle myosin, phosphorylated smooth muscle myosin, and *Dictyostelium myosin* result in different velocities of bead movement and that these velocities are proportional to their actin-activated ATPase activities. The velocities of beads coated with smooth or skeletal muscle myosin correlate with the known in vivo rates of myosin movement along actin filaments in these muscle-cell types. Kohama and Shimmen[156] studied the movement of latex beads coated with *Physarum* myosin along the exposed actin bundles in *Chara,* a close relative of *Nitella.* They found that the presence of calcium inhibited the movement of the beads coated with *Physarum* myosin but was necessary for the movement of beads coated with scallop-muscle myosin. Kachar[157] observed ATP-dependent and calcium-independent movements of organelles along actin bundles in disrupted cytoplasm extruded from *Chara* internode cells; two discrete populations of organelles were observed to move at very different velocities (11 μm/sec vs. 62 μm/sec average velocities).

More recently, Spudich et al.[119] have reported the movement of myosin-coated polystyrene beads in a biochemically reconstituted system. F-actin filaments, polymerized from purified muscle actin, were attached to a transparent solid substrate by their barbed ends and oriented by flowing buffer across the surface; myosin-coated beads attached to and moved unidirectionally along the oriented F-actin filaments toward the barbed end, as would be expected. The bead movement occurred in the absence of calcium or accessory proteins. The velocity of the myosin-coated microsphere is a function of the source of the myosin (skeletal muscle vs. *Dictyostelium*) and agrees closely with those measurements obtained using the *Nitella* system. These exciting observations represent the first fully defined system for visualizing the movement of myosin along an actin filament and focus attention on the fact that at least some cytoplasmic organelle movements may be associated with myosin as the energy transducer and that cytoplasmic membranous organelles may carry their own motor in the form of myosin molecules associated with the cytoplasmic face of their membranes. Workers are now searching for myosin in association with organelle membranes and are seeking purified organelle populations that will move along oriented actin bundles in vitro without addition of exogenous myosin.

Cytoplasmic organelles (membrane-limited) are rapidly moved within the cytoplasm of invertebrate and vertebrate axons in both the anterograde (away from the cell body) and retrograde (toward the cell body) directions and these movements are thought to constitute rapid axoplasmic transport;[120] these movements have been reactivated in permeabilized axons.[121,152] Adams and Bray[122] injected a variety of particulate objects, including polystyrene microspheres derivatized in various ways, polyacrolein beads, paraffin oil, and glass fragments, into the giant axons of the shore crab. In many cases, these objects were observed to move uniformly in the anterograde direction (away from the cell body) with a velocity distribution profile very similar to that of the endogenous organelles. The authors argue that there are few restrictions on the chemical nature of particles that will move in axons although the presence of negative charges appears to be essential (perhaps in order to bind a protein component present in the axoplasm that is essential for expression of motility). These observations are interesting in light of the fact that a nonmembrane bounded object can interact with the motor for organelle transport within the axon. Katz et al.[123] demonstrated that rhodamine-labeled acrylic microspheres (negatively charged; 0.02 to 0.20 μm in diameter) were taken up at axonal endings and retrogradely transported through the axonal cytoplasm to the cell body. No anterograde axoplasmic transport was observed; polystyrene microspheres of comparable size and acrylic microspheres greater than 0.2 μm in diameter were not transported. Although the mechanism by which these microspheres were taken up into the cell and transported along the axon is unknown, the authors hypothesize that they are taken up by endocytosis and the membrane-enclosed microspheres are transported by rapid axoplasmic transport. This hypothesis derives, in part, from the observation of Adams and Bray[122] that negatively charged polystyrene microspheres (0.37 to 0.50 μm diameter) move only in the anterograde direction when injected directly into axons. They observed that these microsphere movements were ATP-dependent and exhibited velocities in the range of 0.4 to 0.6 μm/sec. Allen and coworkers[125] were the first to show that cytoplasm extruded from the squid giant axon exhibited ATP-dependent movement of axonal organelles along microtubules and gliding-type movements of microtubules along a glass surface.[125] Gilbert and Sloboda[128] isolated a population of small membrane-bounded organelles from the axoplasm, labeled these organelles with a fluorescent lipid probe, and demonstrated that they were actively transported after being injected back into extruded squid axoplasm; this motility was abolished if the organelles were protease treated prior to injection into the axoplasm. Allen et al.[125] added fluorescent polystyrene microspheres (0.537 μm in diameter) to extruded squid axoplasm and observed linear movements at 0.135 μm/sec using the fluorescence microscope.

Recent light microscopic observations in combination with electron microscopy have demonstrated that organelles can move bidirectionally along a single microtubule in keratinocytes,[126] in the squid giant axon,[124,125] and in the reticulopodia of certain amoebae.[114] Gilbert and Sloboda[129] have demonstrated movement of membrane-bounded organelles isolated from squid axoplasm in both directions along isolated sea urchin sperm-tail axonemes; these ATP-dependent movements occur in the absence of any dynein arms on the axonemes. This observation suggests that cytoplasmic organelles that move along microtubule tracts possess their own motor attached to the organelle membrane. Vale et al.[130] observed ATP-dependent movement of membranous organelles obtained from squid axoplasm (1.6 μm/sec) and carboxylate-derivatized polystyrene microspheres (0.5 μm/sec) along taxol-stabilized microtubules obtained from squid optic lobes (Figure 7). For both organelles and microspheres, they found that there was a requirement for the soluble supernatant from squid axoplasm; pretreatment of the microspheres or the microtubules with the supernatant was sufficient to obtain movement. The critical factor or factors in the axoplasmic supernatant were nondialyzable, inactivated by heat and trypsin, and inhibited by 100 μ*M* vanadate. This partially reconstituted system provided an excellent bioassay for purification of the molecules necessary for organelle and microsphere movement. It is not clear why the observations made by Gilbert and Sloboda[129] described above did not require any additional axoplasmic components since they appear to be working with the same population of organelles used by Vale et al.;[130] the buffer conditions utilized for the vesicle purification were different and it appears that, in one case, the isolation procedure resulted in loss of the molecules responsible for motility while in the other case they were retained in association with the vesicle membrane. The observations cited above suggest that, at least in the case of axoplasmic transport, there may be a soluble pool of the molecules responsible for organelle movement within the axoplasm and these components may be able to associate initially with either the organelle membrane (or microsphere surface) or the surface of the microtubule.

Building on the observation of Lasek and Brady,[158] Vale et al.[159] used microtubule-affinity chromatography to purify a protein preparation (called kinesin) from squid axoplasm that promoted the movement of microtubules along glass surfaces and the movement of polystyrene microspheres along the surface of brain microtubules in an ATP-dependent manner. The purified preparation contained a major protein component with molecular weight of 110 to 120,000 and a minor doublet component in the 65,000 mol wt range. These workers were not able to demonstrate ATPase activity in the purified kinesin preparation and the purified kinesin was unable to promote endogenous organelle movement along microtubules.

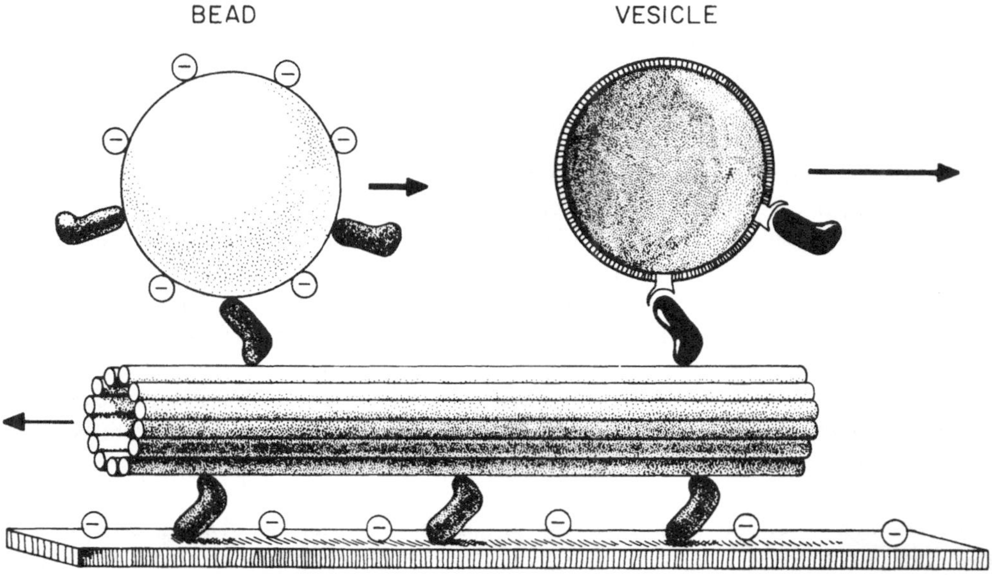

FIGURE 7. Hypothetical scheme showing how certain energy-transducing molecules from the cytoplasm of the squid giant axon may be supporting translocation of membranous organelles and carboxylate microspheres along axonal microtubules, and the movement of axonal microtubules along a solid substrate. The axonal organelle membrane is hypothesized to have specific membrane receptors for the translocator molecule while the carboxylate microsphere and the glass surface nonspecifically adsorb the translocator molecule due to their net negative charge. (From Vale, R. D., Schnapp, B. J., Reese, T. S., and Sheetz, M. P., *Cell*, 40, 559, 1985. With permission.)

This microsphere and microtubule-translocator protein preparation promoted movement in what is equivalent to the anterograde (orthograde) direction of rapid axoplasmic transport only and has been proposed as the mechanism for one direction of rapid axoplasmic transport.[160,163] Kinesin-like proteins have also been reported from sea urchin egg cytoplasm[161] and chick brain.[162] Thus far, no one has been successful in utilizing monoclonal or polyclonal antibodies to kinesin to inhibit organelle movements. Utilizing a monoclonal antibody to the 110 to 120,000 mol wt component of kinesin, Vale et al.[163] have shown that a preparation of squid optic lobe depleted of kinesin contains an activity that promotes movement of polystyrene micropheres along microtubules in what is equivalent to the axonal-retrograde transport direction. Gilbert and Sloboda[164] examined the ATP-binding proteins associated with squid axonal vesticles that are capable of translocating along ciliary axonemal microtubules in the absence of any added soluble axoplasmic proteins. These workers failed to detect kinesin in their vesicle preparations; instead, they observed a high molecular weight (292,000 mol wt) ATP-binding protein that cross-reacted with antibodies specific for mammalian brain MAP2. Microtubule-activated ATPase activity of bovine brain kinesin has recently been reported by Kuznetsov and Gelfand.[167]

Organelle transport within axons is currently the most rapidly moving area in cell motility. Although much remains to be done to sort out the specific molecules and mechanisms involved, it is already clear that rapid axoplasmic transport involves at least one (and possibly two) new molecular mechanisms for cell motility distinct from those previously described, which have been based on myosin interactions with actin filaments or dynein interactions with microtubules. The use of polystyrene microspheres as a means of visualizing force transduction occurring within axons and in reconstituted in vitro systems has played a large role in opening up this new area of cell motility.

Inspired by the work of Adams and Bray,[122] Beckerle[131] microinjected 0.26 μm fluorescent carboxylate-derivatized polystyrene microspheres into cultured cells (BS-C-1 and PtK1) and observed that they exhibited saltatory movements with velocities up to 4.7 μm/sec. High-voltage electron microscopy of the microinjected cells revealed no membrane around the microspheres. Coating the beads with whole-serum proteins or with purified BSA prior to microinjection had no effect on their ability to exhibit motility. An important feature of Beckerle's observations is that the movements of adjacent microspheres within the cytoplasm of the same cell were totally independent, ruling out certain bulk-flow models for organelle movement within these cells. It is possible that the microinjected microspheres adsorbed specific cytoplasmic proteins that then allowed the microspheres to interact with the machinery for movement of endogenous organelles. The motility of the injected microspheres was reversibly inhibited by treatment of the cells with the microtubule-depolymerizing agent nocodazole, suggesting that microtubules are necessary for the expression of this motility. Detection of the movements of these relatively small markers was greatly aided by their fluorescence.

C. Developmental Movements

Polystyrene microspheres and other particulate markers have been utilized in a variety of developmental systems in order to follow the movements of individual cells or sheets of cells. Weston[132] has reviewed the early literature on the use of various cell-marking procedures for developmental studies. In particular, minute particles of carbon, chalk, and carmine have been utilized as cell-surface markers in order to visualize morphogenetic movements during development in meroblastic embryos such as the chick.[133-135] This cell-marking technique has certain potential pitfalls; surface-attached particles can become detached, transferred to the surface of a neighboring cell, or internalized by endocytosis.

Recently, Shimizu[165] utilized carbon particles to mark locations on the surface of the *Tubifex* egg. Using this approach, he was able to demonstrate spiral movements of the egg surface occurring in concert with the movement of organelles in the egg cortex during the process of ooplasmic segregation.

A novel and exciting approach to utilizing polystyrene microspheres in developmental studies has been pioneered by Bronner-Fraser[136-139] who has been studying the mechanism by which cells move (or are moved) along the ventral neural-crest pathway in avian embryos. Untreated polystyrene microspheres or those coated with BSA, Type I collagen, or polytyrosine, when injected into the ventral pathway, were transported along the same pathway taken by neural-crest cells. This translocation of polystyrene microspheres occurred even if the endogenous neural crest was ablated by laser irradiation.[138] However, microspheres coated with fibronectin, laminin, or polylysine were not transported and remained at the site of injection.[137] These observations could not be due to simple charge effects since polylysine is positively charged while fibronectin and laminin are negatively charged. In an effort to understand why fibronectin-coated beads failed to be translocated, Bronner-Fraser[140] coated polystyrene microspheres separately with three different domains of the fibronectin molecule. Beads coated with the heparin-binding and the collagen-binding fragments were translocated; only the beads coated with the fragment containing the cell-binding domain failed to be translocated.

Recent work by Newman and colleagues[154] using a reconstituted extracellular matrix preparation support the concept that the extracellular matrix may play an active role in moving cells through tissues. These workers prepared and joined two gels of pure Type I collagen, one of which contained precartilage mesenchyme cells or polystyrene microspheres and one of which contained human plasma fibronectin; the cells or polystyrene microspheres were observed to migrate rapidly into the gel containing the fibronectin. Both individual 6 μm diameter microspheres and clumps of as many as six of these beads were observed to be moved; heparin-coated microspheres and microspheres derivatized with negative charges (sulfated or carboxylated) were also moved. Small (0.2 μm diameter) polystyrene microspheres, positively charged microspheres, dextran beads, and polyacrylamide beads were not moved. When fragments containing the various domains of the fibronectin molecule were examined for their activity in attracting the microspheres into the second gel the activity was found associated with the heparin-binding domain. Soluble heparin and monoclonal antibodies recognizing the heparin-binding domain of the fibronectin molecule inhibited the migration of the cells and microspheres when added to the fibronectin-containing region of the gel.[155] This remarkable phenomenon is referred to as "matrix-driven translocation" and may have enormous significance for our understanding of the role of the extracellular matrix in cell migration and tissue morphogenesis.

Observations in the literature indicate that labeled cells seeded onto the surface of a solid tissue mass will become internalized into the tissue mass. It has been assumed that this occurs by active "burrowing" of the indicator cells into the tissue[141] and the experiments have been interpreted to suggest that at least some cells within solid tissue masses are motile. Use of inert marker particles has shown that this interpretation is not necessarily correct. Wiseman[142] incubated solid fragments of embryonic chick heart and liver with microspheres made of polystyrene (9.8 μm diameter), aluminum (7 to 15 μm diameter), glass (10 to 15 μm diameter), and stainless steel (8 to 18 μm diameter). After periods of 3 to 4 days in culture, microspheres were observed in sectioned preparations to be located many cell diameters beneath the tissue surface. Dorie et al.[143] examined the uptake of polystyrene microspheres (15 μm diameter) into solid tumor cell spheroids. Within 96 hr, microspheres were observed to be located as far as 280 μm deep into the tissue mass. These two studies suggest that the cells within the solid tissue were able to exert forces upon the inert markers thereby moving them within the tissue.

An interesting use of microspheres has been as ovum surrogates for studies on the mechanism of ovum transport within the oviduct (fallopian or uterine tube).[144-148] A variety of radioactive (^{125}iodine labeled) and nonradioactive microspheres (dextran, glass, copper, silver, and plastic) have been inserted directly into the oviduct or introduced into the peritoneal cavity of humans and rabbits under various physiological conditions. Three major scientific issues have been addressed: (1) Does the behavior of surrogate ova mimic that of natural ova?; (2) What is the effect of size on the transport characteristics?; and (3) What is the effect of the hormonal environment on transport properties? Observations indicated that microspheres are a reasonably good model for transport of natural ova[146], that 200 μm diameter surrogate ova most nearly mimic the natural state,[147] and that the transport characteristics are influenced by estrogen, progesterone, and human chorionic gonatotropin.[146-148]

IV. SUMMARY AND CONCLUSION

A great deal remains to be learned about the mechanisms of cell motility, and it seems apparent that the use of inert microspheres and other particulate objects will continue to play an important role, especially in studies of eukaryotic cell-surface dynamics and the movements associated with cytoplasmic organelles. Methods exist for derivatizing various types of microspheres with a variety of proteins (enzymes, adhesion proteins, contractile proteins); this opens up many as yet unexplored vistas for utilizing microspheres in the study of cell motility. In the words of Richard Harlan;[149]

> "The manor of living nature is so ample, that all may be allowed to sport on it freely; the most jealous proprietor cannot entertain any apprehension that the game will be exhausted, or even perceptibly thinned."

ACKNOWLEDGMENTS

Figures reproduced in this chapter were kindly provided by Dr. H. C. Berg. Dr. S. S.

Bowser, Dr. I. R. Lapidus, Dr. M. P. Sheetz, and Dr. J. A. Spudich, and are reproduced with the permission of the publishers.

Permission to use the quotation from the book *Meet My Maker The Mad Molecule* was provided by the author, J. P. Donleavy.

The author is grateful to the following individuals for providing preprints of their unpublished research: Dr. R. D. Allen, Dr. M. Bonner-Fraser, Dr. S. Gilbert, Dr. M. P. Sheetz, Dr. R. D. Sloboda, and Dr. J. A. Spudich.

The development of this chapter benefited greatly from the comments of a number of the major workers in this field including: Dr. H. C. Berg, Dr. S. S. Bowser, Dr. M. Bronner-Fraser, Dr. R. Burchard, Dr. N. Charon, Dr. S. Gilbert, Dr. A. K. Harris, Dr. I. R. Lapidus, Dr. R. Palazzo, Dr. M. P. Sheetz, Dr. J. A. Spudich, and Dr. J. Travis

The author's original research described here was supported by research grants from the National Institutes of Health (GM28766), the National Science Foundation (DCB-85-02980), and the American Cancer Society (CD-114).

REFERENCES

1. **Doetsch, R. N.,** Motility in procaryotic organisms: problems, points of view, and perspectives, *Biol. Rev.*, 43, 317, 1968.
2. **Silverman, M. and Simon, M.,** Bacterial flagella, *Ann. Rev. Microbiol.*, 31, 397, 1977.
3. **Berg, H. C.,** How bacteria swim, *Sci. Am.*, 233, 36, 1975.
4. **Berg, H. C. and Anderson, R. A.,** Bacteria swim by rotating their flagellar filaments, *Nature (London)*, 245, 380, 1973.
5. **Silverman, M. and Simon, M.,** Flagellar rotation and the mechanism of bacterial motility, *Nature (London)*, 249, 73, 1974.
6. **Berg, H. C., Manson, M. D., and Conley, M. P.,** Dynamics and energetics of flagellar rotation in bacteria, *Symp. Soc. Exp. Biol.*, 35, 1, 1982.
7. **Canale-Parola, E.,** Motility and chemotaxis of spirochetes, *Ann. Rev. Microbiol.*, 32, 69, 1978.
8. **Holt, S. C.,** Anatomy and chemistry of spirochaetes, *Microbiol. Rev.*, 42, 114, 1978.
9. **Berg, H. C.,** How spirochetes may swim, *J. Theor. Biol.*, 56, 269, 1976.
10. **Berg, H. C., Bromley, D. B., and Charon, N. W.,** Leptospiral motility, *Symp. Soc. Gen. Microbiol.*, 28, 285, 1978.
11. **Charon, N. W., Lawrence, C. W., and O'Brien, S.,** Movement of antibody-coated latex beads attached to the spirochete *Leptospira interrogans, Proc. Natl. Acad. Sci. U.S.A.*, 78, 7166, 1981.
12. **Charon, N. W., Daughtry, G. R., McCuskey, R. S., and Franz, G. N.,** Microcinematographic analysis of tethered *Leptospira illini, J. Bacteriol.*, 160, 1067, 1984.
13. **Jarosch, R.,** Gliding, in *Physiology and Biochemistry of Algae*, Lewin, R., Ed., Academic Press, New York, 1962, chap. 36.
14. **Henrichsen, J.,** Bacterial surface translocation: a survey and a classification, *Bacteriol. Rev.*, 36, 478, 1972.
15. **Halfen, L. N.,** Gliding movements, *Encycl. Plant Physiol.*, 7, 250, 1979.
16. **Burchard, R. P.,** Gliding motility of prokaryotes: ultrastructure, physiology, and genetics, *Ann. Rev. Microbiol.*, 35, 497, 1981.
17. **Pringsheim, E. G.,** The relationship between bacteria and myxophyceae, *Bacteriol. Rev.*, 13, 47, 1949.
18. **Schulz, G.,** Bewegungsstudien Sowie Elektronenmikroskopische Membranunterschungen an Cyanophyceen, *Arch. Microbiol.*, 21, 335, 1955.
19. **Dodd, J. D.,** Filament movement in *Oscillatoria sancta* (Kuetz.) gomont, *Trans. Microscop. Soc.*, 79, 480, 1960.
20. **Weibull, C.,** Movement, in *The Bacteria*, Vol. 1, Gunsalus, I. C. and Stanier, R. Y., Eds., Academic Press, New York, 1960, chap. 4.
21. **Pate, J. L. and Chang, L.-Y. E.,** Evidence that gliding motility in prokaryotic cells is driven by rotary assemblies in the cell envelopes, *Curr. Microbiol.*, 2, 59, 1979.
22. **Ridgway, H. F., Wagner, R. M., Dawsey, W. T., and Lewin, R. A.,** Fine structure of the cell envelope layers of *Flexibacter polymorphus, Can. J. Microbiol.*, 21, 1733, 1975.

23. **Lapidus, I. R. and Berg, H. C.,** Gliding motility of *Cytophaga* sp. strain U67, *J. Bacteriol.*, 151, 384, 1982.
24. **Dworkin, M., Keller, K. H., and Weisberg, D.,** Experimental observations consistent with a surface tension model of gliding motility of *Myxococcus xanthus*, *J. Bacteriol.*, 155, 1367, 1983.
25. **de Petris, S.,** Distribution and mobility of plasma membrane components on lymphocytes, in *Dynamic Aspects of Cell Surface Organization*, Vol. 3, Poste, G. and Nicolson, G. L., Eds., North Holland Publishing, Amsterdam, 1977, 643.
26. **Bourguignon, L. Y. W. and Bourguignon, G. J.,** Capping and the cytoskeleton, *Int. Rev. Cytol.*, 87, 195, 1984.
27. **Harris, A. K.,** Recycling of dissolved plasma membrane components as an explanation of the capping phenomenon, *Nature (London)*, 263, 781, 1976.
28. **Bretscher, M. S.,** Directed lipid flow in cell membranes, *Nature (London)*, 260, 21, 1976.
29. **Bretscher, M. S.,** Endocytosis: relation to capping and cell locomotion, *Science*, 224, 681, 1984.
30. **Mast, S. O.,** Structure, movement, locomotion, and stimulation in amoeba, *J. Morphol. Physiol.*, 41, 347, 1926.
31. **Jennings, H. S.,** Contribution to the study of the behavior of lower organisms, *Pub. Carnegie Inst. Washington*, 257, 1906.
32. **Griffin, J. L. and Allen, R. D.,** The movement of particles attached to the surface of amebae in relation to current theories of ameboid movement, *Exp. Cell Res.*, 20, 619, 1960.
33. **Abe, T. H.,** Morpho-physiological study of ameboid movement. II. Ameboid movement and the organization pattern in a striate ameba, *Cytologia*, 27, 111, 1962.
34. **Shaffer, B. M.,** Behavior of particles adhering to amoebae of the slime mold *Polyspondylium violaceum* and the fate of the cell surface during locomotion, *Exp. Cell Res.*, 32, 603, 1963.
35. **Shaffer, B. M.,** Intracellular movement and locomotion of cellular slime-mold amebae, in *Primative motile systems in cell biology*, Allen, R. D. and Kamiya, N., Eds., Academic Press, New York, 1964, 387.
36. **Goldacre, R. J.,** The role of the cell membrane in the locomotion of amoebae, and the source of the motive force and its control by feedback, *Exp. Cell Res.*, Suppl 8, 1, 1961.
37. **Ingram, V. M.,** A side view of moving fibroblasts, *Nature (London)*, 222, 641, 1969.
38. **Abercrombie, M., Heaysman, J. E. M., and Pegrum, S. M.,** The locomotion of fibroblasts in culture. III. Movements of particles on the dorsal surface of the leading lamella, *Exp. Cell Res.*, 62, 389, 1970.
39. **Harris, A. and Dunn, G.,** Centripetal transport of attached particles on both surfaces of moving fibroblasts, *Exp. Cell Res.*, 73, 519, 1972.
40. **Harris, A. K.,** Cell surface movements related to cell locomotion, *Ciba Found. Symp.*, 14, 3, 1973.
41. **Weatherbee, J. A.,** Membranes and cell movement: interactions of membranes with the proteins of the cytoskeleton, *Int. Rev. Cytol*, 12, 113, 1981.
42. **Geiger, B.,** Membrane-cytoskeleton interaction, *Biochim. Biophys. Acta*, 737, 305, 1983.
43. **Dembo, M. and Harris, A. K.,** Motion of particles adhering to the leading lamella of crawling cells, *J. Cell Biol.*, 91, 528, 1981.
44. **Dembo, M., Tuckerman, L., and Goad, W.,** Motion of polymorphonuclear leukocytes: theory of receptor redistribution and the frictional force on a moving cell, *Cell Motility*, 1, 205, 1981.
45. **Middleton, C. A.,** Cell-surface labeling reveals no evidence for membrane assembly and disassembly during fibroblast locomotion, *Nature (London)*, 282, 203, 1979.
46. **Albrecht-Buhler, G. and Goldman, R. D.,** Microspike-mediated particle transport towards the cell body during early spreading of 3T3 cells, *Exp. Cell Res.*, 97, 329, 1976.
47. **Isenberg, G. and Small, J. V.,** Particle movement on microspikes, *Exp. Cell Res.*, 121, 406, 1979.
48. **DiPasquale, A. and Bell, P. B., Jr.,** The upper cell surface: its inability to support active cell movement in culture, *J. Cell Biol.*, 62, 198, 1974.
49. **DiPasquale, A.,** Epithelial cell locomotion: factors involved in extension of the leading edge, *J. Cell Biol.*, 59, 82a, 1973.
50. **Kubota, H. Y.,** Creeping locomotion of the endodermal cells dissociated from gastrulae of the Japanese newt, *Cynops pyrrhogaster*, *Exp. Cell Res.*, 133, 137, 1981.
51. **Godman, G. C., Miranda, A. F., Deitch, A. D., and Tanenbaum, S. W.,** Action of cytocholasin D on cells of established lines. III. Zeiosis and movements at the cell surface, *J. Cell Biol.*, 64, 644, 1975.
52. **Albrecht-Buhler, G.,** A quantitative difference in the movement of marker particles in the plasma membrane of 3T3 mouse fibroblasts and their polyoma transformants, *Exp. Cell Res.*, 78, 67, 1973.
53. **Albrecht-Buhler, G. and Yarnell, M. M.,** A quantitation of movement of marker particles in the plasma membrane of 3T3 mouse fibroblasts, *Exp. Cell Res.*, 78, 59, 1973.
54. **Albrecht-Buhler, G., and Solomon, F.,** Properties of particle movement in the plasma membrane of 3T3 mouse fibroblasts, *Exp. Cell Res.*, 85, 225, 1974.
55. **Marcus, P. I.,** Dynamics of surface modifications in myxovirus-infected cells, *Cold Spring Harbor Symp. Quant. Biol.*, 27, 351, 1962.

56. **Holtzman, E. and Mercurio, A. M.,** Membrane circulation in neurons and photoreceptors: some unresolved issues, *Int. Rev. Cytol.,* 67, 1, 1980.
57. **Bray, D.,** Surface movements during the growth of single explanted neurons, *Proc. Natl. Acad. Sci., U.S.A.,* 65, 905, 1970.
58. **Koda, L. Y. and Partlow, L. M.,** Membrane marker movement on sympathetic axons in tissue culture, *J. Neurobiol.,* 7, 157, 1976.
59. **Roberts, T. M. and Ward, S.,** Membrane flow during nematode spermiogenesis, *J. Cell Biol.,* 92, 113, 1982.
60. **Roberts, T. M. and Ward, S.,** Centripetal flow of pseudopodial surface components could propel the amoeboid movement of *Caenorhabditis elegans* spermatozoa, *J. Cell Biol.,* 92, 132, 1982.
61. **Maher, P. and Molday, R. S.,** Differences in the redistribution of cancanavalin A and wheat germ agglutinin binding sites on mouse neuroblastoma cells, *J. Supramol. Struct.,* 10, 61, 1979.
62. **Smith, C. W. and Hollers, J. C.,** Motility and adhesiveness in human neutrophils. Redistribution of chemotactic factor-induced adhesion sites, *J. Clin. Invest.,* 65, 804, 1980.
63. **Anderson, D. C., Mace, M. L., Brinkley, B. R., Martin, R. R., and Smith, C. W.,** Recurrent infection in glycogenosis type Ib: abnormal neutrophil motility related to impaired redistribution of adhesion sites, *J. Infect. Dis.,* 143, 447, 1981.
64. **Anderson, D. C., Hughes, B. J., and Smith, C. W.,** Abnormal motility of neonatal polymorphonuclear leukocytes. Relationship to impaired redistribution of surface adhesion sites by chemotactic factor or colchicine, *J. Clin. Invest.,* 68, 863, 1981.
65. **Berlin, R. D. and Oliver, J. M.,** Analogous ultrastructure and surface properties during capping and phagocytosis in leukocytes, *J. Cell Biol.,* 77, 789, 1978.
66. **Yahara, I. and Edelman, G. M.,** Modulation of lymphocyte receptor mobility by locally bound concanavalin A, *Proc. Natl. Acad. Sci. U.S.A.,* 72, 1579, 1975.
67. **Peng, H. B., Cheng, P.-C., and Luther, P. W.,** Formation of ACh receptor clusters induced by positively charged latex beads, *Nature (London),* 292, 831, 1981.
68. **Peng, H. B., and Cheng, P.-C.,** Formation of postsynaptic specializations induced by latex beads in cultured muscle cells, *J. Neurosci.,* 1, 1760, 1982.
69. **Peng, H. B. and Phelan, K. A.,** Early cytoplasmic specialization at the presumptive acetylcholine receptor cluster: a meshwork of thin filaments, *J. Cell Biol.,* 99, 344, 1984.
70. **Bloodgood, R. A.,** The flagellum as a model system for studying dynamic cell surface events, *Cold Spring Harbor Symp. Quant. Biol.,* 46, 683, 1981.
71. **Bloodgood, R. A.,** Dynamic properties of the flagellar surface, *Symp. Soc. Exp. Biol.,* 35, 353, 1982.
72. **Bloodgood, R. A.,** Rapid motility occurring in association with the *Chlamydomonas* flagellar membrane, *J. Cell Biol.,* 75, 983, 1977.
73. **Bloodgood, R. A., Leffler, E. M., and Bojczuk, A. T.,** Reversible inhibition of *Chlamydomonas* flagellar surface motility, *J. Cell Biol.,* 82, 664, 1979.
74. **Hoffman, J. L. and Goodenough, U. W.,** Experimental dissection of flagellar surface motility in *Chlamydomonas, J. Cell Biol.,* 86, 656, 1980.
75. **Lewin, R. A.,** Studies on the flagella of algae. I. General observations on *Chlamydomonas moewusii* gerloff, *Biol. Bull.,* 103, 74, 1952.
76. **Bloodgood, R. A.,** Flagella-dependent gliding motility in *Chlamydomonas, Protoplasma,* 106, 183, 1981.
77. **Mesland, D. A. M.,** Mating in *Chlamydomonas eugametos.* A scanning electron microscopical study, *Arch. Microbiol.,* 109, 31, 1976.
78. **Lewin, R. A.,** A new kind of motility mutant (non-gliding) in *Chlamydomonas, Experientia,* 38, 348, 1982.
79. **Snell, W. J., Buchanan, M., and Clausell, A.,** Lidocaine reversibly inhibits fertilization in *Chlamydomonas:* a possible role in sexual signaling, *J. Cell Biol.,* 94, 607, 1982.
80. **Detmers, P. A. and Condeelis, J.,** Trifluoperazine and W-7 inhibit mating in *Chlamydomonas* at an early stage of gametic interaction, *Exp. Cell Res.,* 163, 317, 1986.
81. **Bloodgood, R. A. and May, G. S.,** Functional modification of the *Chlamydomonas* flagellar surface, *J. Cell Biol.,* 93, 88, 1982.
82. **Bloodgood, R. A. and Workman, L. J.,** A flagellar surface glycoprotein mediating cell-substrate interaction in *Chlamydomonas, Cell Motil.,* 4, 77, 1984.
83. **Piperno, G. and Luck, D. J. L.,** An actin-like protein is a component of axomenes from *Chlamydomonas* flagella, *J. Biol. Chem.,* 254, 2187, 1979.
84. **Watanabe, T. and Flavin, M.,** Nucleotide-metabolizing enzymes in *Chlamydomonas* flagella, *J. Biol. Chem.,* 251, 182, 1976.
85. **Bessen, M., Fay, R. B., and Witman, G. B.,** Calcium control of waveform in isolated flagellar axonemes of *Chlamydomonas, J. Cell Biol.,* 86, 446, 1980.
86. **Bloodgood, R. A.,** Direct visualization of dynamic membrane events in cilia, *J. Exp. Zool.,* 213, 293, 1980.

87. **Fitzharris, T. P., Bloodgood, R. A., and McIntosh, J. R.,** Particle movement in the axopodia of *Echinosphaerium:* evidence concerning the role of the axoneme, *J. Mechanochem. Cell. Motil.*, 1, 117, 1972.
88. **Edds, K. T.,** Motility in *Echinosphaerium nucleofilum*. I. An analysis of particle motions in the axopodia and a direct test of the involvement of the axoneme, *J. Cell Biol.*, 66, 145, 1975.
89. **Troyer, D.,** Possible involvement of the plasma membrane in saltatory particle movement in Heliozoan axopods, *Nature (London)*, 254, 696, 1975.
90. **Jahn, T. L. and Rinaldi, R. A.,** Protoplasmic movement in the foraminiferan, *Allogromia laticollaris;* and a theory of its mechanism, *Biol. Bull.*, 117, 100, 1959.
91. **Bloodgood, R. A.,** Unidirectional motility occurring in association with the axopodial membrane of *Echinosphaerium nucleofilum, Cell Biol. Int. Rep.*, 2, 171, 1978.
92. **Kanno, F. and Ishii, K.,** Movement of the surface layer on axopodium and its protoplasm in *Actinosphaerium, Bull. Fac. Liberal Arts, Hosei Univ.*, 31, 1, 1979.
93. **Suzaki, T. and Shigenaka, Y.,** Intra-axopodial particle movement and axopodial surface motility, in *Biological Functions of Microtubules and Related Structures*, Sakai, H., Mohri, H., and Borisy, G. G., Eds., Academic Press, New York, 1982, chap. 10.
94. **Bowser, S. S. and Bloodgood, R. A.,** Evidence against surf-riding as a general mechanism for surface motility, *Cell Motil.*, 4, 305, 1984.
95. **Bowser, S. S., Israel, H. A., McGee-Russell, S. M., and Rieder, C. L.,** Surface transport properties of Reticulopodia: do intracellular and extracellular motility share a common mechanism?, *Cell Biol. Int. Rep.*, 8, 1051, 1984.
96. **Bowser, S. S. and Rieder, C. L.,** Evidence that cell surface motility in *Allogromia* is mediated by cytoplasmic microtubules, *Can. J. Biochem. Cell Biol.*, 63, 608, 1985.
97. **Travis, J. L., Kenealy, J. F. X., Allen, R. D.,** Studies on the motility of the foraminifera. II. The dynamic microtubular cytoskeleton of the reticulopodial network of *Allogromia laticollaris, J. Cell Biol.*, 97, 1668, 1983.
98. **Bowser, S. S. and Rieder, C. L.,** Microtubule-dependent reticulopodial surface motility: reversible inhibition on plasma membrane blebs, *Ann. N.Y. Acad. Sci.*, 466, 933, 1986.
99. **Bowser, S., McGee-Russell, S. M., and Rieder, C.,** EHNA inhibits particle movement in and on the reticulopods of *Allogromia* sp. strain NF (Lee), *J. Protozool.*, 30, 12a, 1983.
100. **Troyer, D.,** Mechanism of saltatory particle movement in Heliozoan axopods, *J. Cell Biol.*, 70, 406a, 1976.
101. **Grell, K. G.,** *Protozoology*, Springer-Verlag, New York, 1973, 333.
102. **Leidy, J.,** *Fresh-water Rhizopods of North America*, U.S. Geological Survey, Government Printing Office, Washington, D. C., 1879, 238.
103. **Kitching, J. A.,** Responses of the Heliozoan *Actinophrys sol* to prey, to mechanical stimulation, and to solutions of proteins and certain other substances, *J. Exp. Biol.*, 37, 407, 1960.
104. **Suzaki, T., Shigenaka, Y., Watanabe, S., and Toyohara, A.,** Food capture and ingestion in the large Heliozoan, *Echinosphaerium nucleofilum, J. Cell Sci.*, 42, 61, 1980.
105. **Cachon, J. and Cachon-Enjumet, M.,** Les movements de cyclose dans les axopodes d'Acanthaires. Leur role lors de la nutrition et de la locomotion. *Bull. Inst. Oceanogr.*, 61, 1, 1964.
106. **Ellis, W. N.,** Recent researches on the Choanoflagellata, *Ann. Soc. R. Zool. Belg.*, 60, 49, 1929.
107. **King, C. A.,** Cell surface interaction of the protozoan *Gregarina* with Concanavalin A beads—implications for models of gregarine gliding, *Cell Biol. Int. Rep.*, 5, 297, 1981.
108. **King, C. A.,** A cell surface linear motor — studies on the motility of gregarine protozoans, *J. Cell Biol.*, 99, 182a, 1984.
109. **Hewitt, J. A.,** Surf-riding model for cell capping, *J. Theor. Biol.*, 80, 115, 1979.
110. **Berlin, R. D. and Oliver, J. M.,** The movement of bound ligands over cell surfaces, *J. Theor. Biol.*, 99, 69, 1982.
111. **Sebba, F.,** A surface-chemical basis for cell motility, *J. Theor. Biol.*, 78, 375, 1979.
112. **McLaughlin, S. and Poo, M.,** The role of electro-osmosis in the electric field-induced movement of charged macromolecules on the surfaces of cells, *Biophys. J.*, 34, 85, 1981.
113. **Keller, K. H., Grady, M., and Dworkin, M.,** Surface tension gradients: feasible model for gliding motility of *Myxococcus xanthus, J. Bacteriol.*, 155, 1358, 1983.
114. **Koonce, M. P. and Schliwa, M.,** Bidirectional organelle transport can occur in cell processes that contain single microtubules, *J. Cell Biol.*, 100, 322, 1985.
115. **Rebhun, L. I.,** Polarized intracellular particle transport: saltatory movements and cytoplasmic streaming, *Int. Rev. Cytol.*, 32, 93, 1972.
116. **Sheetz, M. P. and Spudich, J. A.,** Movement of myosin-coated fluorescent beads on actin cables *in vitro, Nature (London)*, 303, 31, 1983.
117. **Sheetz, M. P., Chasan, R., and Spudich, J. A.,** ATP-dependent movement of myosin *in vitro:* characterization of a quantitative assay, *J. Cell Biol.*, 99, 1867, 1984.

118. **Kersey, Y. M., Hepler, P. K., Palevitz, B. A., and Wessells, N. K.,** Polarity of actin filaments in Characean algae, *Proc. Natl. Acad. Sci. U.S.A.*, 73, 165, 1976.
119. **Spudich, J. A., Kron, S. J., and Sheetz, M. P.,** Movement of myosin-coated beads on oriented filaments reconstituted from purified actin, *Nature (London)*, 315, 584, 1985.
120. **Forman, D. S.,** Saltatory organelle movement and the mechanism of fast axonal transport, in *Axoplasmic Transport*, Vol. 1, Weiss, D. G., Ed., Springer-Verlag, Heidelberg, 1982, 424.
121. **Forman, D. S., Brown, K. J., and Livengood, D. R.,** Fast axonal transport in permeabilized lobster giant axons is inhibited by vanadate, *J. Neurosci.*, 3, 1279, 1983.
122. **Adams, R. J. and Bray, D.,** Rapid transport of foreign particles microinjected into crab axons, *Nature (London)*, 303, 718, 1983.
123. **Katz, L. C., Burkhalter, A., and Dreyer, W. J.,** Fluorescent latex microspheres as a retrograde neuronal marker for *in vivo* and *in vitro* studies of visual cortex, *Nature (London)*, 310, 498, 1984.
124. **Schnapp, B. J., Sheetz, M. P., Vale, R. D., and Reese, T. S.,** Transport filaments from squid axoplasm consist primarily of single microtubules, *Soc. Neurosci. Abstr.*, 10, 1087, 1984.
125. **Allen, R. D., Weiss, D. G., Hayden, J. H., Brown, D. T., Fujiwake, H., and Simpson, M.,** Gliding movement of and bidirectional transport along single native microtubules from squid axoplasm: evidence for an active role of microtubules in cytoplasmic transport, *J. Cell Biol.*, 100, 1736, 1985.
126. **Hayden, J. H. and Allen, R. D.,** Detection of single microtubules in living cells: particle transport can occur in both directions along the same microtubule, *J. Cell Biol.*, 99, 1785, 1984.
127. **Koonce, M. and Schliwa, M.,** Intracellular organelles can move in both directions along a single microtubule, *J. Cell Biol.*, 99, 48a, 1984.
128. **Gilbert, S. P. and Sloboda, R. D.,** Bidirectional transport of fluorescently labeled vesicles introduced into extruded axoplasm of squid *Loligo pealei*, *J. Cell Biol.*, 99, 445, 1984.
129. **Gilbert, S. P., Allen, R. D., and Sloboda, R. D.,** Translocation of vesicles from squid axoplasm on flagellar microtubules, *Nature (London)*, 315, 245, 1985.
130. **Vale, R. D., Schnapp, B. J., Reese, T. S., and Sheetz, M. P.,** Organelle, bead and microtubule translocations promoted by soluble factors from the squid giant axon, *Cell*, 40, 559, 1985.
131. **Beckerle, M. C.,** Microinjected fluorescent polystyrene beads exhibit saltatory motion in tissue culture cells, *J. Cell Biol.*, 98, 2126, 1984.
132. **Weston, J. A.,** Cell marking, in *Methods in Developmental Biology*, Wilt, F. H. and Wessells, N, K., Eds., Thomas Y. Croswell, New York, 1967, 723.
133. **Spratt, N. T., Jr.,** Formation of the primitive streak in the explanted chick blastoderm marked with carbon particles, *J. Exp. Zool.*, 103, 259, 1946.
134. **Spratt, N. T., Jr.,** Regression and shortenings of the primitive streak in the explanted chick blastoderm, *J. Exp. Zool.*, 104, 69, 1947.
135. **Spratt, N. T., Jr. and Haas, H.,** Germ layer formation and the role of the primitive streak in the explanted chick blastoderm, *J. Exp. Zool.*, 158, 9, 1965.
136. **Bronner-Fraser, M.,** Distribution of latex beads and retinal pigment epithelial cells along the ventral neural crest pathway, *Dev. Biol.*, 91, 50, 1982.
137. **Bronner-Fraser, M.,** Latex beads as probes of a neural crest pathway: effects of laminin, collagen, and surface charge on bead translocation, *J. Cell Biol.*, 98, 1947, 1984.
138. **Coulombe, J. N. and Bronner-Fraser, M.,** Translocation of latex beads after laser ablation of the avian neural crest, *Dev. Biol.*, 106, 121, 1984.
139. **Bronner-Fraser, M.,** Latex beads with defined surface coats as probes of neural crest migratory pathways, in *The Role of the Extracellular Matrix in Development*, R. L. Trelstad, Ed., Alan R. Liss, New York, 1984, 399.
140. **Bronner-Fraser, M.,** Effects of different fragments of the fibronectin molecule on latex bead translocation along neural crest migratory pathways, *Dev. Biol.*, 108, 131, 1985.
141. **Wiseman, L. L. and Steinberg, M. S.,** The movement of single cells within solid tissue masses, *Exp. Cell Res.*, 79, 468, 1973.
142. **Wiseman, L. L.,** Contact inhibition and the movement of metal, glass and plastic beads within solid tissues, *Experientia*, 33, 734, 1977.
143. **Dorie, M. J., Kallman, R. F., Rapacchietta, D. F., Van Antwerp, D., Huang, Y. R.,** Migration and internalization of cells and polystyrene microspheres in tumor cell spheroids, *Exp. Cell Res.*, 141, 201, 1982.
144. **Diaz, J., Vasquez, J., Diaz, S., Diaz, F., and Croxatto, H. B.,** Transport of ovum surrogates by the human oviduct, in *Ovum Transport and Fertility Regulation*. Harper, M. J. K., Pauerstein, C. J., Adams, C. E., Coutinho, E. M., Croxatto, H. B., and Paton, D. M., Eds., Scriptor, Copenhagen, 1976, 404.
145. **Crosby, R. J., Chatkoff, M. L., and Pauerstein, C. J.,** Methods for studying ovum transport rates, in *Ovum Transport and Fertility Regulation*, Harper, M. J. K., Pauerstein, C. J., Adams C. E., Coutinho, E. M., Croxatto, H. B., and Paton, D. M., Eds., Scriptor, Copenhagen, 1976, 99.

146. **Pauerstein, C. J., Hodgson, F. J., Young, R. J., Chatkoff, M. L., and Eddy, C. A.,** Use of radioactive microspheres for studies of tubal ovum transport, *Am. J. Obstet. Gynecol.*, 122, 655, 1975.
147. **Hodgson, B. J., Croxatto, H. B., Vargas, M. I., and Pauerstein, C. J.,** Effect of particle size on time course of transport of surrogate ova through the rabbit oviduct, *Obstet. Gynecol.*, 47, 213, 1976.
148. **Overstrom, E. W., Bigsby, R. M., and Black, D. L.,** Effects of physiological levels of estradiol-17beta and progesterone on oviduct edema and ovum transport in the rabbit, *Biol. Reprod.*, 23, 100, 1980.
149. **Harlan, R.,** *Fauna Americana: Being a Description of the Mammiferous Animals Inhabiting North America,* Anthony Finley, Philadelphia, 1835; reproduced in *Science,* 225, 160, 1984.
150. **Czarska, L. and Grebecki, A.,** Membrane folding and plasma-membrane ratio in the movement and shape transformation in *Amoeba proteus, Acta Protozool.*, 4, 201, 1966.
151. **Chapman-Andresen, C.,** Surface renewal in *Amoeba proteus, J. Protozool.*, Suppl. 11, 11, 1964.
152. **Adams, R. J.,** Organelle movement in axons depends on ATP, *Nature (London),* 297, 327, 1982.
153. **Castenholz, R. W.,** Motility and taxes, in *The Biology of the Cyanobacteria,* Carr, N. G. and Whitton, B. A., Eds., University of California, 19, 413, 1982.
154. **Newman, S. A., Frenx, D. A., Tomasek, J. J., and Rabuzzi, D. D.,** Matrix-driven translocation of cells and nonliving particles, *Science,* 228, 885, 1985.
155. **Newman, S. A., Frenz, D. A., Hasegawa, E., and Akiyama, S. K.,** Matrix-driven translocation: dependence on the heparin binding domain of fibronectin, *J. Cell Biol.*, 99, 167a, 1984.
156. **Kohama, K., and Shimmen, T.,** Inhibitory Ca^{2+}-control of movement of beads coated with *Physarum* along actin-cables in *Chara* internode cells, *Protoplasma,* 129, 88, 1985.
157. **Kachar, B.,** Direct visualization of organelle movement along actin filaments dissociated from Characean algae, *Nature (London),* 227, 1355, 1985.
158. **Lasek, R. J. and Brady, S. T.,** Attachment of transported vesicles to microtubules in axoplasm is facilitated by AMP-PNP, *Nature (London),* 316, 645, 1985.
159. **Vale, R. D., Reese, T. S., and Sheetz, M. P.,** Identification of a novel force-generating protein, kinesin, involved in microtubule-based motility, *Cell,* 42, 39, 1985.
160. **Schnapp, B. J. and Reese, T. S.,** New developments in understanding rapid axonal transport, *Trends Neurosci.* 9, 155, 1986.
161. **Scholey, J. M., Porter, M. E., Grissom, P. M., and McIntosh, J. R.** Identification of kinesin in sea urchin eggs, and evidence of its localization in the mitotic spindle, *Nature (London),* 318, 483, 1985.
162. **Brady, S. T.,** A novel brain ATPase with properties expected for the fast axonal transport motor, *Nature (London),* 317, 73, 1985.
163. **Vale, R. D., Schnapp, B. J., Mitchison, T., Steuer, E., Reese, T. S., and Sheetz, M. P.,** Different axoplasmic proteins generate movement in opposite directions along microtubules in vitro, *Cell,* 43, 623, 1985.
164. **Gilbert, S. P. and Sloboda, R. D.,** Identification of a MAP 2-like ATP binding protein associated with axoplasmic vesicles that translocate on isolated microtubules, *J. Cell Biol.*, 103, 947, 1986.
165. **Shimizu, T.,** Bipolar segregation of mitochondria, actin network, and surface in the *Tubifex* egg: role of cortical polarity, *Dev. Biol.*, 116, 241, 1986.
166. **Rosenbaum, J. R.,** personal communication.
167. **Kuznetsov, S. A. and Gelfand, V. I.,** Bovine brain kinesin is a microtubule-activated ATPase, *Proc. Nat. Acad. Sci. U.S.A.*, 83, 8530, 1986.

Chapter 11

APPLICATIONS AND CHARACTERIZATION OF RADIOLABELED OR MAGNETIZABLE NANO- AND MICROPARTICLES FOR RES, LYMPH, AND BLOOD FLOW STUDIES

Sven-Erik Strand, Lena Andersson, and Lennart Bergqvist

TABLE OF CONTENTS

I. INTRODUCTION

For diagnostic purpose, particles can be labeled with radionuclides or contain magnetizable materials making them suitable for external measurements of their biokinetics in vivo. In therapy, particles can be used as targetable drug-delivery systems or, if containing radioactivity, for internal radiation therapy. Loaded with, e.g., stable iodine, particles may also be used for photon activation therapy.

The particle nomenclature is mainly based on the size. Particles in the submicrometer size range are referred to as nanoparticles whereas larger particles will be called microparticles. The denomination colloids is in most literature used for both nano- and microparticles less than a few micrometers. Figure 1 shows the main biological behavior after intravascular injection for particles of different size ranges. The sizes of some cellular components and molecules are also indicated in this figure.

The biological behavior of particles depends strongly on their size and the mode of delivery to the organism. After interstitial injection, particles in the size range of a few nanometers up to about 100 nm enter the lymphatic capillaries, where they are transported with the lymph-to-lymph nodes, where phagocytosis occur. Thus, studies of the lymphatic drainage and localization of lymph nodes are possible. After intravenous/intraarterial injection, particles less than about 2 μm will rapidly be cleared from the blood stream by the reticuloendothelial system (RES). Particles larger than 7 μm will, after intravenous injection, be mechanically entrapped in the lung capillaries. After intraarterial injection, the entrapment will be in the first capillary bed reached. If inhaled, the particles will be trapped in the alveolar macrophages.

II. CHARACTERIZATION OF RADIOLABELED NANOPARTICLES

A. Radiocolloids

Many different radiolabeled colloids have been used for about 30 years in nuclear medicine for liver-spleen imaging. Nanoparticles in liver scintigraphy have been used frequently in

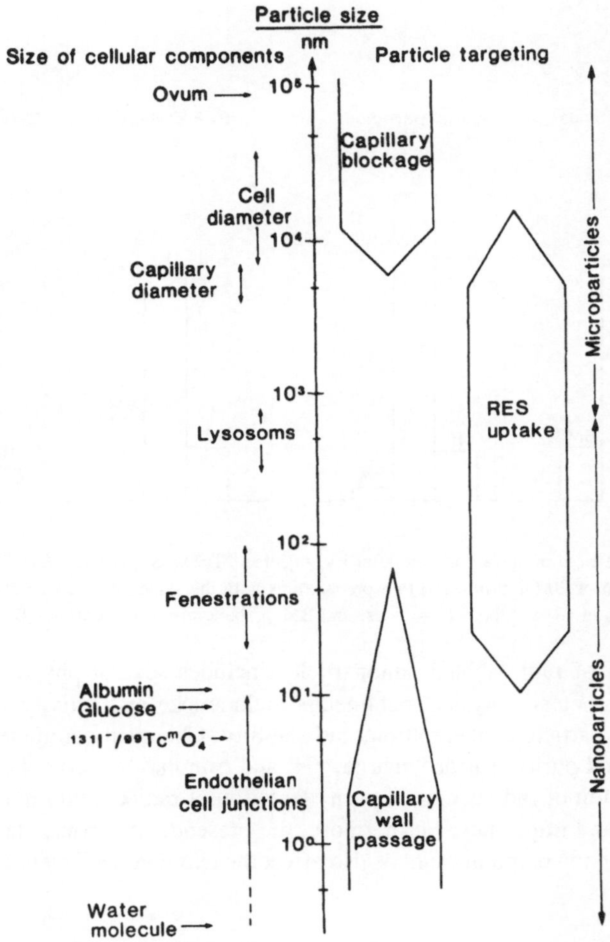

FIGURE 1. Size dependence on particle targeting after intravascular injection. Also indicated are the commonly used size intervals for the definitions of nano- and microparticles. Some sizes of important cellular components and molecules are included.

medical diagnosis since the 1950s, when colloidal gold (^{198}Au) was introduced. Today there are many radiopharmaceuticals for liver scintigraphy, most of them consisting of nanoparticles labeled with ^{99}Tc^m. The clinical work has led to an extensive number of published papers. During the last decade lymphoscintigraphy with radiolabeled colloids has also been proved to be useful in several clinical situations.

The radiocolloids used at present can be divided in two categories: (a) inert colloids such as ^{99}Tc^m-labeled sulfur, antimony sulfide, tin, rhenium, and phytate colloids, and (b) biodegradable colloids such as ^{99}Tc^m-labeled microaggregated albumin and liposomes.

The development of radiocolloids has often been made with emphasis on rapidity and efficiency of labeling, stability, and reasonable particle size.[1] Thus, considerations of physicochemical factors of the colloids have been less stressed.

B. Physicochemical Factors

Very few articles concerning the physicochemical properties of different colloids have been published. This fact is a considerable drawback, because a good knowledge of the physicochemical properties will ease the search for optimal radiocolloids, as well as explain differences in biological behavior of colloids.

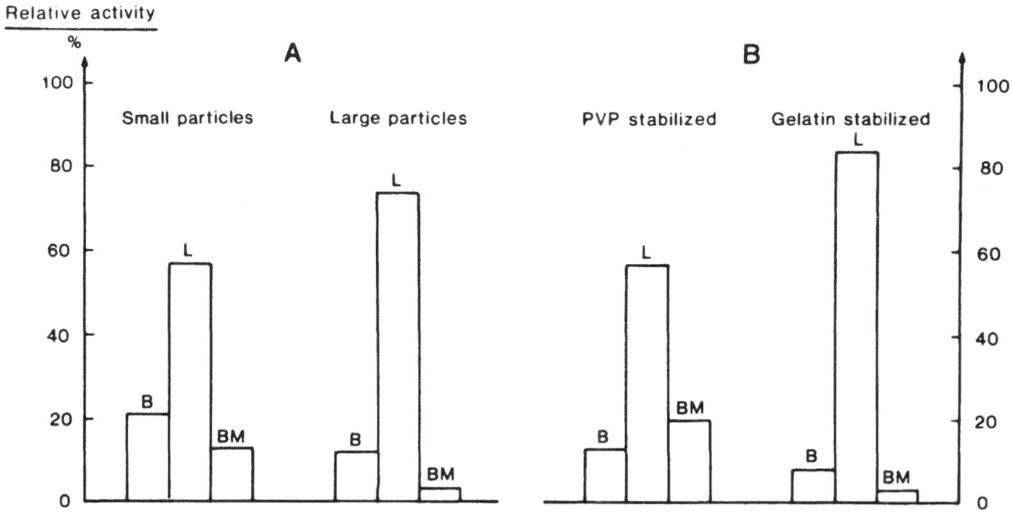

FIGURE 2. Organ distribution in rats 15 min after i.v. injected $^{99}Tc^{m}Sb_2S_3$ particles. (A) Two preparations with particles either <10 nm or 200-400 nm; (B) two preparations with the same particle size distribution; stablilized either with PVP or gelatin. B = blood, L = liver, and BM = bone marrow. (Redrawn from Frier.[2])

Characterization of radiolabeled nanoparticles includes several physical, chemical, and radiochemical parameters. Physical parameters include size and activity-size distributions, shape of particles, particle concentration, and stability. Chemical parameters include chemical composition and purity, surface charge, pH, and osmolarity. Radiochemical parameters include quantification of radiochemical purity (fraction of radioactivity in the wanted chemical composition) and impurities. Furthermore, the presence of carrier, stabilizer, and competing substances in the preparation may also affect the extraction efficiency from the blood.[2]

1. Particle Size

The particle size of an intravenously injected radiocolloid is generally considered to influence the clearance rate from the blood and biodistribution.[2-4] Small particles will generally be cleared more slowly, and to a larger extent accumulate in the bone marrow in comparison with larger particles (Figure 2A). The results in this figure may, however, also be influenced by differences in the number of injected particles. No studies on particle distribution in vivo where the particle size is the only variable, have to our knowledge been made.

The behavior of colloids injected interstitially is very dependent on their size.[5-10] Particles with a diameter of up to a few nanometers will mainly leak to the blood capillaries. Larger particles with diameters around 50 to 100 nm will slowly migrate to the lymph capillaries. Still larger particles will be trapped for a longer time in the interstitial space. The number of injected colloid particles has also been reported to influence the migration rate from the injection site[11] and the phagocytosis in lymph nodes.[12]

2. Number of Particles

The distribution and uptake rate within the RES will also be affected by the number of particles injected, which may be due to an increased occupation of available receptor sites on macrophages, or of an exhaustion of available opsonins.[2,3]

Atkins et al.[13] demonstrated in mice that the liver uptake decreased and the bone marrow uptake increased with increasing doses of colloid above a certain dose.

In a study at our department,[3] the number of nanoparticles in a $^{99}Tc^{m}{}_2S_7$ preparation was increased by increasing the amount of sodium thiosulphate, from 12 to 900 mg. The average

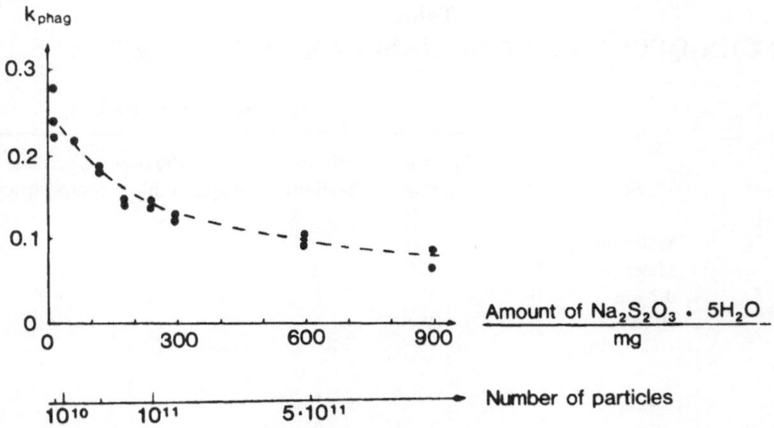

FIGURE 3. The dependence on the phagocytic index, k_{phag}, on the amount of i.v. injected number of particles.

particle size of the preparations was found to be relatively uninfluenced by the amount of sodiumthiosulphate obtained with a Coulter Nano-Sizer® (range 230 to 390 nm). After intravenous injection of 0.5 mℓ in rats the phagocytic index (k_{phag} — see below), was calculated for the different experiments, and the results are shown in Figure 3. It is obvious that the phagocytic index decreases with increasing number of particles injected.

3. *Surface Characteristics*

The influence of surface characteristics of particles on the biodistribution and kinetics has been demonstrated by Frier,[2] who injected i.v. in rats $^{99}Tc^{m}$-antimony sulfide colloid stabilized with different agents (Figure 2B). In another study by Artursson et al.,[14] their results indicate that the phagocytosis might be mediated by a receptor mechanism which makes it possible for the macrophages in the RES both to discriminate between different surface characteristics and to estimate the extent of the foreign nature of the particles.

If the particles and phagocytes have surface charges, there will be either electrostatic repulsion or attraction between them. Wilkins and Myers[15] have reported that negatively charged particles are mainly taken up by the liver, while positively charged particles show an initial appreciable accumulation in the lungs and a later accumulation in the spleen.

4. *Effects In Vivo*

When characterizing radiolabeled nanoparticles for biokinetic studies, it is also important to know what happens to the particles in vivo, before making contact with fixed macrophages of the RES. However, very few studies have been carried out in this field and much still remains to be done.

It has been found in some studies that particles may increase or decrease in size after incubation in serum. An increase in particle size may be due to the fact that some materials are coated by opsonin proteins before they can be recognized by the phagocytes.[16] For sulfur colloids, it has been documented that deficiencies in certain opsonins will affect the rate of phagacytosis in, e.g., patients with carcinoma. Another reason for increased particle size may be aggregation of particles. This was found by Dornfest et al.[16] when incubating a gelatin-stabilized sulfur colloid in normal rat sera. The mean activity size, obtained with microfiltration, increased from 0.30 to 0.70 μm. A decrease in particle size has, however, also been observed by Frier et al.[17] for another gelatin-stabilized $^{99}Tc^{m}$-sulfur colloid when incubated in normal rat serum. By incorporating ^{35}S in the colloid, they were able to show that sulfur was dissolved from the particle surface.

Table 1
TECHNIQUES FOR SIZING NANO- AND MICROPARTICLES

		Possibilities to measure					
Technique	**Size interval**	**Size distribution**	**Activity-size distribution**	**Shape**	**Particle number**	**Chemical composition**	**Charge**
Electrophoresis	Molecules	—[a]					Y
Gel filtration	Molecules — 100 nm		Y				
TEM	0.2 nm	Y		Y	Y	—[b]	
SEM	10 nm	Y		Y		—[b]	
Light-scattering spectroscopy	1 nm—3 μm	Y		—[c]			
Ultrafiltration	1—60 nm		Y				
Microfiltration	10 nm—12 μm		Y				
Ultracentrifugation	0.1 μm		Y				
Phase-contrast microscopy	0.2 μm	Y		Y	Y		
Coulter®Counter	0.4—800 μm	Y					

[a] In combination with light-scattering spectroscopy.
[b] In combination with X-ray fluorescence analysis.
[c] Only in some advanced systems.

In the study of Dornfest et al.,[16] a decrease in particle size of sulfur colloid was observed when incubated in saline to which albumin had been added, resulting in a mean activity size of 0.14 μm. A decrease was also obtained after incubation in sera from some of their rats with leukemia. These findings indicate that the protein content of sera may have an important role in regulating the particle size.

C. Physicochemical Characterization

Most of the available techniques for physical characterization of particles are given in Table 1. Some of the techniques in the table will be further presented below.

It should be noted that the total radioactivity content in a particle is generally considered to be proportional to either the volume or the surface area of a particle.[18] Frier et al.[17] have, however, shown that one $^{99}Tc^{m}$-sulfur colloid had the radioactivity localized in the central part of the particle.

1. Techniques

Electrophoresis — Colloid particles bearing charges may be separated by electrophoresis in, e.g., sucrose-density gradients. Lim et al.[19] have developed an elaborate technique to measure the size and charge distributions of colloid particles by combining electrophoresis and laser light-scattering measurements.

Gel Filtration — Colloid preparations may be eluted through a chromatographic bed in a column in order to separate particles of different sizes. Billinghurst and Jette[18] have demonstrated a technique for determining the size-activity distribution of colloid solutions by fraction collection of calibrated gel columns. Persson et al.[20] suggested a filtration technique where the column is developed with 10 mℓ of physiological saline and then sealed. The column, which still retains all the radioactivity, is then scanned with a slit–collimated NaI(T1) detector. The obtained scanning profile gives a qualitative information on the size distribution and also indicates the presence of radiolabeled impurities.

Transmission Electron Microscopy (TEM) — The size distribution, particle shape, number, and chemical composition of particles can be determined by TEM, equipped with X-ray fluorescence analysis and an image-analyzer system. The sample is spotted or nebulized

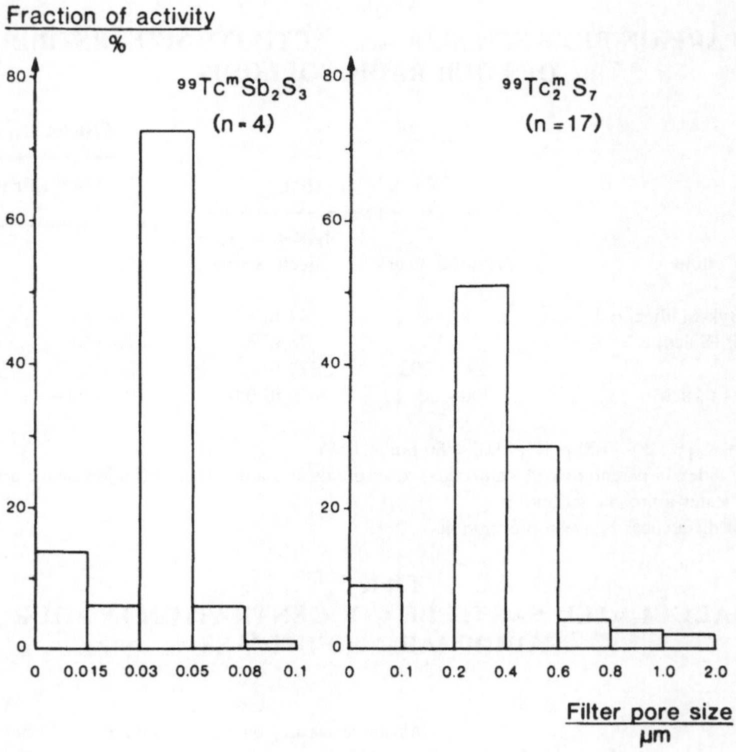

FIGURE 4. Activity-size distributions of $^{99}Tc^mSb_2S_3$ and $^{99}Tc_2^mS_7$ nanoparticle preparations obtained with microfiltration through Nuclepore® filters.

on a plastic-coated grid, allowed to dry or partially dry, and then analyzed.[21] Unfortunately the particles may change or sublime due to the vacuum or heat of the electron beam. Furthermore, it is difficult to analyze preparations containing stabilizer and contaminants.[21]

Scanning Electron Microscopy (SEM) — As for TEM, SEM may also determine size distribution, particle shape and number, and chemical composition. The sample is spread on a polycarbonate filter and allowed to dry. Low-Z particles are coated with a thin layer of, e.g., gold-palladium alloy by sputtering to increase the contrast. The layer prevents heat effects but may influence the surface structure of small particles. With a freeze fracture technique, it is also possible to eliminate the risk of volatilization of particles.[18]

Microfiltration — The proportion of radioactivity in different particle size ranges may be studied with filtration through, e.g., polycarbonate capillary pore membranes (Nuclepore®).[18,22-24] Membranes are available with 17 different sizes, ranging from 0.01 to 12 μm. Small aliquots of particle preparations are passed through the membranes, held in filter holders, followed by flushing with 2 mℓ of physiological saline and a positive air pressure. The filter and filtrate are then measured for activity. The activity-size distributions of two nanoparticle preparations obtained with this technique are given in Figure 4.

Laser Light-Scattering Spectroscopy — This technique involves illumination of a particle solution with the light from a laser, Some of the light scattered at 90° is detected by a photomultiplier, As the particles move and diffuse in the solution due to Brownian motion, the scattered light will give rise to a diffraction pattern. The rate at which this intensity pattern changes is inversely proportional to particle size. A computer will give the average particle size and a polydispersity index which is an indication of the width of the size distribution. If the size distribution is wide, the computed size will have a tendency of being larger than the actual mean size. In Table 2, the results of measurements of four different

Table 2
A COMPARISON BETWEEN SIZE AND ACTIVITY-SIZE DISTRIBUTION OF FOUR RADIOCOLLOIDS

	Particle size (nm)		Activity size (nm) Microfiltration	
Colloid	Manufacturer	Light-scattering spectroscopy	Range	Average
$^{99}Tc^{m}Sb_2S_3$ (Byk-Mallinckrodt)	3—30	44 (0.11)[b]	30—50	40
$^{99}Tc^{m}$Nanocoll (Solco)	—[a]	75 (0.42)	10—50	40
$^{99}Tc^{m}{}_2S_7$ (25)	600 ± 200	270 (0.03)	200—600	350
$^{99}Tc^{m}$Albu-Res (Solco)	200—2000	455 (0.08)	100—2000	—[c]

[a] > 100 nm < 1%; 80—100 nm < 4%; < 80 nm > 95%.
[b] Dispersity index in parenthesis. A value close to zero indicates a narrow distribution, while a value close to one indicates a broad distribution.
[c] Significant differences between preparations.

Table 3
CALCULATED PARTICLE CONCENTRATION IN FOUR RADIOPHARMACEUTICALS

Colloid	Amount of colloid forming substance	Assumed mean particle size (nm)	Assumed particle density (g·cm^{-3})	Volume of suspension (mℓ)	Estimated number of particles per milliliter
$^{99}Tc^{m}Sb_2S_3$	1.3 mg Sb_2S_3	40	4	6.0	10^{12}
$^{99}Tc^{m}$ Nanocoll	0.5 mg HSA	50	1	5.0	10^{12}
$^{99}Tc^{m}{}_2S_7$	12 mg $Na_2S_2O_3 \cdot 5H_2O$	350	2	11.5	10^{10}
$^{99}Tc^{m}$ Albu-Res	2.5 mg HSA	500	1	5.0	10^{10}

nanoparticle preparations with a Malvern® System 4600 are compared with microfiltration and manufacturers' data.

Coulter Counter — In a Coulter® Counter, particles suspended in an electrolyte are made to pass through a small aperture across which an electric current has been established. Each particle displaces electrolyte in the aperture, thus producing a pulse equal to its displaced volume. Each pulse is counted and sized in order to obtain the size distribution. The method is fast, simple, and well suited for sizing large nanoparticles and spherical microparticles.[21]

2. Number of Particles

The number of particles in a preparation may be determined by several different methods. Particles with diameters above 0.2 μm can be determined with a phase-contrast microscope if the sample is placed on a hemocytometer grid or with autoradiography.[17] For smaller particles TEM may be used.[17] It is also possible to calculate the number of particles if the amount of particle-forming material, particle density, and diameter are known.[17,24] Calculated number of particles for four different preparations are given in Table 3.

3. Stability

Since the particle size of a radiopharmaceutical is of great importance for the biological behavior, the preparation must be stable until the time of administration. The long-term stability for various nanoparticles regarding particle size has been checked in some studies with gel filtration,[24] light-scattering spectroscopy,[23] and microfiltration.[23]

The radiochemical stability may be controlled with paper chromatography, thin-layer chromatography, (TLC), high-pressure liquid chromatography (HPLC), gel filtration, and electrophoresis.

III. CHARACTERIZATION OF RADIOLABELED MICROPARTICLES

A. Introduction

Different types of microparticles have been used in different biomedical applications during the last 40 years. In the late 1940s, Prinzmetal et al.[26] started to make blood flow measurements with glass beads. Larger particles of macroaggregated albumin (MAA, 20 to 50 μm) were used for lung perfusion studies. Later on, in the 1960s, cheramic and plastic microspheres were available in different sizes and labeled with many different radionuclides.[27] Today microspheres of albumin, macroaggregated albumin, and plastic microspheres are the most common microparticles in biomedical use.

In vitro characterization of radiolabeled microparticles includes the same parameters as mentioned for radiolabeled nanoparticles above. The microparticles are, however, easier to test due to their greater size.

B. Physicochemical Characterization

Lung perfusion studies in man are mostly performed with MAA or microspheres of albumin, labeled with $^{99}Tc^{m}$. The MAA consists of albumin denatured by heat or by chemical treatment in an agitated aqueous solution. The microspheres are albumin molecules, spherized by homogenization in oil and solidified by heating.[28]

Radiolabeled plastic microparticles are used for blood flow measurements in animals. The particles consist of polystyrene spheres labeled with a radionuclide and annealed at over 400°C. To prevent leaching, the microspheres are coated with a polymeric resin. The density of the particles is about 1.2 to 1.4 g/cm^3 which is close to that of red blood cells (1.1 g/cm^3).

1. Particle Size In Vitro and In Vivo

The size range and the possibility of fragmentation of the microparticles should be controlled with either phase-contrast microscopy, microfiltration, or Coulter Counter measurements. In Figure 5, photomicrographs of MAA and microspheres of some commonly used commercial preparations are presented.

Particles 7 to 15 μm in diameter are comparable in size with red blood cells and because no sedimentation in vivo has been found[29] there should be a close agreement between the behavior of such particles and red blood cells concerning the rheology. With larger particles the error in the measurement of blood flow to small regions will be magnified with increased particle size.[30]

2. Number of Particles

The number of particles in a preparation can easily be determined with phase-contrast microscopy. Concerning plastic microparticles, the manufacturers state the number of particles per gram and the specific activity. A presentation of parameters of some preparations is given in Table 4A and B.

If too few particles are injected in lung perfusion studies the image quality will decrease. If blood flow measurements are performed, the statistical error will increase. Too many microparticles, on the other hand, may cause hemodynamic changes.[31,32]

3. Stability

Loss of Particles — Loss of microparticles in vivo after entrapment in the capillaries

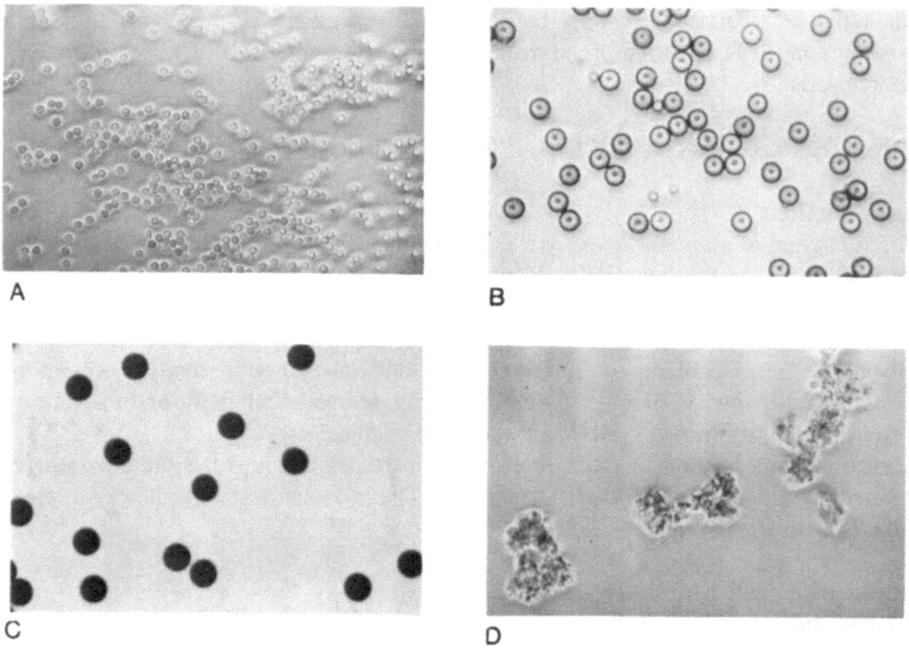

FIGURE 5. Photomicrographs of (A) and (B) magnetizable microparticles 3 μm and 10 μm, respectively, and (C) radiolabeled plastic microparticles, 15 μm in diameter, for experimental blood flow measurements. (D) MAA used for lung perfusion studies in clinical nuclear medicine.

Table 4A
MICROPARTICLES FOR EXPERIMENTAL BLOOD FLOW MEASUREMENTS

Microparticles	Content	No. particles per milligram	Particle size (μm)
3M Tracer microspheres	a	$60 \cdot 10^6$	3 ± 1
	a	$2 \cdot 10^6$	9 ± 1
	a	$0.45 \cdot 10^6$	15 ± 3
	a	$0.10 \cdot 10^6$	25 ± 5
	a	$0.012 \cdot 10^6$	50 ± 10
NEN-TRAC ® microspheres	b	$1.4 \cdot 10^6$	10 ± 1.0
	b	$0.40 \cdot 10^6$	15 ± 1.5
	b	$0.087 \cdot 10^6$	25 ± 2.5
	b	$0.011 \cdot 10^6$	50 ± 5.0

Note: (a)polystyrene; (b)styrene-divinyl benzene copolymer.

may occur. Several authors have investigated the loss of microspheres in ischemic and normal tissues for different particle sizes. Reports in the literature are contradictory about the particle loss. Small particles, 7 to 10 μm, have been found to be lost in normal tissue in contrast to particles greater than 10 μm.[33,34] In ischemic myocardium, however, loss may occur for all particle sizes.[34,35]

Murdoch et al.[36] and Reimer et al.,[37] on the contrary, report of no loss of small particles in ischemic myocardium. Because of the possibility of particle loss, it must be recommended in blood flow measurements that the migration of different microparticles has to be determined for each experimental situation.

Table 4B
MICROPARTICLES LABELED WITH $^{99}Tc^m$ FOR CLINICAL BLOOD FLOW MEASUREMENTS

Microparticles	Content	No. particles per milliliter	Particle size
Tecepart® (Hoechst AG)	a	$0.13 \cdot 10^6$—$1.0 \cdot 10^6$	<10 μm; < 5% 10—30 μm; 70—80% 70—80 μm; 5—10%
Solco® MAA (Solco)	a	$0.21 \cdot 10^6$—$2.3 \cdot 10^6$	<5 μm; < 5% 5—40 μm; >95% >50 μm; 0%
TechneScan® (Byk-Mallinckrodt)	a	$0.40 \cdot 10^6$—$1.8 \cdot 10^6$	8—60 μm; none >150 μm
Amerscan® (Amersham)	a	$0.13 \cdot 10^6$—$0.67 \cdot 10^6$	10—100 μm
TCK-8 MONO (International CIS)	a	$0.20 \cdot 10^6$—$0.70 \cdot 10^6$	10—70 μm; none >150 μm
TCK-5 (International CIS)	b	$0.05 \cdot 10^6$—$0.40 \cdot 10^6$	23—45 μm; about 95%
TCK-5S (International CIS)	b	$0.50 \cdot 10^6$—$4.0 \cdot 10^6$	7—25 μm; about 95%

Note: (a) Macroaggregated albumin; (b) microspheres of albunin.

Aggregation — Aggregation of microparticles can alter the biodistribution and cause false results. The occurence of any aggregation can be studied in detail with a scanning electron microscope. By vigorous shaking or by using an ultrasonicator, the aggregates can be broken.[38]

Leaching — Activity can leach from the microspheres into the suspending fluid. This can be controlled by centrifugation of a small volume of the particle suspension, followed by microfiltration of the supernatant. This procedure ought to be repeated every 2 to 4 weeks to check leaching.[38] Leaching has been reported to occur from ^{125}I-labeled microparticles.[38]

Biodegradability in vivo — The blood flow measurements are based on the assumption that the in vivo distribution of the microparticles is not altered until detection is performed. In lung perfusion studies it is essential that the particles are not metabolized, and thus the distribution not significantly change before and during the measurement/imaging.

The clearance of albumin particles from the lung capillaries is explained either by mechanical breakup of the larger aggregates, making them capable of passing through the capillaries or by enzymatic degradation of the denatured albumin.[39] The radionuclide may also be cleared from the intact albumin particles.

The clearance rate is found to be dependent on particle size. Larger particles remain in the capillaries longer than smaller ones. This gives clearance curves that are not monoexponential, but have several components.[40] Also, parameters such as temperature and duration of heating at the microsphere production have an effect on the biodegradability. Biological half-lives of 2 to 8 hr[40] have been reported, but clearance may be even faster depending upon the preparation.

If, e.g., for $^{99}Tc^m$-MAA, the radionuclide is cleared from the intact albumin aggregates, there will be an uptake of activity in other organs, i.e., the thyroid and the stomach (Figure 6) together with an increased activity concentration in the circulating blood.

Concerning plastic microspheres, biodegradability is no problem. These particles are not metabolized and the biodistribution can thus not be altered due to this. The particles may,

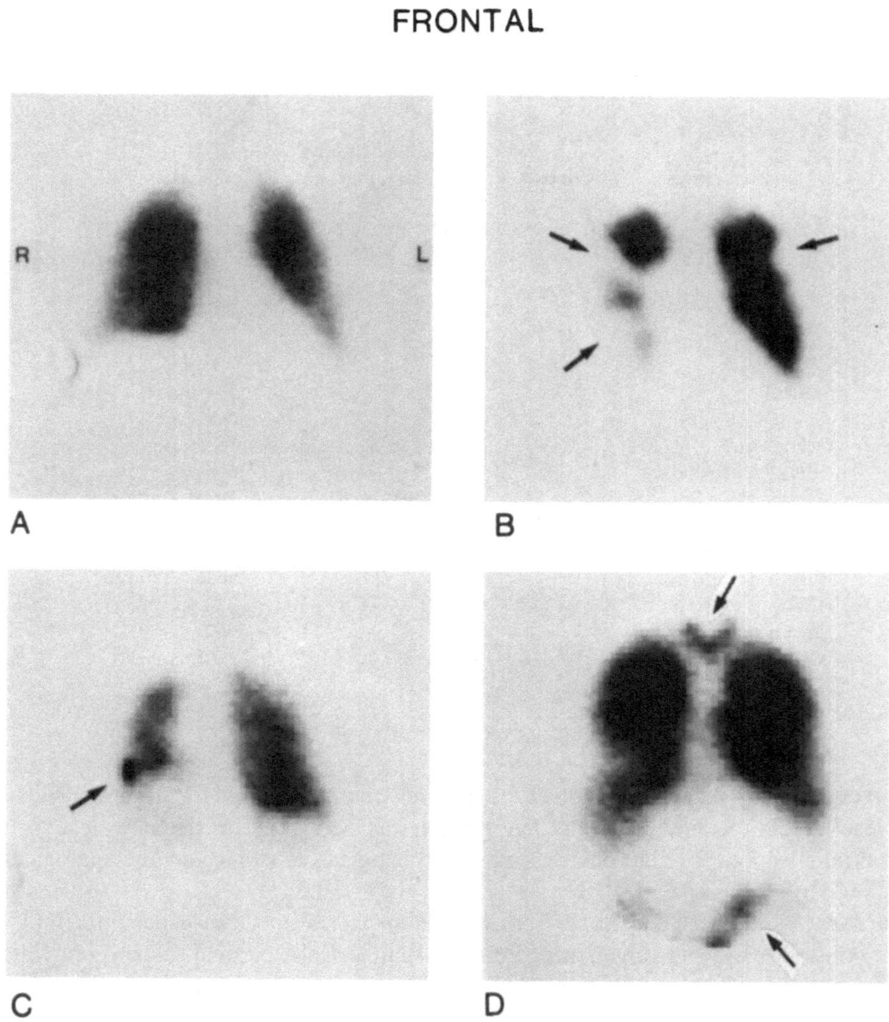

FIGURE 6. Four different lung perfusion scintigrams performed with $^{99}Tc^{m}$-MAA. (A) a normal scintigram showing left and right lungs and (B) a patient with multiple pulmonary embolisms (some are indicated with arrows in the figure); (C) the arrow indicates a "hot spot" obtained due to a bad preparation; (D) a perfusion scintigram with activity uptake in other organs, i.e., the thyroid and the stomach due to biodegradability of $^{99}Tc^{m}$-MAA.

however, be shunted[33] away, and the radionuclide may leach out from the microspheres as mentioned above.

IV. BIOLOGICAL BEHAVIOR OF RADIOLABELED NANOPARTICLES AFTER INTRAVENOUS INJECTION

A. Reticuloendothelial System (RES)

The RES is a cell system, scattered in the body, which has the ability of phagocytosis.[41] It was Ilja Metchnikoff's discovery in the 1880s that laid the groundwork for the theory of phagocytosis.[42] Aschoff[41] regarded these cells as specialized endothelial cells and named these scattered fixed macrophages "reticuloendothelial cells".

The clearance of material from the blood stream is confined to the macrophages which line blood vascular channels.[43] The dominance of the liver in RE blood clearance has led to the incorrect belief that 90% of the body macrophages are situated in the liver. A large population of macrophages in the body, not lining blood channels, cannot contribute to blood clearance.

Partition of an intravenous particulate substance between liver, spleen, and bone marrow depends on its relative blood flow, the local kinetics of uptake at each site, and the phagocytic capacity of each site. This is dramatically revealed if the normally predominant hepatic clearance is depressed using an intravenous blocking agent. In this case there is a slower rate of clearance of a subsequent test substance, but organ uptake studies reveal that, far from being decreased, the total uptake into spleen and bone marrow is greatly increased. This is assumed to be because the blood sinus-lining macrophages in these latter sites now have better access to blood-borne material, previously denied them by overwhelming competition from the liver.[43]

In the exchange process between the blood and the hepatic cells, endogenous products are taken up, metabolized, stored, synthesized, and excreted. The liver lobule is defined as a unit of parenchymal tissue, peripherally characterized by branches of the portal vein and hepatic artery, and centrolobularly by a branch of the hepatic vein, i.e., the central vein.

The normal transit time of indicator solutions through the liver lobule seems to be in the order of 5 to 10 sec[44] but great differences in flow rate within the sinusoids have been demonstrated.[44] Direct connections between the portal and central veins do not exist. Vasoactive substances or extensive endocytosis by sinusoidal cells directly influence the sinusoidal blood flow.[44]

The cells forming a liver sinusoid are endothelial cells and underlying fat-storing cells, both in contact with the microvilli of the parenchymal cells keeping open the space of Disse. Due to the difference in sinusoidal orientation, blood flow will vary widely in the sinusoids. Differences in red blood cell speed of 270 to 410 μm/sec have been observed.[44] The sinusoids have a length of 250 μm with a diameter of 3 to 12 μm.[44]

Fresh red blood cells have a diameter of about 8μm and thus the sinusoids should be too small for them to pass. The red blood cells can, however, adapt within 0.06 sec to different shapes, resulting in a row of single cells in the sinusoids separated by small volumes of plasma.[44] The size of white blood cells are between 6 to 8 μm and compared to red blood cells they are more rigid.[44]

Four types of liver sinusoidal cells have been described: endothelial cells, Kupffer cells, fat-storing cells, and pit cells. Endothelial cells have a high pinocytotic capacity and their flat processes have open fenestrations and no basal lamina. Kupffer cells are solitary macrophages with a high phagocytotic and a considerable pinocytotic capacity. They contain a rich population of lysosomes and can phagocytose considerable amounts of material.

Kupffer cells possess three structures that are related to pinocytosis: (a) bristle-coated micropinocytotic vesicles with a diameter of about 100 nm; (b) a thick fuzzy coat of the cell surface; and (c) worm-like structures protruding from the cell surface.[44]

The space of Disse constitutes a unique tissue space, bordered on the lumen side by a fenestrated endothelial lining (fenestrae diameter 105 to 110 nm,[44] and on the other side by the sinusoidal surface of the parenchymal cells bearing numerous microvilli. About 70% of the parenchymal cell membrane is located in the space of Disse, where the microvilli are 70 to 120 nm in diameter.

Red blood cells fill the sinousoidal lumen and touch the wall with the exception of a small slip layer remaining between the surface of the red blood cell and the endothelial cell. Particles, fluids, and solutes in this layer may be influenced to enter the space of Disse through the fenestrations by the constantly passing red blood cells, resulting in a kind of forced sieving (Figure 7).[44] When a white blood cell moves in the sinusoid, by either its own mobility or by the driving forces of the small pressure gradient, impressed parts of the endothelial lining may come back into position and, as a result, fresh fluids will be drawn into the space of Disse. The latter is called endothelial massage,[44] (Figure 7).

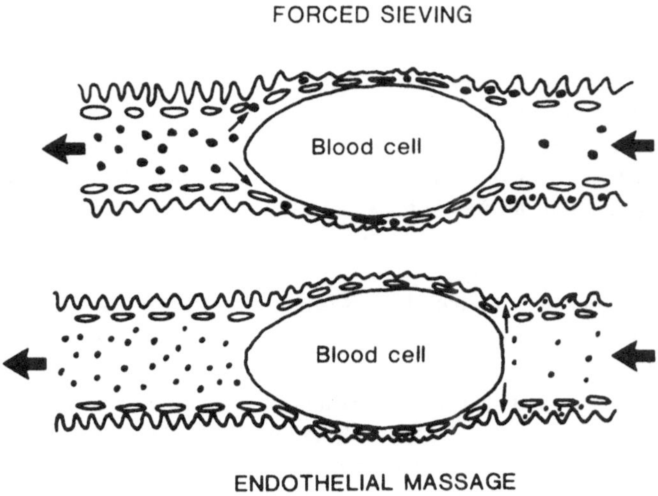

FIGURE 7. Nanoparticles are passed through the endothelial fenestrae to the Disses space by "forced sieving" or "endothelial massage". (Redrawn from Wisse and DeLeeuw.[44])

Parenchymal cells represent about 78 to 83% of the total volume of the liver parenchyma. These cells have a diameter of 18 to 20 μm. Uptake by parenchymal cells can occur by molecular transport mechanisms in the cell membrane or by bristle–coated micropinocytosic vesicles.[44]

The importance of serum factors was already acknowledged by Metchnikoff who suggested the name "stimulins".[45]

A large molecular weight protein (α-2-macro globulin, α-2-opsonic protein) that coated foreign particles and augmented Kupffer cell phagocytosis has been shown to be identical to cold insoluble globulin or fibronectin.[46] It was found that depletion of the protein was correlated with decreased phagocytic activity in the liver.

Endogenous particles, when exposed to blood, were coated with opsonins, particularly fibronectin.[47] The plasma level of fibronectin appeared to correlate directly with Kupffer cell nonimmunospecific phagocytic capacity.[48] Immunospecific, IgG or IgM, mediated phagocytosis was receptor specific via Fc receptors and C3b receptors, respectively.[49]

B. Biological Factors Influencing Phagocytosis

The distribution in the organism of intravenously injected colloids is dependent not only on the physicochemical properties of the colloid but also on some important biological factors such as blood flow to reticuloendothelial (RE) organs, the degree of macrophage activation, and the presence in plasma of mediating factors.

1. Blood Flow

The blood flow to the main RE organs plays a crucial role in determining the distribution of injected particles. This is particularly true when small amounts of colloidal particles are injected. Several investigators have observed that the removal of tracer doses of colloids is almost completely confined to the liver, and the particles are taken up by the liver almost completely during their first passage through the organ.[50,51]

Blood clearance of colloids will in such cases mainly be a measure of liver blood flow and this technique has indeed been used for this purpose.[52] In recent years the technique has been adopted for estimations of the arterial and portal components of total liver blood flow.[53-56] This requires a fast dynamic registration of colloid uptake with a scintillatiom camera and on-line computer system. By increasing the amount of injected colloid particles, the uptake in the liver decreases and uptake in extrahepatic RE organs is observed. This was observed by Biozzi et al. in the early 1950s [51] and these authors noted the importance of a critical colloid dose for studies of RE phagocytosis. Above this dose, blood clearance of colloids is more a measure of total RE phagocytic capacity. However, deviations of blood flow from major RE organs will inevitably affect the colloid uptake of these organs.

2. Macrophage Activation

Macrophages in the body are normally in a resting state and must be activated or elicited to be able to perform their various activities. Macrophages are activated by numerous stimuli such as immunoglobulins, complement factors, endotoxin, bacteria, and foreign material.[3] The degree of activation can be measured by studying secretory products of macrophages.[57]

In certain disease states such as infection and tumor growth, RE clearance of injected particles has been shown to be increased, a finding probably attributable to macrophage activation.[3]

3. Macrophage Depression

Many of the compounds known to activate macrophages will, if administered in larger quantities, suppress RE function. This is also true for colloids. Studies in our department have shown that repeated injections of $^{99}Tc^{m}$-labeled sulfur colloids result in a progressive decrease in liver uptake rate.[3] Infusion of plasma substitutes, which to a variable extent are taken up by the RES, will also decrease the uptake of subsequently administered colloids.[58] This is probably due both to a depression of cellular activity and to a depletion of opsonins in plasma, since the RE depression can partly be reversed by adding fresh plasma to the colloid.[59]

4. Opsonins

Plasma opsonins play a major role in promoting phagocytosis. Inert colloid particles may be phagocytosed by Kupffer cells in vitro even in the absence of plasma factors. Phagocytosis is, however, greatly enhanced by the addition of opsonin.[59]

Depression of RE function has been associated with a decrease in serum opsonins.[59,60] A technique of evaluating opsonin activity was described by Saba and DiLuzio[61] studying the in vitro phagocytosis of a gelatinized lipid emulsion by rat liver slices. Kupffer cells of normal rat liver in an incubation medium consisting of normal rat serum and heparin manifested a 17-fold enhancement in colloid uptake as compared to the degree of phagocytosis manifested in buffer medium.[62] Pisano et al.[63] demonstrated that human serum was as effective as rat serum in promoting rat liver-slice phagocytosis.

The plasma level of fibronectin, an opsonic protein, appears to correlate directly with Kupffer cell nonimmunospecific-phagocytic capacity.[47] Depletion of plasma opsonins have been registered in trauma, sepsis, and advanced malignancies, all of which have been associated with decreased Re phagocytosis.

5. Organ-Specific Factors

The above-mentioned biological factors may affect total RE uptake. There are, however, numerous conditions which may affect only some parts of the RES. In cases of liver disease or after suppression of liver-colloid uptake, a compensatory increase of colloid uptake has been noted in the extrahepatic RES.[3] It thus seems that the extrahepatic RES has a reserve capacity which may partly compensate for a decreased phagocytic capacity of the liver. This fact emphasized the importance of techniques for measuring RE function, which allows for discrimination between various RE organs. The development of scintigraphic techniques for studying colloid uptake has given such possibilities.

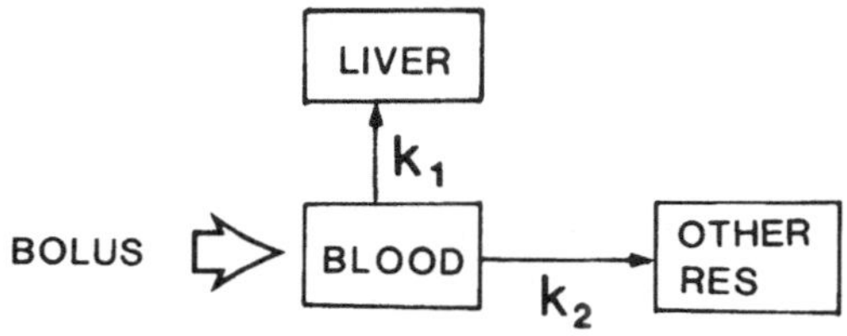

FIGURE 8. An open two-compartment model used for evaluation of RES function.[69]

C. Techniques for Measuring RE Function

1. Blood Clearance

From a physiologic point of view the RES can be characterized in terms of its phagocytic behavior. This has lead to the development of the "colloid clearance technique" for studying RE function.

Quantitative studies of vascular carbon clearance in the rat allowed for the application of mathematical expressions of phagocytosis.[51] The concentration of injected carbon in the blood as function of time (t) after injection could be expressed by

$$C(t) = C(0) \cdot 10^{-k_{phag}t} \tag{1}$$

The constant k_{phag} characterized the rate of clearance and was called the "granulopectic (or phagocytic) index". It was found that, by increasing the dose of carbon, the granulopectic index tended to decrease, and above a certain dose (the critical dose) the granulopectic index was inversely proportional to the dose of carbon injected. Contemporary studies by Dobson and Jones[50] on the blood clearance of radioactive chromic phosphate further emphasized the significance of the "critical colloid dose".

Above the critical dose, colloid particles were not completely absorbed in their first passage through the liver, and the value of k_{phag} was more a measure of phagocytic activity provided blood flow was normal.[64] It was observed that by increasing the dose, the colloid uptake by the extrahepatic RES, mainly the spleen, increased.[51]

2. Compartment Analysis

The kinetics of intravenously injected radiolabeled nanoparticles can be quantitatively measured with a scintillation camera equipped with an on-line computer system.

One reported technique is based on biodegradable radiolabeled human serum albumin (HSA) nanoparticles and dynamic registration of the activity distribution during 90 min.[65,66]Differences have been observed between normal healthy adults and tumor patients regarding accumulation rate (increased in tumor patients) and, particulary, elimination rate, i.e., particle degradation (decreased in tumor patients).

A second method that has been reported is compartment analysis of the biokinetics of inert radiocolloids.[3,67] In this case sequential images can be recorded during a 15-min period from the injection, and regions of interest selected over liver and blood pool. Time-activity curves can then be generated and evaluated.

To estimate the flow rate of the radiocolloid into the liver, k_1, and into the extrahepatic RES, k_2, an open two-compartment model can be used, schematically presented in Figure 8. Using the symbols in the figure, the rate of change of the radiocolloid in the blood (B) and in the liver (L) with time can be expressed by:

$$\frac{dB}{dt} = -B(k_1 + k_2) \tag{2}$$

$$\frac{dL}{dt} = B \cdot k_1 \quad (3)$$

The precentage amount of activity in the blood and the liver at any time (t) can then be expressed by Equations 4 and 5:

$$B(t) = 100 \cdot e^{-(k_1 + k_2)t} \quad (4)$$

$$L(t) = 100 \cdot \frac{k_1}{k_1 + k_2} \{1 - e^{-(k_1 + k_2)t}\} \quad (5)$$

The rate constants k_1 and k_2 can be obtained by fitting these equations to the experimental curves with a nonlinear least-squares fit.[68]

3. Phagocytic Index

The uptake of colloids by the RES are, as mentioned earlier, mostly described by the phagocytic index (k_{phag}). As the first very rapid component of the liver uptake curve (40 to 60 sec after injection) represents both mixing of the test substance with the blood and phagocytosis by the liver, calculations of k_{phag} have to be made in the second part of the curve, mainly representing phagocytosis of the colloid. Blood mixture of the colloid can then be assumed to be complete.

The phagocytic index has been calculated by Rydén et al.[69] in this second part of the curve by using Equation 6:

$$k_{phag} = \log \frac{L(120)}{L(60)} \quad (6)$$

Combining Equations 5 to 6 gives:

$$k_{phag} = \log \frac{1 - e^{-(k_1 + k_2)120}}{1 - e^{-(k_1 + k_2)60}} \quad (7)$$

A slow uptake rate of the liver, low k_1, combined with a fast uptake rate of the extrahepatic RES, high k_2, will thus give the same value of k_{phag} as the opposite situation (high k_1 and low k_2). For very low values of $k_1 + k_2$, k_{phag} approaches the limit log 2 or 0.3010.

It is thus obvious, that selection of a time interval on the uptake curve by measuring the uptake at only a few different times, will not provide a proper estimate of the uptake function in the organ. The same applies to the analysis of blood clearance curves. This emphasizes the inferiority of using Equation 1 uncritically for determination of RE function.

D. Experimental Applications

1. Depression of RE Function

Numerous compounds depressing RE phagocytic activity have been described such as methyl palmitate,[70] dextran sulfate,[71] gadolinium chloride,[72] colloidal gold,[73] and plasma substitutes such as dextran 40 and 70, gelatin (Haemaccel®), and polyvinyl pyrrolidone (PVP).[58] The in vitro depressant effect of the plasma substitutes were more pronounced in combination with burn trauma, and seemed to be mediated through interaction with humoral factors. A transitory depression of RE function has been observed in rats following surgery. This depression was not associated with alterations in liver blood flow.[59] The RE depression in rats after surgery was closely related to a decreased opsonic activity of serum and was followed by a compensatory increase in RE activity 3 to 4 hours after surgery.[74]

Studies at our department have revealed that intravenous administration of gelatin (Haemaccel®) significantly reduced the total colloid uptake in the liver as well as the colloid uptake rate.[69] The same was true for methyl palmitate, which, if administered on two consecutive days before testing RE-function, profoundly depressed hepatic-colloid uptake rate and the total colloid uptake in the liver. These studies of RE function were performed in the rat using $^{99}Tc^{m}$-sulfur colloid and dynamic scintigraphic measurement.[75]

2. Stimulation of RE Function

Administration of endotoxin to animals decreases the phagocytic function of the RES as measured by carbon clearance according to Benacerraf and Sebestyen.[64] These investigators, however, showed that the prior administration of a small dose of endotoxin rendered the macrophages resistant to the effect of subsequently given endotoxin. A yeast polysaccharide, zymosan, derived from the cell wall of Saccaromyces Cerevisiae, was also found to have stimulatory effects on the macrophage system, increasing the rate of clearance of colloidal carbon.[64] Histological observations revealed that zymosan was phagocytized by the RES and that the RE stimulation was accompained by proliferation of RE elements[76] and with the reversible formation of macrophage-containing granulomas in the liver.[77]

The active RE stimulatory component of zymosan was identified as glucan, a β1,3-polyglucose.[78] The effects of zymosan in activating the C_3 part of the complement has also been observed.[79]

The stimulatory effects of zymosan have been confirmed in our studies which have revealed proliferation of RE cells and granuloma formation after zymosan injection in the rat.[80] The secretion of a lysosomal enzyme, *N*-acetyl-β-glucosaminidase, was significantly increased from zymosan-stimulated macrophages.[81]

3. RE Function in Tumor Growth

With the introduction of the colloid clearance technique, experimental tumor growth was found to be associated with enhancement of clearance of injected particles from the blood.[82-84] In subsequent experiments in mice, Old et al.[85] observed that enhanced phagocytic activity, as reflected by the intravascular removal of colloidal carbon, increased during the early phase of sarcoma and carcinoma tumors. Prior to the death of animals, the disappearance rate of colloidal carbon returned to normal. An increase in tumor size correlated with a decline of RE function. Studies in humans with gastric cancer revealed a more depressed RE function in patients with locally advanced disease than in patients where radical surgery was possible.[86] Other observations have shown an increased RE function, measured by particle blood clearance, in patients with cancer.[87,88]

The influence of RE depression on tumor growth was evaluated by Fischer and Fischer, who found an augmented growth of intrahepatic tumor after portal administration of tumor cells in the rat, following administration of colloidal carbon or Thorotrast[89] or after surgical trauma.[90] A decreased resistance to intravenous tumor cell challenge during RE depression following surgery was also reported by Saba and Antikatzides.[91] The important role of the RES as a major host defense system has been reviewed recently.[92]

By use of a $^{99}Tc^{m}$-sulfur colloid, we have studied the effects on colloid uptake rate and colloid distribution in rats with experimental tumors of different size and location. Animals with small liver or subcutaneous tumors showed an increased activity of both hepatic and extrahepatic RES. Animals with larger tumors showed, however, a significant depression of hepatic RE function.[93] These findings are in accordance with the above-mentioned earlier observations. It has also been shown in our studies that depression of hepatic RE function on the time of inoculation in the liver of an experimental tumor resulted in accelerated growth of the tumor and an increased mortality in RE-depressed rats.[75]

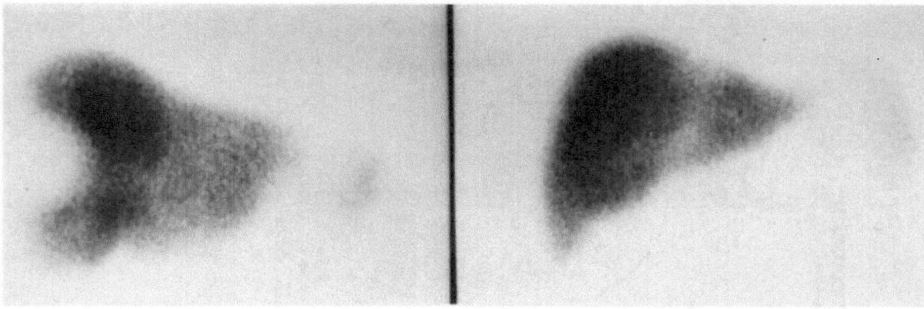

A B

FIGURE 9. Liver-spleen scintillation camera images of two patients (anterior views) (A) showing a focal defect in the right liver lobe and (B) normal conditions.

E. Clinical Applications

Liver-spleen imaging is normally performed with a scintillation camera about 15 min after an intravenous injection of a radiocolloid. By taking anterior, posterior, and lateral images it is possible to assess the size, shape, and position of the liver and the spleen, which — together with the observed activity distribution in and between the organs — are suitable in medical diagonosis in oncology and for infectious diseases.[94] Figure 9 gives two examples of liver-spleen scintigrams — one with normal findings and one with a solitary "space-occupying" lesion. Unfortunately, lesions with a diameter less than 2 to 3 cm in diameter (depending on the localization) are difficult to detect with planar imaging, due to surrounding radioactivity and patient motion mainly because of breathing. A somewhat better detectability can be obtained with single-photon emission-computed tomography (SPECT)[95] and motion correction in the images.

In recent years, the number of liver-spleen scintillation camera studies performed has declined due to development of other diagnostic techniques such as transmission-computed tomography, magnetic resonance tomography ultrasound, and better blood serum analysis.

V. BIOLOGICAL BEHAVIOR OF RADIOLABELED NANOPARTICLES AFTER INTERSTITIAL INJECTION

A. Lymphatic System

As early as 300 B.C., Herophilus and Erasistratus in Alexandria observed that the vessels of the small intestine appeared white after a fatty meal. In 1653, after Pequet's discovery in 1651 of the thoracic duct, Bartholin found serous vessels in many parts of the body and called them "lymphatics". A comprehensive review of the lymphatic system is given by Yoffey and Courtice.[96] In brief, the lymphatic system provides a mechanism, other than the blood vascular system, clearing the tissue of substances not readily absorbed by the blood. In Figure 10 an outline of the lymphatic system is given, showing the drainage of tissue by lymphatic capillaries, transporting lymph in the afferent lymph vessels to lymph nodes and further via efferent lymph vessels to the thoracic duct.

The ultrastructure of lymphatic capillaries resembles that of blood capillaries, but are usually wider in their lumina. The endothelial wall varies in thickness from 0.1 to 6 μm.

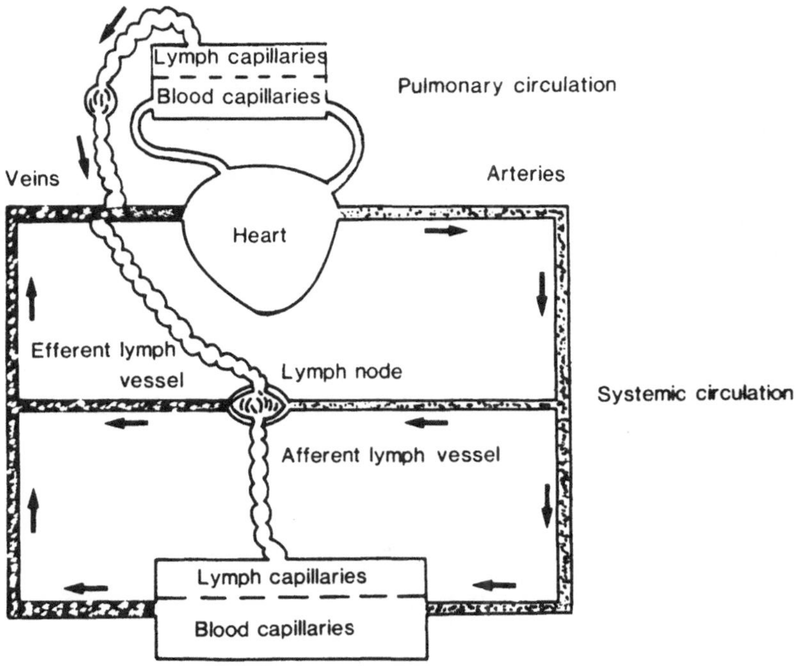

FIGURE 10. Schematic presentation of the blood and lymph circulation in man. (Redrawn from Yoffey and Courtice.[96])

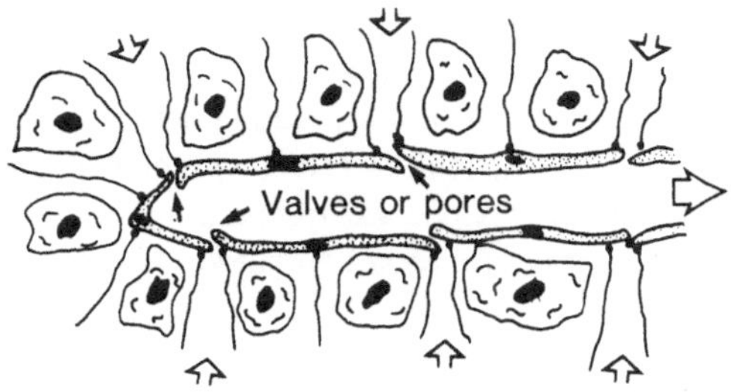

FIGURE 11. Schematics of the structure of an end lymphatic capillary. The direction of inflow from the surrounding tissue is indicated by open arrows.

All the endothelial cells contain vesicles and caveolae intracellulares. The endothelial junctions vary considerably. In some tissues intercellular gaps are not seen, but these gaps or cellular open junctions appear frequently in others. Endothelial junctions are probably continuously undergoing changes, depending on the local circumstances at the time. In Figure 11 an end-lymphatic capillary is schematically drawn.

The bulk of materials absorbed enter the lymphatics through endothelial gaps; some of the smaller complexes such as proteins might also be transported to some extent by the vesicular route.

The thicker-walled collecting ducts, thickness 15 to 50 μm, consist of elastic and muscular elements. The muscle cells enable the vessels to contract rhythmically, providing a continous lymph flow. In man the total lymph flow is about 1 to 2 ℓ per day. The endothelium in the ducts contains fewer open junctions. In the lymphatic vessels valves are prominent.

Lymph passes through one or more lymph nodes, although in some tissues lymph may bypass the nodes. The walls of the lymphatic sinuses are composed of thin, flattened, greatly elongated RE cells.

The lymphatic trunks join the subclavian or jugular veins near their junctions — on the left side, the deep cervical duct draining the head and neck, the subclavian duct draining the arm, and the thoracic duct draining the abdominal viscera and lower extremities, the ducts enter the venous system in close association with one another.

Besides the communication between the thoracic and right lymph ducts, lymphaticovenous communications, other than those at the base of the neck, have long been known to exist. Another possible re-entry of lymph into the blood vascular system is during its passage through the lymph nodes.

An important intrinsic factor for the lymph flow is the fact that the lymphatic vessels rhythmically undergo contractions and relaxations for the propulsion of lymph due to the smooth-muscle cells in the capillary wall.

Extrinsic factors such as muscular activity will increase the lymph pressure and move lymph in the direction determined by the anatomical disposition of the valves. Respiratory movement is another important extrinsic factor in the flow of lymph along the main lymph channels from the abdominal and thoracic cavities. Other factors affecting lymph flow are passive movement, pulsation of blood vessels, mobility of intestinal tract, venous pressure, and gravity. Anesthesia may affect the formation of tissue fluid as well as being the force responsible for lymph propulsion. The effect may be variable depending partly on the anesthetic used.

Lymph flows in a lymph node first into the cortical or subcapsular sinus, and from there through the medullary sinuses into the hilus and efferent vessels. The sinuses consist of cells which can hold particulate matter mechanically but also act actively phagocytic. Phagocytic cells are found both in the sinuses, where they are either fixed or free, and scattered throughout the lymphoid tissue. In the sinuses they occur: (1) as network of cells, (2) as the endothelium lining the walls of the sinuses where they are sometimes known as the littorial cells, or (3) as rounded macrophages floating freely in the lymph.

The lymph nodes appear to provide two main types of filtration. The first is of a simple mechanical type. The second involves, in addition, a biological reaction on the part of the phagocytic elements in the node; the RE cells seem to be the first to be involved.

B. Interstitial Particle Absorption

The behavior of nanoparticles injected interstitially is strongly dependent on their size. The size dependence for interstitial particle absorption has thoroughly been investigated at our department in animal studies.[5,24,97] In these studies different nanoparticle preparations have been characterized both for particle size and lymph node accumulation.

Experiments were carried out in rabbits, anesthetized and fixed in supine position under a scintillation camera. The nanoparticles (0.2 to 0.4 mℓ) were injected subcutaneously, bilaterally just below xiphoid process. Images were taken at 2 to 4 hr after injection.

In Figure 12, a summary of our results for the uptake in the parasternal lymph nodes 2 hr after injection for different nanoparticle preparations are given. For most of the preparations, the size ranges obtained with microfiltration are given. As discussed above for RES, different parameters will influence the particle phagocytosis. This might explain the heterogeneity of the results for different preparations with about the same particle sizes. It is, however, obvious that there is a dependence on the uptake values of the particle size. In the figure the shaded area indicates the correlation showing a maximum uptake for particles around 10 to 50 nm in diameter.

The results can be interpreted as small particles; less than a few nanometers will mostly be exchanged through the blood capillaries. Larger particles, with diameters up to tens of nanometers, will preferably be absorbed into the lymph capillaries and phagocytosed in the

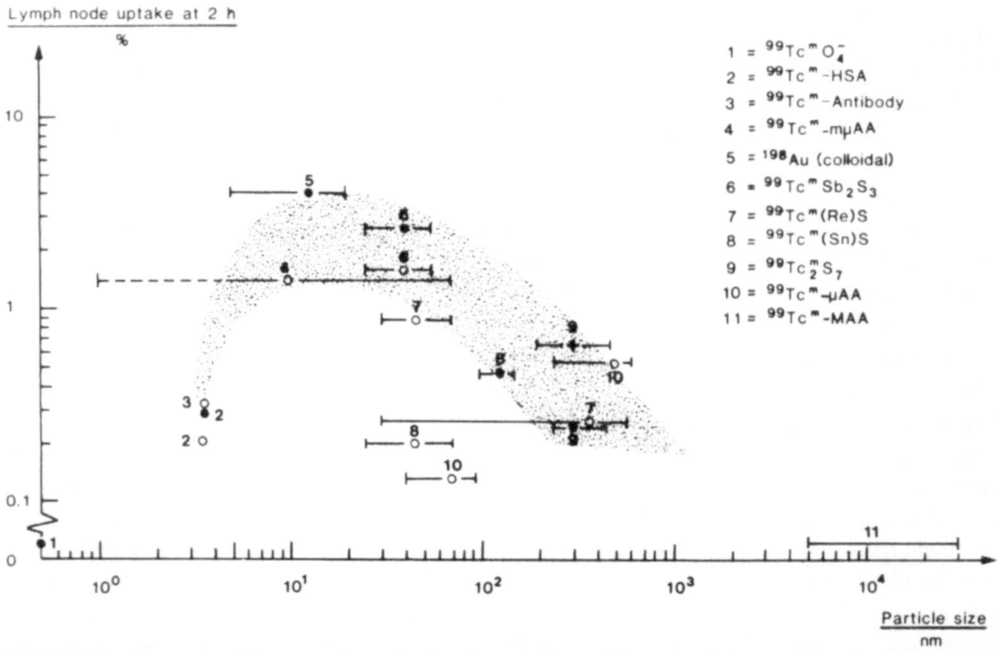

FIGURE 12. Summary of three experimental animal investigations of the accumulation of radioactive nanoparticles in the lymph nodes in rabbits after subcutaneous injection. Size ranges for the preparations are given. The symbols (●), (○), and (+) represent results from references 5, 24, and 97, respectively. The shaded area, giving the correlation of uptake vs. particle size, indicates that there is a maximal lymph node uptake for particles around 10

lymph nodes. Still larger particles, hundreds of nanometers, will for a long time be trapped in the interstitial space.

The above size dependence for particle absorption has been verified both in animal and human studies.[6-10]

C. Techniques for Studying Lymphatic Function

Radiolabeled colloids with suitable physical and chemical properties have the potential to provide functional, anatomical, and morphological information about the lymphatic system when injected interstitially.

The optimal size of the radiocolloid has been found to be about 5 nm.[5,24] Colloidal ^{198}Au could be made with this size and was used for many years to study the lymphatic system. Two disadvantages with ^{198}Au, which has made this colloid less used today, are a high absorbed dose in the injection site that sometimes causes local tissue necrosis[98] and an unfavorable photon energy for scintillation camera measurements. Strand and Persson[5] found that a $^{99}Tc^{m}$-labeled antimony sulfide colloid ($^{99}Tc^{m}Sb_2S_3$, particle size 3 to 30 nm) was better suited for lymphoscintigraphy. This radiocolloid has been the most frequently used in the last decade.

The injection site is chosen either at the origin of the lymphatics under study or, as for malignant melanoma, close to the site of the primary lesion. After interstitial injection of 0.1 to 0.5 mℓ, dynamic scintillation camera images can be registered to study the lymphatic drainage rate and/or static images 2 to 4 hr later for optimal visualization of regional lymph nodes. Muscular motion or massage on the injection sites is sometimes carried out in order to improve the rate of colloid outflow from the injection site.[98]

D. Applications

A safe, reproducible, and noninvasive technique to image the lymphatic system is desirable as an alternative technique to contrast lymphangiography which has several well known drawbacks.[99,100]

Table 5
THE REPORTED VALUES OF DIFFERENT CLINICAL LYMPHOSCINTIGRAPHIC PROCEDURES

Application	Type of disease	Injection site(s)	Reported value(s)	Ref.
Staging	Breast carcinoma	Bilateral subcostal	Valuable	109
	Breast carcinoma	Periareolar areas	No value—possible value	110, 112, 113
	Breast carcinoma	Periareolar and interdigital	Possible value	111
	Malignant melanoma	Bipedal	No value	123
	Cervical cancer	Labium majus	No value	128
	Pelvic neoplasms	Bilateral perianal	Possible value—valuable	99, 114
	Pelvic neoplasms	Bipedal	Valuable	98, 129
Before lymphadenectomy	Malignant melanoma	Peritumoral	Valuable	105, 119—123
	Esophageal carcinoma	Submucosal layer	Valuable	117
During lymphadenectomy	Breast carcinoma	Periareolar areas	Valuable	115
	Cervical cancer	Bipedal	Valuable	116
After lymphadenectomy	Breast carcinoma	Bilateral subcostal and hands	Valuable	118
	Malignant melanoma	Peritumoral	No value	130
Follow-up	Breast carcinoma	Bilateral subcostal	Valuable	131
Radiotherapy planning	Breast carcinoma	Bilateral subcostal	Valuable	124—127
Lymphatic function	Edema	Bipedal	Valuable	132, 133

Lymphoscintigraphy with radiocolloids has been considered for a long time to have the potential to partly replace lymphography, but so far its clinical use has been restricted to rather few medical centers.[101]

In the early 1950s, Sherman and Ter-Pogossian[102] reported migration of an interstitally injected ^{198}Au colloid to regional lymph nodes. It was soon found that cancerous lymph nodes also contained radioactivity. These findings initiated studies on whether or not colloidal ^{198}Au had therapeutic effects on lymph node metastases. Hultborn et al.[103] and Seaman and Powers[104] found, however, that not all cancer-containing nodes had colloidal uptake. Furthermore, they showed by autoradiography that the radioactivity was restricted to the residual normal tissue in the cancerous nodes. These observations have recently been confirmed by Sullivan et al.[105] when using $^{99}Tc^{m}Sb_2S_3$.

During the 1950s and 1960s, techniques for visualizing lymph node regions were developed, e.g., for axillary and parasternal lymph nodes,[103] cervical nodes,[106] ilio-inguinal and para-aortic nodes,[98] and internal mammary nodes.[107]

In later years, more thorough evaluations of the clinical usefulness of lymphoscintigraphy have been carried out (Table 5).

As a staging procedure, i.e., determining presence or absense of regional lymph node metastases, there is a controversy of how to interpret abnormal scintigrams.[108] Furthermore, there are different opinions about how to carry out the injection technique. Ege has performed internal mammary lymphoscintigraphy by bilateral subcostal injections in more than 1000 patients and found this technique of value as a staging procedure.[109] Axillary lymphoscintigraphy with periareolar subcutaneous injections for demonstrating axillary lymph node

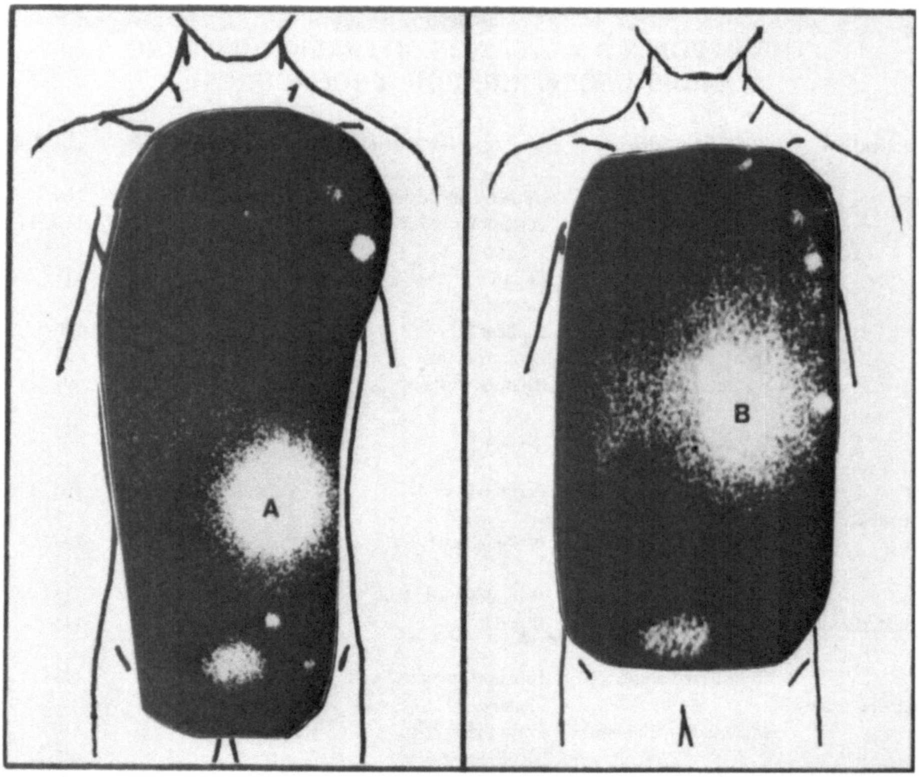

FIGURE 13. Lymphoscintigraphy with peritumoral injections in two patients with malignant melanoma. (A) injection 6 cm left and 2 cm above the umbilicus demonstrating lymph drainage to left axilla and left groin; (B) injection 6 cm left of midline of L2 showing lymph flow to left axilla and flank.

metastases has been made in rather limited patient studies. Most authors find this technique either to be of possible value[110,111] or of no value.[112,113] Iliopelvic lymphoscintigraphy with bilateral perianal injections in initial patient trials suggests that the technique is clinically valuable in staging genitourinary cancers.[114]

Lymphoscintigraphy can readily be used as guidance for localizing regional lymph nodes when lymphadenectomy is,[115,116] or will be[117] performed. It can also be carried out postoperatively to verify that no residual nodes are present.[118]

The clinical value of lymphoscintigraphy in patients with malignant melanoma is of value to demonstrate the lymphatic drainage from the primary tumor site when injections are made peritumorally,[119-123] as shown in Figure 13. Lymphoscintigraphy with bilateral dorsopedal s.c. injections in patients with malignant melanoma on the lower extremities has been found by Bergqvist et al.[123] to be of no value for demonstrating ilio-inguinal lymph-node metastases.

When planning radiation therapy of patients with breast carcinoma, internal mammary lymphoscintigraphy has been found to be valuable for establishing the exact localization of the lymph nodes.[124-127]

VI. BLOOD FLOW MEASUREMENTS WITH RADIOLABELED MICROPARTICLES

A. Intravenous Injection

To make a study of the lung blood flow the capillaries are microembolized by the intravenously injected particles in proportion to the blood supply. Particles must therefore be greater than the smallest lung capillaries. Particles greater than 100 μm may, however, alter

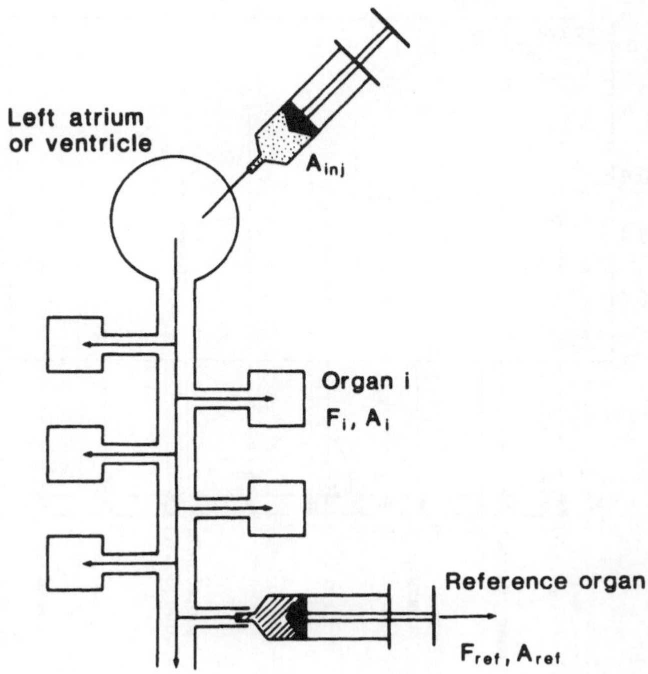

FIGURE 14. The principle of the microparticle technique for experimental blood flow measurements using microparticles injected intraarterially.

the perfusion pattern and if a large number of particles are injected it may cause pulmonary hypertension. The ideal particle size is 15 μm.[28]

Albumin particles are, due to their metabolism in vivo, commonly used to study the human circulation. The wide variation in size distribution (5 to 100 μm) of the MAA particles is, however, a disadvantage in these preparations.

A critical parameter in lung perfusion studies is the number of particles injected. Too few particles trapped in the lung results in decreased image quality and insufficient data, thus irradiating the patient without benefit. Overdosing is unusual, but the risk of toxicity increases with particle number.[31] Dworkin et al.[31] recommends at least 60 radiolabeled particles per gram of lung. In a 70-kg man this means 60,000 particles. An upper limit of 250,000 particles is suggested; then very little is gained when a greater number is injected. The number of lung capillaries[134] is approximately 10^{11}. If the injection consists of 10^5 particles and the major part of the microspheres are so small that they only get trapped in the capillaries, then one per million capillaries will be blocked. Larger particles cause obstruction of the arterioles. The number of pulmonary arterioles are approximately 10^8 and if 10^5 particles larger than 20 μm are injected, one out of thousand arterioles will be occluded in a normal subject.[134] Even in a patient with a severe cardiopulmonary disease there will be a wide safety margin.

B. Intraarterial Injection

When microspheres are injected intraarterially, measurements of organ blood flow and cardiac output can be made. A known amount of radiolabeled microspheres is injected into the left heart to distribute in the cardiac output.[135] For calculation of cardiac output, an artificial organ is created, i.e., via a catheter in an artery, from which blood is drawn at a constant rate into a syringe (Figure 14). The relation between the flow rate and the activity

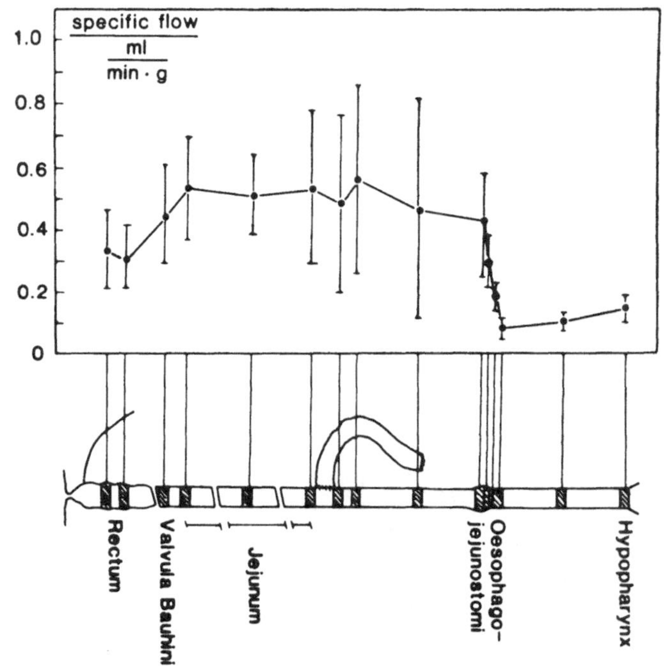

FIGURE 15. Example of the results from blood flow measurement in an oesophago-jejunostomi in pigs, obtained with the microsphere method.

is thus obtained. The blood flow to any organ, represented by a tissue sample, can then be measured and calculated by:[136]

$$F_i = A_i \cdot \frac{F_{ref}}{A_{ref}} \tag{8}$$

where F_i and F_{ref} are the blood flow to an organ, i, and to the reference organ respectively. A_i and A_{ref} are similary the activity in organ, i, and in the reference organ. If an assumption is made that the flow is equally distributed in the organ, then any tissue sample from this organ can be used for the activity measurement. Cardiac output (C.O.) is calculated in the corresponding way;[136]

$$C.O. = A_{inj} \cdot \frac{F_{ref}}{A_{ref}} \tag{9}$$

The reference sample is drawn during 1 to 2 min with start prior to the microsphere injection. According to Buchberg et al.[137] 1% of cardiac output could be safely drawn within 1 min without disturbing cardiac output and blood pressure. An example of a blood flow measurement is given in Figure 15.

Several factors have to be considered when the microsphere technique is used. After injection, microspheres must be well mixed and distributed in proportion to the regional blood flow. When the particles reach an organ they should get trapped in the first circulation and remain there until the activity in the tissue samples is measured.[38]

To make an accurate investigation, the injection site and the number of spheres injected are two parameters of great importance. An adequate mixing is best obtained by an injection in the left atrium or left ventricle. To study cardiac output and coronary blood flow, the particles have to be injected into the left atrium to get complete mixing with the blood.[27] A

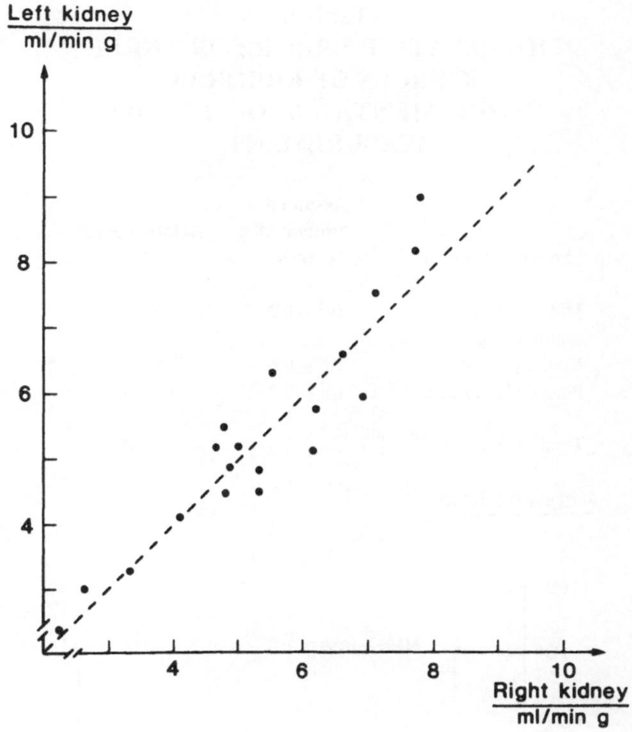

FIGURE 16. Correlation between the blood flow in the left and the right kidney in experimental blood flow measurements, to assure adequate mixing.

control of the mixing can be done by measuring the blood flow to the left and right kidney or the left and right cerebral hemisphere, where only small differences should occur (Figure 16). Mixing can also be assessed by injection of microspheres labeled with two (or more) different radionuclides. If injected only a few minutes apart they should present the same pattern of distribution.[38]

The number of particles in the reference and the tissue samples, together with the quantification of the activity of the radionuclide in the samples, influence the accuracy in the measurement of the blood flow.[138] If the particles are shunted away from the capillaries, then the number of trapped and measured particles will not be proportional to the blood flow. This error is a function of sphere and vessel size distribution. The total amount of radioactive nuclei in the particle can be calculated if it is assumed that each particle contains nuclei in proportion to its volume, and if the size distribution of the particles is known. In Table 6 the three sources of statistical error are presented. The most critical parameters are the number of microspheres and the number of recorded pulses. As given in Table 6, the total relative error can vary between 1 to 14% for the examples given.[139]

VII. BIOMEDICAL APPLICATIONS OF MAGNETIZABLE MICROPARTICLES

An alternative for using radiolabeled nano- and microparticles is to have magnetizable spheres which, after magnetization in vivo, could be measured externally with a magnetometer.

The body is a source of magnetic fields produced either by naturally occuring electric currents in the body or by magnetic materials in the body. Fluctuating magnetic fields are produced by organs in the body containing muscles or nerves. Peak values in the order of 10^{-10}, 10^{-11} and 10^{-12} T can be measured from heart, skeletal muscle, and brain, respectively.

Table 6
THE RELATIVE ERROR FOR THREE SOURCES OF ERROR IN EXPERIMENTAL BLOOD FLOW MEASUREMENTS

Sources of error	Assumed number of errors	Relative error (%)
Microparticles in sample (1g)	10^2—10^4	10—1
Nuclei per particle	10^6—10^8	0.1—0.01
Pulses per sample	10^2—10^6	10—0.1
Total error		14.1—1.0

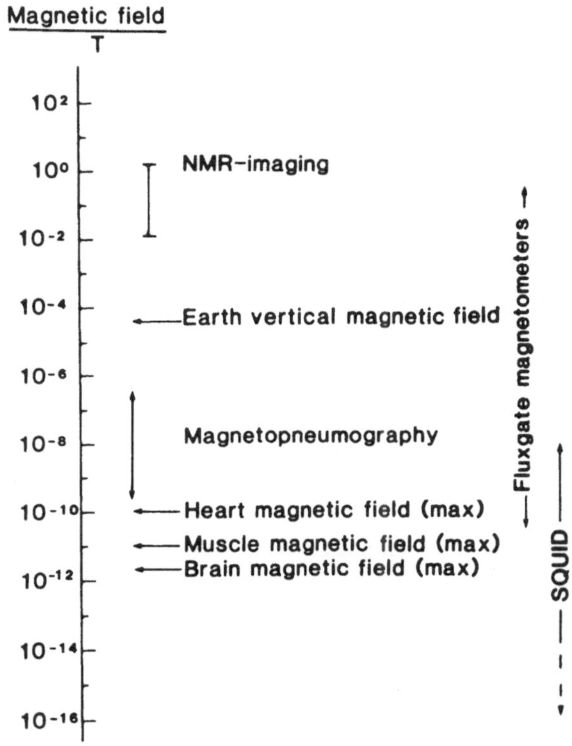

FIGURE 17. Magnitude of magnetic field strengths, applicable in magnetometry. Detector sensitivities are indicated.

Steady magnetic fields are produced by magnetic particles in the body. For example, γ-Fe_2O_3 particles in the lungs will produce a field between 0.1 to 100 nT. The later measurements are called magnetopneumography. A summary of different field strengths are given in Figure 17.

To measure these very weak fields, very sensitive magnetic detectors are needed. For fields down to about 10^{-10} T, ordinary fluxgate magnetometers can be used. For weaker fields a SQUID (superconducting quantum interference device) has to be used. In Figure 17 these detector sensitivites are indicated.

Cohen pioneered biomedical applications of magnetometry with measurements of magnetic

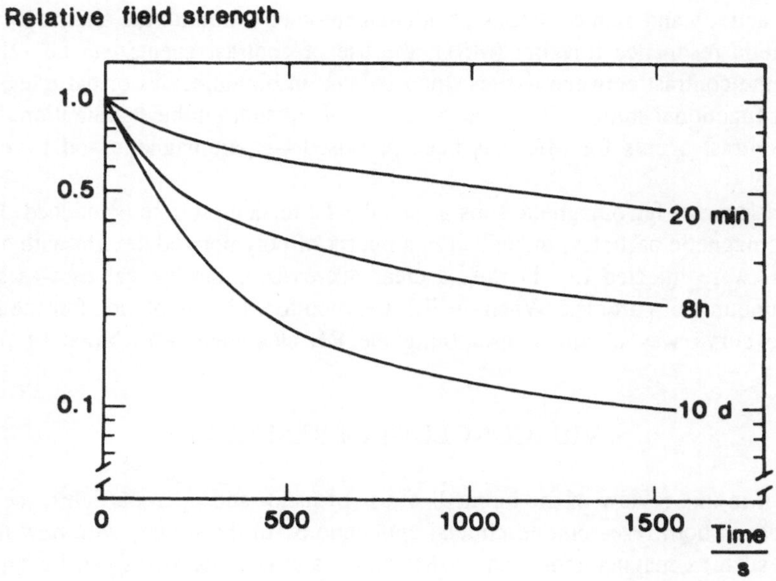

FIGURE 18. Three relaxation curves for lung macrophages, obtained at 20 min, 8 hr, and 10 days after inhalation of magnetic aerosols. (Data redrawn from Brain et al.[142])

contamination in human lungs in 1973.[140] The particles in the lungs were magnetized in an external homogenous magnetic field, whereafter the remanent field caused by the magnetized particles was measured with a sensitive magnetometer.

After inhalation of magnetic aerosols, application of an external magnetic field produces a remanent magnetic field in the direction of the magnetizing field, proportional to the external field and the density of the contaminant particles. This field decays rapidly after the magnetization with half-times in the order of 2 to 3 min.[141,142] A graph of this decreasing field is called a relaxation curve. Figure 18 gives examples, taken from Brain et al.[142] of three relaxation curves measured 20 min, 8 hr and 10 days after inhalation. The rate of relaxation is very slow just after the inhalation, but increases with time, when assumed phagocytosis of particles by macrophages takes place.[143] The relaxation rate then decreases again after a few days.[144]

The decrease of field strength after a magnetization, may be due to cell movement, intracellular processes such as organelle motion caused by microfilaments or Brownian motion. Because the particles are found in the phagosomes, the relaxation measurements can serve as a probe for external measurements of the motion of the cytoskeleton causing phagosomal or secondary lysosomal motion, which reflects the phagocytic process in the cells.[145] Changes in the location of particles from extracellular to intracellular sites, and movements from ectoplasmic to endoplasmic sites within cells may be responsible for the observed changes in relaxation rates with time.[144]

In an in vitro experiment, Valberg[146] has shown that pulmonary macrophages containing maghemite (γ-Fe_2O_3) aerosol rotate the particles and thus the relaxation curve reflects the macrophage cytoplasmic movement.

After intravenous injection of magnetic Fe_2O_3 particles, these are phagocytosed by the RES macrophages, predominantly in the liver. As for the lung macrophages after magnetization, a relaxation curve can be measured. Thus, with this method, the phagocytic process in the liver may also be possible to study.[145]

Because the magnetic particles can be labeled with radionuclides[147,148] it might be possible both to measure the initial trapping of the particles in the liver by external measurements of the radioactivity and then continously with magnetometry monitor the phagocytic process.

In magnetic resonance imaging (MRI), the use of contrast agents can be valuable for enhancing the contrast between tissues. Incorporated in biomolecules or particles, they can be used for functional studies. It may also be possible to shorten the examination time. Two kinds of contrast agents for MRI has been proposed — paramagnetic and ferromagnetic materials.[149]

In the presence of ferromagnetic substances, the T2 relaxation time is reduced. In a study where ferromagnetic particles, imbedded in a matrix of polymerized dextrin with a diameter of 800 nm, were injected i.v. in rats, a clear decrease of the T2 values was seen with increased amounts of particles. When the T2 was monitored in the blood after the injection, a clearance curve was obtained resembling the RE clearance from blood of particulate materials.[149]

VIII. CONCLUDING REMARKS

As shown in this review of the biomedical use of nano- and microparticles, they play an essential role in both research and clinical applications. In the future, with new measuring techniques such as magnetometry and MRI, nano- and microparticles can be supposed to be as important as earlier. In order to increase our knowledge of their biological behavior, further research concerning their basic biological behavior ought to be encouraged.

ACKNOWLEDGMENTS

This study was supported by grants from Medical Research Council grant No. B85-17X-6530-02; John and Augusta Perssons Foundation for Medical Research, Lund; Åke Wibergs Foundation, Stockholm; and Research Funds of the Medical Faculty, Lund. We acknowledge the clinical suggestions by Assistant Professor Per-Ebbe Jönsson and Assistant Professor Stefan Rydén and their kindness in permitting us to use their results.

REFERENCES

1. **Nelp, W. B.,** An evaluation of colloids for RES function studies, in *Radiopharmaceuticals*, Subramanian, G., Rhodes, B. A., Cooper, J. F., and Sodd, V. J., Ed., Society of Nuclear Medicine, New York, 1975, 349.
2. **Frier, M.,** Phagocytosis, in *Progress in Radiopharmacology*, Vol. 2, Cox, P. H., Ed., Elsevier, Amsterdam, 1981, 249.
3. **Rydén, S.,** Reticuloendothelial Function. Methods to Evaluate Function and Correlation between Function and Tumour Growth — Experimental Studies in the Rat, Thesis, University of Lund, Lund, Sweden, 1983.
4. **Scott, B. G. D., Williams, H. S., and Marriott, P. M.,** The phagocytosis of colloidal particles of different sizes, *Br. J. Exp. Pathol.*, 48, 411, 1967.
5. **Strand, S. E. and Persson, B. R. R.,** Quantitative lymphoscintigraphy. I. Basic concepts for optimal uptake of radiocolloids in the parasternal lymph nodes of rabbits, *J. Nucl. Med.*, 20, 1038, 1979.
6. **Aspegren, K., Strand, S. E., and Persson, B. R. R.,** Quantitative lymphoscintigraphy for dectection of metastases to the internal mammary lymph nodes. Biokinetics of $^{99}Tc^{m}$-sulfur colloid uptake and correlation with microscopy. *Acta Radiol. Oncol.*, 17, 17, 1977.
7. **Ege, G. N. and Warbick, A.,** Lymphoscintigraphy: a comparison of 99m-Tc-antimony sulphide colloid and 99m-Tc-stannous phytate, *Br. J. Radiol.*, 32, 124, 1979.
8. **Kaplan, W. D., Davis, M. A., and Rose, C. M.,** A comparison of two technetium-99m labeled radiopharmaceuticals for lymphoscintigraphy, *J. Nucl. Med.*, 20, 933, 1979.

9. **Nagai, K., Ito, Y., Otsuka, N., Muranaka, A., Kaji, T., and Kajihara, Y.,** Experimental studies on uptake of ^{99m}Tc antimony sulfide colloid in RES. A comparison with various ^{99m}Tc-colloids, *Int. J. Nucl. Med. Biol.*, 8, 85, 1980.
10. **Nagai, K., Ito, Y., Otsuka, N., and Muranaka, A.,** Deposition of small ^{99m}Tc labeled colloids in bone marrow and lymph nodes., *Eur. J. Nucl. Med.*, 7, 66, 1982.
11. **Griffin, F. M. Jr., Griffin, J. A., and Leider, J. E.,** Studies on the mechanism of phagocytosis. I. Requirements of circumferential attachment of particle-bound ligands to specific receptors on the macrophage plasma membrane, *J. Exp. Med.*, 142, 1263, 1975.
12. **Zum Winkel, K.,** Lymphologie mit Radionukliden, Verlag Hildegard Hoffman, 1972.
13. **Atkins, H. L., Hauser, W., and Richards, P.,** Factors affecting distribution of technetium-sulfur colloid, *RES J. Reticuloendothel. Soc.*, 8, 176, 1970.
14. **Arturson, P., Laakso, T., and Edman, P.,** Acrylic microspheres in vivo IX: blood elimination kinetic and organ distribution of microparticles with different surface characteristics, *J. Pharm. Sci.*, 72, 1415, 1983.
15. **Wilkins, D. J. and Myers, P. A.,** Studies on the relationship between electrophoretic properties of colloids and their blood clearance and organ distribution in the rat, *Br. J. Exp. Pathol.*, 47, 568, 1966.
16. **Dornfest, B. S., Lenehan, P. F., Reilly, T. M., Mestler, G. E., Steigman, J., and Solomon, N. A.,** Effects of sera of normal, anemic and leukemic rats on particle size distribution of ^{99m}Technetium-sulfur colloid in vitro, *RES. J. Reticuloendothel. Soc.*, 21, 317, 1977.
17. **Frier, M., Griffiths, P., and Ramsay, A.,** The physical and chemical characteristics of sulphur colloids, *Eur. J. Nucl. Med.*, 6, 255, 1981.
18. **Billinghurst, M. W., and Jette, D.,** Collodial particle-size determination by gel filtration, *J. Nucl. Med.*, 20, 133, 1979.
19. **Lim, T. K., Bloomfield, A., and Krejcarek, G.,** Size and charge distributions of radiocolloid particles, *Int. J. Appl. Radiat. Isot.*, 30, 531, 1979.
20. **Persson, B. R. R., Strand, S. E. and Knöös, T.,** Radioanalytical studies of ^{99m}Tc-labelled colloids and macromolecules with gel chromatography column scanning technique, *J. Radiol. Chem.*, 43, 275, 1978.
21. **Warbick, A., Ege, G. N., Henkelman, R. M., Maier, G., and Lyster, D. M.,** An evaluation of radiocolloid sizing techniques, *J. Nucl. Med.*, 18, 827, 1977.
22. **Davis, M. A., Jones, A. G., and Trindade, H.,** A rapid and accurate method for sizing radiocolloids, *J. Nucl. Med.*, 15, 923, 1974.
23. **Pedersen, B. and Kristensen, K.,** Evaluation of methods for sizing of colloidal radiopharmaceuticals, *Eur. J. Nucl. Med.*, 6, 521, 1981.
24. **Bergqvist, L., Strand, S. E., and Persson, B. R. R.,** Particle sizing and biokinetics of interstitial lymphoscintigraphic agents, *Semin. Nucl. Med.*, 13, 9, 1983.
25. **Persson, B. R. and Naversten, Y.,** Technetium-99m sulfide colloid preparation for scintigraphy of the reticuloendothelial system, *Acta Radiol. Ther. Phys. Biol.*, 9, 567, 1970.
26. **Prinzmetal, M., Simkin, B., Bergman, H. C., and Kruger, H. E.,** Studies on the coronary circulation. II. The collateral circulation of the normal human heart by coronary perfusion with radioactive erythrocytes and glass spheres, *Am. Heart J.*, 33, 420, 1947.
27. **Wagner, H. N., Jr., Rhodes, B. A., Sasaki, Y., and Ryan, J. P.,** Studies of the circulation with radioactive microspheres, *Invest. Radiol.*, 4, 374, 1969.
28. **Rhodes, B. A. and Croft, B. Y.,** *Basics of Radiopharmacy*, C. V. Mosby, St. Louis, Mo., 1978, chap. 3.
29. **Hales, J. R. S.,** Radioactive microsphere techniques for studies of the circulation, *Clin. Exp. Pharm. Physiol.*, 1, 31, 1974.
30. **Phibbs, R. H., Wyler, F., and Neutze, J.,** Rheology of microspheres injected into circulation of rabbits, *Nature (London)*, 216, 1339, 1967.
31. **Dworkin, H. J., Gutkowski, R. F., Porter, W., and Potter, M.,** Effect of particle number on lung perfusion images: concise communication, *J. Nucl. Med.*, 18, 260, 1977.
32. **Stanek, K. A., Smith, T. L., Murphy, W. R., and Coleman, T. G.,** Hemodynamic distrubances in the rat as a function of the number of microspheres injected, *Am. J. Physiol.*, 245, H920, 1983.
33. **Fan, F. C., Schuessler, G. B., Chen, R. Y. Z., and Chien, S.,** Determinations of blood flow and shunting of 9- and 15 μm spheres in regional beds, *Am. J. Physiol.*, 237, H25, 1979.
34. **Andersen, K. S., Skjerven, R., and Lekven, J.,** Stability of 8-, 15-, and 26 μm microspheres entrapped in feline myocardium, *Am. J. Physiol.*, 244, H121, 1983.
35. **Lekven, L. and Andersen, K. S.,** Migration of 15 micron microspheres from infarcted myocardium, *Cardiovasc. Res.* 14, 280, 1980.
36. **Murdoch, R. H., Jr. and Cobb, F. R.,** Effects of infarcted myocardium on regional blood flow measurements to ischemic regions in canine heart, *Circ. Res.*, 47, 701, 1980.
37. **Reimer, K. A. and Jennings, R. B.,** The changing anatomic reference base of evolving myocardial infarction, *Circulation*, 60, 866, 1979.

38. **Heymann, M. A., Payne, B. D., Hoffman, J. I. E., and Rudolph, A. M.,** Blood flow measurements with radionuclide-labelled particles, *Prog. Cardiovasc. Dis.*, 20, 55, 1977.
39. **Davis, M. A.,** Particulate radiopharmaceuticals for pulmonary studies, in *Radiopharmaceuticals*, Subramanian, G., Rhodes, B. A., Cooper, J. F., and Sodd, V. J., Eds., Society of Nuclear Medicine, New York, 1975, chap. 29.
40. **Davis, M. A. and Holman, B. L.,** Radiopharmaceuticals for perfusion scanning, in *Progress in Nuclear Medicine: Regional Pulmonary Function in Health and Disease*, Holman, B. L., Lindeman, J. F., and Karger, S., Eds., Basel, A. G., 1973, chap. 2.
41. **Aschoff, L. I.,** Morphologie des reticulo-endothelialen systems, in *Ergebnisse der Inneren Medizin und der Kinderheilkunde*, Kraus, F., Meyer, E., Minkowski, O., Muller, F. R., Sahli, H., Schittenhelm, A., Czerny, A., Heubner, O., and Langstein, L., Eds., Julius Springer, Berlin, 1924.
42. **Metchnikoff, I.,** *Lectures on the Comparative Pathology of Inflammation*, Kegan Paul, Trench, Trubner & Co, London, 1893; reprinted, Dover Publishers, New York, 1968, 137.
43. **Bradfield, J. W. B.,** The reticulo-endothelial system and blood clearence, in *Microspheres and Drug Therapy*, Davis, S. S., Illum, L., McVie, J. G., and Tomlinson, E., Eds., Elsevier, Amsterdam, 1984, chap. 2.
44. **Wisse, E. and De Leeuw A. M.,** Structural elements determining transport and exchange processes in the liver, in *Microspheres and Drug Therapy*, Davis, S. S., Illum, L., McVie, J. G., and Tomlinson, E., Eds., Elsevier, Amsterdam, 1984, chap. 1.
45. **Metchnikoff, I.,** *Immunity of Infective Disease* (transl.), Binnie, F. G., Ed., University Press, New York, 1905.
46. **Munthe-Kaas, A. C. and Kaplan, G.,** Endocytosis by macrophages, in *The Reticuloendothelial System: A Comprehensive Treatise*, Carr, I. and Daems, W. T., Eds., Plenum Press, New York, 1981, 19.
47. **Saba, T. M., Blumenstock, F. A., Weber, P., and Kaplan, J. E.,** Physiologic role for cold-insoluble globulin in systemic host defense: implications of its characterization as the opsonic α_2-SB-glycoprotein, *Ann. N.Y. Acad. Sci.*, 312, 43, 1978.
48. **Saba, T. M.,** Aspecific opsonins, in *The Immune System in Infections and Diseases*, Nater, E. and Milgram, F., Eds. Karger, Basel, 1975, 489.
49. **Munthe-Kaas A. C.,** Phagocytosis in rat Kupffer cells in vitro, *Exp. Cell. Res.*, 99, 319, 1976.
50. **Dobson, E. L. and Jones, H. B.,** The behaviour of intravenously injected particulate material, *Acta Med. Scand.*, 273, 1, 1952.
51. **Biozzi, G., Benacerraf, B., and Halpern, B. N.,** Quantitative study of the granulopectic activity of the reticuloendothelial system. II. A study of the kinetics of the granulopectic activity of the RES in relation to the dose of carbone injected. Relationship between the weight of the organs and their activity. *Br. J. Exp. Pathol.*, 34, 441, 1953.
52. **Pirttiaho, H. I. and Pitkänen, U.,** Size and blood flow of the liver estimated with $^{99}Tc^{m}$ scanning, *Acta Radiol. Ther.*, 16, 497, 1977.
53. **Sarper, R., Fajman, W. A. Tarcan, Y. A., and Nixon, D. W.,** Enhanced detection of metastatic liver disease by computerized flow scintigrams: concise communication, *J. Nucl. Med.*, 22, 318, 1981.
54. **Fleming, J. S., Humphries, N. L. M., Karran, S. J., Goddard, B. A., and Ackery, D. M.,** In vivo assessment of hepatic arterial and portal venous components of liver perfusion: concise communication, *J. Nucl. Med.*, 22, 18, 1981.
55. **Izzo, G., DiLuzio, S., Guerrisi, M., Favella, A., and Magrini, A.,** On the interpretation of the early part of the liver time-activity curve: double tracer experiments, *Eur. J. Nucl. Med.*, 8, 101, 1983.
56. **Fleming, J. S., Ackery, D. M., Walmsley, B. H., and Karran, S. J.,** Scintigraphic estimation of arterial and portal blood supplies to the liver, *J. Nucl. Med.*, 24, 1108, 1983.
57. **Davies, P. and Booney, R. J.,** Secretory products of mononuclear phagocytoses: a brief review, *RES. J. Reticuloendothel. Soc.*, 31, 37, 1979,
58. **Schildt, B., Bouveng, R., and Sollenberg, M.,** Plasma substitute induced impairment of the reticuloendothelial system function, *Acta Chir. Scand.*, 141, 7, 1975.
59. **Saba, T. M. and DiLuzio, N. R.,** Reticuloendothelial blockade and recovery as a function of opsonic activity, *Am. J. Physiol.*, 216, 197, 1969.
60. **Jenkin, C. R. and Rowley, D.,** The role of opsonins in the clearance of living and inert particles by the cells of the reticuloendothelial system, *J. Exp. Med.*, 114, 363, 1961.
61. **Saba, T. M. and DiLuzio, N. R.,** Kupffer cell phagocytosis and metabolism of a variety of particles as a function of opsonization, *RES. J. Reticuloendothel. Soc.*, 2, 437, 1965.
62. **DiLuzio, N. R., Miller, E., McNamee, R., and Pisano J. C.,** Alterations in plasma recognition factor activity in experimental leukemia, *RES. J. Reticuloendothel. Soc.*, 11, 186, 1972.
63. **Pisano, J. C., DiLuzio, N. R., and Salky, N. K.,** Absence of macrophage humoral recognition factor(s) in patients with carcinoma, *J. Lab. Clin. Med.*, 76, 141, 1970.
64. **Benacerraf, B. and Sebestyen, M. M.,** Effect of bacterial endotoxin on the reticuloendothelial system, *Fed. Proc.*, 16, 860, 1957.

65. **Munz, D., Standke, R., and Hörr, G.,** Measurement of phagocytic and proteolytic function of macrophages in liver, spleen and bone marrow, in Progress in Radiopharmacology, Vol. 2, Cox, P. H., Ed., Elsevier, Amsterdam, 1981, 261.
66. **Reske, S. N., Vyska, K., and Feinendegen, L. E.,** In vivo assessment of phagocytic properties of Kupffer cells, *J. Nucl. Med.*, 22, 405, 1981.
67. **DeNardo, S. J., Bell, G. B., DeNardo, G. L., Carretta, R. F., Scheibe, P. O., Imperator, T. J., and Jackson, P. E.,** Diagnosis of chirrosis and hepatitis by quantitative hepatic and other reticuloendothelial clearence rates, *J. Nucl. Med.*, 17, 449, 1976.
68. **Palmer, J. G., Persson, B. R. R., Strand, S. E., and Naversten Y.,** Computer assisted biokinetic studies of radiopharmaceuticals in animals and humans. I. A method for on-line biokinetic studies with computerized scintillation camera, in Proceedings of Third International Radiopharmaceutical Dosimetry Symposium, Oak Ridge, Tenn., 1980, 157.
69. **Rydén, S., Strand, S. E., Palmer, J., Stenram, U., Hafström, L. O., and Persson, B.,** A scintillation camera technique for measurements of the reticuloendothelial function. Comparison of different methods for measuring RES function, *Eur. J. Nucl. Med.*, 7, 16, 1982.
70. **Saba, T. M. and DiLuzio, N. R.,** Evaluation of humoral and cellular mechanisms of methyl palmitate induced reticuloendothelial depression., *Life Sci.*, 7 (Suppl. 2), 337, 1968.
71. **Bradfield, J. W. B. and Wagner, H.,** Th relative importance of blood flow and liver phagocytic function in the distribution of technetium-99m sulphur colloid, *J. Nucl. Med.*, 18, 620, 1977.
72. **Husztik, E., Lázár, G., and Szilágyi, S.,** Study on the mechanism of Kupffer cell phagocytosis blockade induced by gadolinium chloride, in *Kupffer Cells and Other Liver Sinusoidal Cells*, Wiesse, E. and Knook, D. L., Eds., Elsevier, Amsterdam, 1977, 387.
73. **Wagner, H. N. and Iio, M.,** Studies of the reticuloendothelial system (RES). III. Blockade of the RES in man, *J. Clin Invest.*, 43, 1525, 1964.
74. **Saba, T. M.,** Mechanism mediating reticuloendothelial system depression after surgery, *Proc. Soc. Exp. Biol. Med.*, 133, 1132, 1970.
75. **Rydén, S., Bergqvist, L., Hafström, L. O., Hultberg, B., Stenram, U., and Strand, S. E.,** Influence of reticuloendothelial suppressing agent on liver tumor growth in the rat, *J. Surg. Oncol.*, 26, 245, 1984.
76. **Heller, J. H.,** Non toxic RES stimulatory lipids, *Ann. N.Y. Acad. Sci.*, 88, 116, 1960.
77. **DiLuzio, N. R.,** Influence of glucan on hepatic macrophage structure and function. Relation to inhibition of hepatic metastases, in *Kupffer Cells and Other Liver Sinusoidal Cells*, Wisse, E. and Knook, D. L., Eds., Elsevier, Amsterdam, 1977, 397.
78. **Diller, I. C., Mankowski, Z. T., and Fischer, M. E.,** The effect of yeast polysaccarides on mouse tumors, *Cancer Res.*, 23, 201, 1963.
79. **Bitter-Sauermann, D.,** Aktivierung des Komplementsystems — ein Monopol des Immunokomplexes, *Klin. Wochenschr.*, 50, 277, 1972.
80. **Rydén, S., Bergqvist, L., Hafström, L. O., Hultberg, B., and Stenram, U.,** Release of beta-hexosaminidase after administration of different agents affecting the reticuloendothelial system. An experimental study of rats, *Enzyme*, 31, 104, 1984.
81. **Rydén, S., Hultberg, B., Hafström, L. O., and Isaksson, A.,** Influence on plasma beta-hexosaminidase of reticuloendothelial stimulation and depression. An experimental study in rats, *J. Clin. Chem. Clin. Biochem.*, 22, 219, 1984.
82. **Stern, K. and Duvelius, W. A.,** Hepatic uptake of colloidal radio-gold (^{198}Au) in tumor-bearing rats, *Proc. Am. Assoc. Cancer Res.*, 2, 348, 1958.
83. **Biozzi, G., Stiffel, C., Halpern, B. N., and Mouton, D.,** Etude de la fonction phagocytaire du S.R.E. au course de développement de tumeurs malignes expérimentales chez le rat et la souris, *Ann. Inst. Pasteur*, 94, 681, 1958.
84. **Old, L. J., Clarke, D. A., and Goldsmith, M.,** Reticuloendothelial function during experimental tumor growth, *Proc. Am. Assoc. Cancer Res.*, 3, 49, 1958.
85. **Old, L. J., Clarke, D. A., Benacerraf, B., and Goldsmith, M.,** The reticuloendothelial system and the neoplastic process, *Ann. N.Y. Acad. Sci,*. 88, 264, 1960.
86. **Omori, Y.,** The relation between the reticuloendothelial system and the spread of cancer, especially of gastric carcinoma, *Tohuku J. Exp. Med.*, 81, 315, 1964.
87. **Salky, N. K., DiLuzio, N. R., Levin, A. G., and Goldsmith, H. S.,** Phagocytic activity of the reticuloendothelial system in neoplastic disease, *J. Lab. Clin. Med.*, 70, 393, 1967.
88. **Margarey, C. J. and Baum, M.,** Reticuloendothelial activity in humans with cancer, *Br. J. Surg.*, 57, 748, 1970.
89. **Fischer, E. R. and Fischer, B.,** Experimental studies of factors influencing hepatic metastases. VII. Effect of reticuloendothelial interference, *Cancer Res.*, 21, 275, 1961.
90. **Fischer, B. and Fischer, E. R.,** Experimental factors influencing hepatic metastases. III. Effect of surgical trauma with special reference to liver injury, *Ann. Surg.*, 150, 731, 1959.

91. **Saba, T. M. and Antikatzides, T. G.,** Decreased resistance to intravenous tumor-cell challenge during reticuloendothelial depression following surgery, *Br. J. Cancer,* 34, 381, 1976.
92. **Mackaness, G. B.,** Role of macrophages in host defense mechanisms, in *The Macrophage in Neoplasia,* Fink, M. A., Ed., Academic Press, New York, 1976, 3.
93. **Rydén, S., Bergqvist, L., Hafström, L. O., and Strand, S. E.,** Reticuloendothelial function in normal and tumor-bearing rats. Measurements with a scintillation camera technique, *Eur. J. Cancer Clin. Oncol.,* 19, 965, 1983.
94. **McAfee, J. G., Ause, R. G., and Wagner, H. N.,** Diagnostic value of scintillation scanning of the liver, *Arch. Intern. Med.,* 116, 95, 1965.
95. **Strauss, L., Bostel, F., Clorius, J. H., Raptou, E., Wellman, H., and Georgi, P.,** Single-photon emission computed tomography (SPECT) for assessment of hepatic lesions, *J. Nucl. Med.,* 23, 1059, 1982.
96. **Yoffey, J. M. and Courtice, F. C.,** *Lymphatics, Lymph and the Lymphomyeloid Complex,* Academic Press, London, 1970.
97. **Bergqvist, L., Strand, S. E., and Jönsson, P. E.,** The characterization of radiocolloids used for administration to the lymphatic system, in *Microspheres and Drug Therapy,* Davis, S. S., Illum, L., McVie, J. G., and Tomlinson, E., Eds., Elsevier, Amsterdam, 1984, 263.
98. **zum Winkel, K. and Scheer, K. E.,** Scintigraphic and dynamic studies of the lymphatic system with radio-colloids, *Minerva Nucl.,* 9, 390, 1965.
99. **Kaplan, W. D.,** Iliopelvic lymphoscintigraphy, *Semin. Nucl. Med.,* 13, 42, 1983.
100. **Dworkin, H. J.,** Teaching editorial. Potentinal for lymphoscintigraphy, *J. Nucl. Med.,* 23, 936, 1982.
101. **Dworkin, H. J.,** Re: Teaching editorial. Potential for lymphoscintigraphy, *J. Nucl. Med.,* 24, 371, 1983.
102. **Sherman, A. I. and Ter-Pogossian, M.,** Lymph-node concentration of radioactive colloidal gold following interstitial injection, *Cancer,* 6, 1238, 1953.
103. **Hultborn, K. A., Larsson, L. G., and Ragnhult, I.,** The lymph drainage from the breast to axillary and parasternal lymph nodes, studied with the aid of colloidal Au-198, *Acta Radiol.,* 43, 52, 1955.
104. **Seaman, W. B. and Powers, W. E.,** Studies on the distribution of radioactive colloidal gold in regional lymph nodes containing cancer, *Cancer,* 8, 1044, 1955.
105. **Sullivan, D. C., Croker, B. P., Harris, C. C., Deery, P., and Seigler, H. F.,** Lymphoscintigraphy in malignant melanoma, *Am. J. Roentgenol.,* 137, 847, 1981.
106. **Schwab, W., Scheer, K. E., and zum Winkel, K.,** Scintigraphie des Zervikalen Lymphsystems, *Arch. Ohren. Nasen. Kehlkopfheilkd.,* 183, 382, 1965.
107. **Schenk, P.,** Scintigraphische Darstellung des Parasternalen Lymphsystems, *Strahlentherapie,* 130, 504, 1966.
108. **Osborne, M. P., Meijer, W. S., and DeCrosse, J. J.,** Lymphoscintigraphy in the staging of solid tumors, *Surg. Gynecol. Obstet.,* 156, 384, 1983.
109. **Ege, G. N.,** Internal mammary lymphoscintigraphy in breast carcinoma: a study of 1072 patients, *Int. J. Radiat. Oncol. Biol. Phys.,* 2, 755, 1977.
110. **Agwunobi, T. C. and Boak, J. L.,** Diagnosis of malignant breast disease by axillary lymphoscintigraphy: a preliminary report, *Br. J. Surg.,* 65, 379, 1978.
111. **Hill, N. S., Ege, G, N., Greyson, N. D., Mahoney, L. J., and Jirsch, D. W.,** Predicting nodal metastases in breast cancer by lymphoscintigraphy, *Can. J. Surg.,* 26, 507, 1983.
112. **Christensen, B., Blichert-Toft, M., Siemssen, O. J., and Nielsen, S. L.,** Reliability of axillary lymph node scintiphotography in suspected carcinoma of the breast, *Br. J. Surg.,* 67, 667, 1980.
113. **Peyton, J. W. R., Crosbie, J., Bell, T. K., Roy, A. D., and Odling-Smee, W.,** High colloidal uptake in axillary nodes with metastatic disease, *Br. J. Surg.,* 68, 507, 1981.
114. **Ege, G. N. and Cummings, B. J.,** Interstitial radiocolloid iliopelvic lymphoscintigraphy: technique, anatomy, and clinical application, *Int. J. Rad. Oncol. Biol. Phys.,* 6, 1483, 1980.
115. **Gitsch, E., Philipp, K., and Kubista, E.,** Die Intraoperative Lymphszintigraphie bei der Radikaloperation des Mammakarzinoms, *Geburtshilfe Fraunheilkd.,* 43, 112, 1983.
116. **Gitsch, E. Philipp, K., and Pateisky, N.,** Intraoperative lymph scintigraphy during radical surgery for cervical cancer, *J. Nucl. Med.,* 25, 486, 1984.
117. **Terui, S., Kato, H., Hirashima, T., Iizuka, T., and Oyamada, H.,** An evaluation of the mediastinal lymphoscintigram for carcinoma of the esophagus studied with ^{99m}Tc rhenium sulfur colloid, *Eur. J. Nucl. Med.,* 7, 99, 1982.
118. **Bourgeois, P., Frühling, J., and Henry, J.,** Postoperative axillary lymphoscintigraphy in the management of breast cancer, *Int. J. Rad. Oncol. Biol. Phys.,* 9, 29, 1983.
119. **Fee, H. J., Robinson, D. S., Sample, W. F., Graham, L. S., Holmes, E. C., and Morton, D. L.,** The determination of lymph shed by colloidal gold scanning in patients with malignant melanoma: a preliminary study, *Surgery,* 84, 626, 1978.
120. **Meyer, C. M., Lecklitner, M. L., Logic, J. R., Balch, C. E., Bessey, P. Q., and Tauxe, W. N.,** Technetium-99m sulfur-colloid cutaneous lymphoscintigraphy in the management of truncal melanoma, *Radiology,* 131, 205, 1979.

121. **Sullivan, D. C., Croker, B. P., Harris, C. C., Deery, P., and Seigler, H. F.,** Lymphoscintigraphy in maligant melanoma: ^{99m}Tc antimony sulfur colloid, *Am. J. Roentgenol.*, 137, 847, 1981.
122. **Munz, D. L., Altmeyer, P., Sessler, M. J., and Hör, G.,** Axillary lymph node groups — the center in lymphatic drainage from the truncal skin in man, *Lymphology*, 15, 143, 1982.
123. **Bergqvist, L., Strand, S.-E., Hafström, L., and Jönsson, P.-E.,** Lymphoscintigraphy in patients with malignant melanoma: a quantitative and qualitative evaluation of its usefulness, *Eur. J. Nucl. Med.*, 9, 129, 1984.
124. **Dufresne, E. N., Kaplan, W. D., Zimmerman, R. E., and Rose, C. M.,** The application of internal mammary lymphoscintigraphy to planning of radiation therapy, *J. Nucl. Med.*, 21, 697, 1980.
125. **Bourgeois, P. and Frühling, J.,** Internal mammary lymphoscintigraphy in the diagnosis and management of breast cancer, *J. Belge. Radiol.*, 63, 669, 1980.
126. **Siddon, R. L., Chin, L. M., Zimmerman, R. E., Mendel, J. B., and Kaplan, W. D.,** Utilization of parasternal lymphoscintigraphy in radiation therapy of breast carcinoma, *Int. J. Radiat. Oncol. Biol. Phys.*, 8, 1059, 1982.
127. **Collier, B. D., Palmer, D. W., Wilson, J. F., Greenberg, M., Komaki, R., Cox, J. D., Lawson, T. L., and Lawlor, P. M.,** Internal mammary lymphoscintigraphy in patients with breast cancer, *Radiology*, 147, 845, 1983.
128. **Iversen, T. and Aas, M.,** Pelvic lymphoscintigraphy with ^{99m}Tc-colloid in lymph node metastases, *Eur. J. Nucl. Med.*, 7, 455, 1982.
129. **Voutilainen, A. and Wiljasalo, M.,** On the correlation of lymphography and lymphoscintigraphy in metastases of tumours of the pelvic region, *Ann Chir. Gyneacol.*, 54, 268, 1965.
130. **Rees, W. V., Robinson, D. S., Holmes, E. C., and Morton, D. L.,** Altered lymphatic drainage following lymphadenectomy, *Cancer*, 45, 3045, 1980.
131. **Ege, G. N.,** Radiocolloid lymphoscintigraphy in neoplastic disease, *Cancer Res.*, 40, 3065, 1980.
132. **Vieras, F. and Boyd, C. M.,** Radionuclide lymphangiography in the evaluation of pediatric patients with lower-extremity edema: concise communication, *J. Nucl. Med.*, 18, 441, 1977.
133. **McConnell, R. W., McConnell, B. G., and Kim, E. E.,** Other applications of interstitial lymphoscintigraphy, *Semin. Nucl. Med.*, 13, 70, 1983.
134. **Taplin, G. V. and MacDonald, S.,** Radiochemistry of macroaggregated albumin and newer lung scanning agents, *Semin. Nucl. Med.*, 1, 132, 1971.
135. **Lifson, N.,** Use of microspheres to measure intraorgan distribution of blood flow in the splanchnic circulation, in *Measurement of Blood Flow*, Granger, D. N. and Bulkley, G. B., Eds., Williams & Wilkins, Baltimore, 1981, chap. 10.
136. **Hafström, L. O., Persson, B. and Sundqvist, K.,** Measurement of cardiac output and organ blood flow in rats using ^{99}Tc^{m} labelled microspheres, *Acta Physiol. Scand.*, 106, 123, 1979.
137. **Buckberg, G. D., Luck, J. C., Payne, D. B., Hoffman, J. I. E., Archie, J. P., and Fixler, D. E.,** Some sources of error in measuring regional blood flow with radioactive microspheres, *J. Appl. Physiol.*, 31, 598, 1971.
138. **Dole, W. P., Jackson, D. L., Rosenblatt, J. I., and Thompson, W. L.,** Relative error and variability in blood flow measurements with radiolabelled microspheres, *Am. J. Physiol.*, 243, H371, 1982.
139. **Ståhlberg, F.,** personal communication.
140. **Cohen, D.,** Ferromagnetic contamination in the lungs and other organs of the human body, *Science*, 180, 745, 1973.
141. **Valberg, P. A. and Brain, J. D.,** Generation and use of three types of iron oxide aerosol, *Am. Rev. Respir. Dis.*, 120, 1013, 1979.
142. **Brain, J. D., Bloom, S. B., Valberg, P. A., and Gehr, P.,** Correlation between the behavior of magnetic iron oxide particles in the lungs of rabbits and phagocytosis, *Exp. Lung Res.*, 6, 115, 1984.
143. **Nemoto, I.,** A model of magnetization and relaxation of ferrimagnetic particles in the lung, *IEEE Trans. Biomed. Eng.*, BME-29, 745, 1982.
144. **Gehr, P., Brain, J. D., Nemoto, I., and Bloom, S. B.,** Behavior of magnetic particles in hamster lungs: estimates of clearance and cytoplasmic motility, *J. Appl. Physiol. Respirat. Environ. Exercise Physiol.*, 55, 1196, 1983.
145. **Gehr, P., Brain, J. D., Bloom, S. B., and Valberg, P. A.,** Magnetic particles in the liver: probe for intracellular movement, *Nature (London)*, 302, 336, 1983.
146. **Valberg, P. A.,** Magnetometry of ingested particles in pulmonary macrophages, *Science*, 224, 513, 1984.
147. **Hoffer, P. B. and Huberty, J. P.,** Carbonyl iron (magnetic) suspensions labelled with technetium-99m, in *Proc. Med. Radionuclide Imag.*, IAEA, Vienna, 1977, 71.
148. **Lichtenstein, M., Pojer, P. M., and Spokas, R. J.,** The labelling of magnetically responsive particles with ^{99}Tc^{m} and ^{32}P, *Int. J. Nucl. Med. Biol.*, 11, 153, 1984.
149. **Olsson M., Persson, B., Salford, L. G., and Scröder, U.,** The use of paramagnetic and ferromagnetic contrast agent in NMR-imaging of T1 and T2 respectively, in Proc. 1st Congr. Eur. Soc. Magnetic Resonance Med. Biol., Geneva, Switzerland, 1984.

INDEX

D

E

F

G

H

I